Elliptic Functions

In its first six chapters this text seeks to present the basic theory of the
Jacobi elliptic functions as an historical essay, an attempt to answer the
question: 'what would the treatment of elliptic functions have been like if Abel had
developed the ideas, rather than Jacobi?' Accordingly, it is based on the idea of
inverting integrals which arise in the theory of differential equations and, in particular,
the differential equation that describes the motion of a simple pendulum.

The later chapters present a more conventional approach to the Weierstrass
functions and to elliptic integrals, and then the reader is introduced to the richly varied
applications of the elliptic and related functions. Applications spanning arithmetic
(solution of the general quintic, the representation of an integer as a sum of three
squares, the functional equation of the Riemann zeta function), dynamics (orbits,
Euler's equations, Green's functions), and also probability and statistics, are discussed.

Elliptic Functions

J. V. ARMITAGE

Department of Mathematical Sciences
University of Durham

and the late

W. F. EBERLEIN

CAMBRIDGE
UNIVERSITY PRESS

CAMBRIDGE UNIVERSITY PRESS
Cambridge, New York, Melbourne, Madrid, Cape Town, Singapore, São Paulo

Cambridge University Press
The Edinburgh Building, Cambridge CB2 2RU, UK

Published in the United States of America by Cambridge University Press, New York

www.cambridge.org
Information on this title: www.cambridge.org/9780521780780

© Cambridge University Press 2006

First published 2006

Printed in the United Kingdom at the University Press, Cambridge

A catalogue record for this book is available from the British Library

ISBN-13   978-0-521-78078-0 hardback
ISBN-10   0-521-78078-0 hardback

ISBN-13   978-0-521-78563-1 paperback
ISBN-10   0-521-78563-4 paperback

W. F. Eberlein's original was dedicated to Patrick, Kathryn, Michael, Sarah, Robert, Mary and Kristen;
I should like to add:
To Sarah, Mark and Nicholas
(J. V. Armitage).

For some minutes Alice stood without speaking, looking out in all directions over the country – and a most curious country it was. There were a number of tiny little brooks, running straight across it from side to side, and the ground between was devided up into squares by a number of little green hedges, that reached from brook to brook. 'I declare it's marked out just like a large chess-board!' Alice said at last.

Lewis Carroll, Through the looking Glass

Contents

Contents

Preface

This is essentially a prolegomenon to the Partial Preface and serves primarily to place in a proper context the contents of this book and how they relate to the original six chapters, to which it refers.

Those six chapters, originally by W. F. Eberlein, sought to relate the ideas of Abel to the later work of Jacobi and concluded with the transformation theory of the theta functions. The first chapter began with the differential equation associated with the motion of a simple pendulum, very much in the tradition of Greenhill's *'Applications of Elliptic Functions'* (1892), but much influenced by the spirit of modern analysis. (Greenhill's obituary reads that 'his walls (were) festooned with every variety of pendulum, simple or compound.'[1]) The version given here is inspired by those early chapters and, apart from the addition of illustrative examples and extra exercises, is essentially unchanged.

The present account offers six additional chapters, namely 7 to 12, together with an Appendix, which seek to preserve the essentials and the spirit of the original six, insofar as that is possible, and which include an account of the Weierstrass functions and of the theory of elliptic integrals in Chapters 7 and 8. There follows an account of applications in (mainly classical) geometry (Chapter 9); in algebra and arithmetic – the solution of the quintic in Chapter 10, and sums of three squares, with references to the theory of partitions and other arithmetical applications in Chapter 11); and finally, in classical dynamics and physics, in numerical analysis and statistics and another arithmetic application (Chapter 12). Those chapters (9 to 12), were inspired by Projects offered by Fourth Year M. Math. undergraduates at Durham University, who chose topics on which to work, and they are offered here partly to encourage similar work. Finally the Appendix includes topics from its original version and then an application to the Riemann zeta function. There are extensive references to further reading in topics outside the scope of the present treatment.

[1] J. R. Snape supplied this quotation.

Original partial preface

(Based on W. F. Eberlein's preface to chapters 1 to 6, with some variations and additions)

Our thesis is that on the untimely death of Abel in 1829, at the age of 26, the theory of elliptic functions took a wrong turning, or at any rate failed to follow a very promising path. The field was left (by default) to Jacobi, whose 18th century methods lacked solid foundations until he finally reached the firm ground of theta functions. Thence emerged the notion of pulling theta functions out of the air and then defining the Jacobian elliptic functions as quotients of them.

All that is, of course, rigorous, but perhaps it puts the theta function cart before the elliptic functions horse! For example, the celebrated theta function identity

$$\prod_{n=1}^{\infty}(1 - q^{2n-1})^8 + 16q \prod_{n=1}^{\infty}(1 + q^{2n})^8 = \prod_{n=1}^{\infty}(1 + q^{2n-1})^8$$

looks impressive and was described by Jacobi as 'aequatio identica satis abstrusa', but if one starts in the historical order with elliptic functions, it reduces to the relatively trivial, if perhaps more inscrutable, identity $k^2 + (1 - k^2) = 1$. (See Chapter 4.)

In this book we shall apply *Abel's methods*, supplemented by the rudiments of complex variable theory, to *Jacobi's functions* to place the latter's elegance upon a natural and rigorous foundation.

A pedagogical note may be helpful at this point; we prefer to motivate theorems and proofs. Influenced by the writings of George Polya, we have tried to motivate theorems and proofs by 'induction and analogy' and by 'plausible inference' on all possible occasions. For example, the addition formulae for the Jacobian elliptic functions are usually pulled out of a hat, but (cf the book by Bowman [12]) in Chapter 2 we *guess* the basic addition formula for $cn(u + v, k), 0 < k < 1$, by interpolating between the known limiting cases $k = 0$ (when $cn(u + v, 0) = \cos(u + v)$) and $k = 1$ (when $cn(u + v, 1) = \sec h(u + v)$). We have tried to adopt similar patterns of plausible reasoning throughout.

Acknowledgements

With grateful thanks to Sarah, who typed the preliminary version, and to eight fourth year undergraduates at Durham University, whose projects in the applications of elliptic functions I supervised and from whom I learnt more than they did from me.

Acknowledgements are made throughout the book, as appropriate, to sources followed, especially in Chapters 8 to 12, which are devoted to the applications of elliptic functions. The work of the students referred to in the preceding paragraphs was concerned with projects based on such applications and used the sources quoted in the text. For example the work on the solution of the general quintic equation (Chapter 10) relied on the sources quoted and on individual supervisions in which we worked through and interpreted the references cited, with additional comments as appropriate. I should like to acknowledge my indebtedness to those texts and to the students who worked through them wih me. I should also like to acknowledge my indebtedness to Dr Cherry Kearton, with whom I supervised a project on cryptography and elliptic curves.

I lectured on elliptic functions (albeit through a more conventional approach and with different emphases and without most of the applications offered here) since the nineteen-sixties at King's College London and at Durham University. Again, I should like to express my gratitude to students who attended those lectures, from whose interest and perceptive questions I learnt a great deal. Over the years I built up an extensive collection of exercises, based on the standard texts (to which reference is made in what follows) and on questions I set in University examinations. Those earlier courses did not involve the applications offered here, nor did they reflect the originality of the unconventional approach offered here, as in the early Chapters by W. F. Eberlein, but I should like nevertheless to acknowledge my indebtedness to collections of exercises due to my colleagues at Durham, Professor A. J. Scholl and Dr J. R. Parker, though

I should add that neither of them is to be blamed for any shortcomings to be found in what follows.

I would like to express my thanks to Caroline Series of the London Mathematical Society, who invited me to complete Eberlein's six draft chapters for the Student Texts Series. I would also like to thank Roger Astley of Cambridge University Press for his patient encouragement over several years, and Carol Miller, Jo Bottrill and Frances Nex for their most helpful editorial advice.

1

The 'simple' pendulum

1.1 The pit and the pendulum

The inspiration for this introduction is to be found in Edgar Allen Poe's *The Pit and the Pendulum, The Gift* (1843), reprinted in *Tales of Mystery and Imagination*, London and Glasgow, Collins.

A simple pendulum consists of a heavy particle (or 'bob') of mass m attached to one end of a light (that is, to be regarded as weightless) rod of length l, (a constant). The other end of the rod is attached to a fixed point, O. We consider only those motions of the pendulum in which the rod remains in a definite, vertical plane. In Figure 1.1, P is the particle drawn aside from the equilibrium position, A. The problem is to determine the angle, θ, measured in the positive sense, as a function of the time, t.

The length of the arc AP is $l\theta$ and so the velocity of the particle is

$$\frac{d}{dt}(l\theta) = l\frac{d}{dt}\theta,$$

and it is acted upon by a downward force, mg, whose tangential component is $-mg\sin\theta$. Newton's Second Law then reads

$$m\left[\frac{d}{dt}l\frac{d\theta}{dt}\right] = -mg\sin\theta,$$

that is

$$\frac{d^2\theta}{dt^2} + \frac{g}{l}\sin\theta = 0. \tag{1.1}$$

If one sets $x = (g/l)^{1/2}\,t$, (so that x and t both stand for 'time', but measured in different units), then (1.1) becomes

$$\frac{d^2\theta}{dx^2} + \sin\theta = 0. \tag{1.2}$$

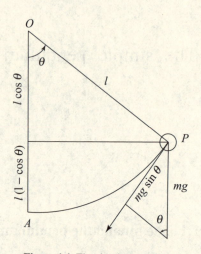

Figure 1.1 The simple pendulum.

Equation (1.2) has a long history (of some 300 years at least), but a rigorous treatment of it is by no means as straightforward as one might suppose (whence the 'pit'!). As one example of the pitfalls we might encounter and must try to avoid in our later development of the subject, let us begin with the familiar linearization of (1.2) obtained by supposing that θ is small enough to permit the replacement of $\sin\theta$ by θ. We obtain

$$\frac{\mathrm{d}^2\theta}{\mathrm{d}x^2} + \theta = 0 \qquad\qquad (1.3)$$

and the general solution of (1.3) is

$$\theta = A\cos x + B\sin x,$$

where A, B are constants, and the motion is periodic with x having period 2π and t having period $2\pi(l/g)^{1/2}$. The unique solution of Equation (1.3) satisfying the initial conditions $\theta(0) = 0$, $\theta'(0) = 1$, where θ' denotes $\frac{\mathrm{d}\theta}{\mathrm{d}x}$, is $\theta(x) = \sin x$.

Equation (1.3) is the familiar equation defining the simple harmonic motion of a unit mass attached to a spring with restoring force $-\theta$, where θ is the displacement from the equilibrium position at the time x. As such it is discussed in elementary calculus and mechanics courses; so what can go wrong?

Since the independent variable, x, is absent from (1.3) the familiar procedure is to put $v = \theta'$ and then

$$\theta'' = \frac{\mathrm{d}v}{\mathrm{d}x} = \frac{\mathrm{d}v}{\mathrm{d}\theta}\frac{\mathrm{d}\theta}{\mathrm{d}x} = v\frac{\mathrm{d}v}{\mathrm{d}\theta};$$

so that (1.3) becomes $v\dfrac{dv}{d\theta} + \theta = 0$, or

$$\frac{d}{d\theta}\left\{\frac{1}{2}v^2 + \frac{1}{2}\theta^2\right\} = 0;$$

that is, $\frac{1}{2}v^2 + \frac{1}{2}\theta^2 \equiv C$, where C is a constant.

Now $\frac{1}{2}v^2$ is the kinetic energy and $\frac{1}{2}\theta^2$ is the potential energy of the system and so we recover the familiar result that the total energy is constant (the sum is the energy integral). On inserting the initial conditions, namely $\theta = 0$ and $v = 1$ when $x = 0$, we obtain $C = 1/2$ and so

$$\left(\frac{d\theta}{dx}\right)^2 = 1 - \theta^2. \tag{1.4}$$

Clearly the solution already found for (1.3), $\theta(x) = \sin x$, satisfies (1.4), but, in passing from (1.3) to (1.4), we have picked up extra solutions. For example, if we write

$$\theta(x) = \begin{cases} \sin x, & x < \dfrac{\pi}{2}, \\[2mm] 1, & x \geq \dfrac{\pi}{2}, \end{cases} \tag{1.5}$$

then (1.5) is a C^1 solution[1] of (1.4) for all x, but not of (1.3), when $x > \frac{\pi}{2}$. Physically, the solution (1.5) corresponds to the linearized pendulum 'sticking' when it reaches maximum displacement at time $x = \pi/2$. (Of course we chose θ to be small and so the remark is not applicable to the pendulum problem, but it is relevant in what follows.)

We shall have to be aware of the pitfalls presented by that phenomenon, when we introduce the elliptic functions in terms of solutions of differential equations. It falls under the heading of 'singular solutions'; for a clear account see, for example, Agnew (1960), pp. 114–117.

Exercise 1.1

1.1.1   Re-write (1.4) in the form $v^2 = 1 - \theta^2$ and observe that $\frac{dv}{d\theta}$ is infinite at $\theta = \pm 1$.

---

[1] Here and in what follows we use the notation $f(x) \in C^k(I)$ to mean that f is a complex valued function of the real variable x defined on the open interval $I : a < x < b$ and having k continuous derivatives in I (with obvious variations for closed or half-open intervals). If $k = 0$, the function is continuous.

1.2   Existence and uniqueness of solutions

So far, we have not proved that Equation (1.2) has any solutions at all, physically obvious though that may be. We now address that question.

Set $\omega = \frac{d\theta}{dx}$ and re-write (1.2) as a first order autonomous[2] system

$$\frac{d\theta}{dx} = \omega \equiv f(\theta, \omega),$$

$$\frac{d\omega}{dx} = -\sin\theta \equiv g(\theta, \omega). \tag{1.6}$$

The functions $f(\theta, \omega)$, $g(\theta, \omega)$ are both in $C^\infty(\mathbb{R}^2)$ and the matrix

$$\begin{pmatrix} \dfrac{\partial f}{\partial \theta} & \dfrac{\partial g}{\partial \theta} \\[2mm] \dfrac{\partial f}{\partial \omega} & \dfrac{\partial g}{\partial \omega} \end{pmatrix} = \begin{pmatrix} 0 & -\cos\theta \\ 1 & 0 \end{pmatrix}$$

is bounded on $\mathbb{R}^2$.

We can now appeal to the theory of ordinary differential equations (see Coddington & Levinson (1955), pp. 15–32, or Coddington (1961)) to obtain the following.

Theorem 1.1 *The system (1.6) corresponding to the differential equation (1.2) has a solution $\theta = \theta(x)$, $(-\infty < x < \infty)$, such that $\theta(a) = A$ and $\theta'(a) = B$, where a, A and B are arbitrary real numbers. Moreover, that solution is unique on any interval containing a.*

Note that Equation (1.2) implies that the solution $\theta(x)$, whose existence is asserted in Theorem 1.1, is in $C^\infty(\mathbb{R})$.

The result of Theorem 1.1 is fundamental in what follows.

1.3   The energy integral

On multiplying Equation (1.1) by $ml^2 \frac{d\theta}{dt}$, we obtain

$$ml^2 \frac{d\theta}{dt}\frac{d^2\theta}{dt^2} + mgl\sin\theta \frac{d\theta}{dt} = 0,$$

that is

$$\frac{d}{dt}\left[\frac{1}{2}m\left(l\frac{d\theta}{dt}\right)^2 + mgl(1 - \cos\theta)\right] = 0.$$

---

[2] Equations (1.6) are the familiar way of writing a second order differential equation as a system of linear differential equations; the system is said to be autonomous when the functions $f(\theta, \omega)$, $g(\theta, \omega)$ do not depend explicitly on x (the time).

Hence,

$$\frac{1}{2}m\left(l\frac{d\theta}{dt}\right)^2 + mgl(1 - \cos\theta) = E, \tag{1.7}$$

where E is a constant. By referring to Figure 1.1, we see that the first term in (1.7) is the kinetic energy of the pendulum bob and the second term is the potential energy, measured from the equilibrium position, A. The energy required to raise the pendulum bob from the lowest position ($\theta = 0$) to the highest possible position ($\theta = \pi$) (though not necessarily attainable – that depends on the velocity at the lowest point) is $2mgl$. So we may write

$$E = k^2(2mgl), \quad k \geq 0. \tag{1.8}$$

Clearly, we can obtain any given $k \geq 0$ by an appropriate choice of the initial conditions. We now assume $0 < k < 1$ (oscillatory motion).

It will be helpful later to look at (1.8), on the assumption $0 < k < 1$, from a slightly different point of view, as follows (see Exercise 1.3.2)

Suppose that $v = v_0$ and $\theta = \theta_0$, when $t = 0$. Then we have

$$\frac{1}{2}\left(v_0^2 - v^2\right) = gl(1 - \cos\theta);$$

that is

$$v^2 = v_0^2 - 4gl\sin^2\frac{\theta}{2}.$$

On writing $v = l\dfrac{d\theta}{dt}$, $h^2 = \dfrac{g}{l}$, we obtain

$$\left(\frac{d\theta}{dt}\right)^2 = 4h^2\left(\frac{v_0^2}{4gl} - \sin^2\frac{\theta}{2}\right). \tag{1.9}$$

On comparing (1.7), (1.8) and (1.9), we see that $k^2 = v_0^2/(4gl)$.

Recall our assumption that $0 < k < 1$, that is that $v_0^2 < 4gl$; so that the bob never reaches the point given by $\theta = \pi$ and the motion is, accordingly, oscillatory (which is what one would expect of a pendulum). Comparison with (1.9) suggests that we write $k = v_0/(2\sqrt{(gl)}) = \sin\alpha/(2)$, where $0 < \alpha < \pi$, and re-introduce the normalized time variable $x = (g/l)^{1/2} t$, to obtain (cf. (1.9)).

$$\left(\frac{d\theta}{dx}\right)^2 = 4\left(k^2 - \sin^2\frac{\theta}{2}\right) = 4\left(\sin^2\frac{\alpha}{2} - \sin^2\frac{\theta}{2}\right). \tag{1.10}$$

It is clear that any solution of (1.2) corresponding to $k, 0 < k < 1$, is a solution of (1.10), but is the converse true? (In other words have we produced a situation similar to that given by (1.5)?)

One sees immediately that $\theta \equiv \pm\alpha + 2\pi n$ is a solution of (1.10), but not of (1.2). Are there other solutions of (1.10) analogous to the solution (1.5) of (1.3)?

Recall that $\theta''\left(\pi/2\right)$ does not exist in (1.5); so consider the $C^2(-\infty, \infty)$ solutions of (1.10). If we reverse the argument that led to (1.10), we obtain

$$(\theta'' + \sin\theta) \cdot \theta' = 0,$$

and so

$$\theta'' + \sin\theta = 0,$$

except possibly on the set Γ defined by

$$\Gamma = \{x \in \mathbb{R} | \theta'(x) = 0\}.$$

Now (1.10) implies $\theta'(x) = 0$ only when

$$\theta(x) = \pm\alpha + 2\pi n, \quad n \in \mathbb{Z}.$$

So there are two cases to consider:

Case 1: Γ contains interior points;
Case 2: Γ contains no interior points.

Theorem 1.2 *In Case 1, $\Gamma = \mathbb{R}$ and then $\theta = \pm\alpha + 2\pi n$. In Case 2, $\theta'' + \sin\theta = 0$ for all x.*

Proof Suppose that Case 1 holds and assume that $\Gamma \neq \mathbb{R}$. We argue indirectly.

By hypothesis, there exists an interval $[a, b]$, $(-\infty < a < b < +\infty)$, on which $\theta' = 0$ and a point, c, such that $\theta'(c) \neq 0$. Then either $c < a$ or $c > b$. Suppose that $c > b$ and write $d = \sup\{x \in \mathbb{R} | b \leq x, \theta'(s) = 0, a \leq s \leq x\}$. Then $d \leq c$ and $\theta'(d) = 0$ by the continuity of θ' (recall that we are considering $C^2(-\infty, +\infty)$ solutions). It follows that $d < c$ and $\theta'(x) = 0$ for $a \leq x \leq d$.

Moreover, given $\varepsilon > 0$, there exists $x \in [d, d + \varepsilon]$, with $\theta'(x) \neq 0$. It follows that there exists a sequence $\{x_n\}$, with $x_n + d \neq 0$ and $\theta'(x_n) \neq 0$ for every n. But then $\theta''(x_n) = -\sin\theta(x_n)$, whence, by the continuity of θ'',

$$\theta''(d) = \lim_{n\to\infty} \theta''(x_n) = -\lim_{n\to\infty} \sin\theta(x_n) = -\sin\theta(d) = -\sin(\pm\alpha) \neq 0.$$

But $\theta'(x) = 0$ for $a \leq x \leq d$ implies $\theta''(x) = 0$ for $a < x < d$, whence $\theta''(d) = 0$, by the continuity of θ''. So we have obtained the contradiction we sought.

The case $c < a$ is similar and can be reduced to the case $c > b$ by making the substitution $x \to -x$, under which both (1.2) and (1.10) are invariant.

The remaining statements when Case 1 holds are trivial.

Finally, in Case 2, we must show that if $a \in \Gamma$, then $\theta''(a) + \sin \theta(a) = 0$. Now since a is a boundary point of Γ, there exists a sequence $\{x_n\} \subset R - \Gamma$ such that $x_n \to a$. Since θ, θ'' are continuous, $\theta''(a) + \sin \theta(a) = \lim_{n \to \infty} \{\theta''(x_n) + \sin \theta(x_n)\} = 0$.

It follows that the only C^2 solutions of (1.10) which are not solutions of (1.2) are the singular solutions $\theta = \pm\alpha + 2\pi n$. Later, we shall construct a 'sticking solution' of (1.10), analogous to (1.5).

That completes the proof of Theorem 1.2.

We conclude this section with a result that is physically obvious.

Proposition 1.1 *Let θ be a solution of (1.10) such that $-\pi \leq \theta(a) \leq \pi$, for some a. Then $-\alpha \leq \theta(x) \leq \alpha$, for $-\infty < x < +\infty$.*

Proof With respect to the variable x, the energy equation (1.7) reads

$$\frac{1}{2}\left(\frac{d\theta}{dx}\right)^2 + (1 - \cos \theta) = \frac{E}{mgl} = 2k^2 = 2\sin^2\frac{\alpha}{2}. \tag{1.11}$$

Let $\theta(a) = A$. Then $2\sin^2(A/2) = 1 - \cos A \leq 2\sin^2(\alpha/2)$ and $-\pi \leq A \leq \pi$ together imply $-\alpha \leq A \leq \alpha$.

Suppose that $\theta(b) > \alpha$, for some b. Then, by the Intermediate Value Theorem, there exists c such that $a < c < b$ and $\alpha < \theta(c) < \pi$. But then (1.10) implies $\theta'(c)^2 < 0$ – a contradiction. Hence $\theta(x) \leq \alpha$ for all x. The proof that $\theta(x) \geq -\alpha$ for all x is similar and is left as an exercise.

The essential content of Proposition 1.1 is that, without loss of generality, we may and shall assume henceforth that all solutions θ of (1.10) satisfy $-\alpha \leq \theta(x) \leq \alpha$, for all x, since θ and $\theta + 2\pi n$ are simultaneously solutions of (1.2) and (1.10). Note that $0 < k \ll 1$ ('k is very much less than 1') implies $\alpha \ll 1$, whence $|\theta| \leq \alpha \ll 1$ and then (1.3) is a good approximation to (1.2).

Exercises 1.3

1.3.1 Show that the changes of variable

$$\omega = \frac{d\theta}{dt}, \quad \frac{d^2\theta}{dt^2} = \frac{d\omega}{dt} = \frac{d\omega}{d\theta}\frac{d\theta}{dt} = \omega\frac{d\omega}{d\theta}$$

applied to Equation (1.1) yield the energy integral (1.7).

1.3.2 Starting from the equation of energy for the simple pendulum, namely

$$\dot{\theta}^2 = -4g \sin^2 \frac{\theta}{2} + \text{constant},$$

suppose that when the pendulum bob is at its lowest point, the velocity v_0 satisfies

$$\frac{v_0^2}{2g} = \frac{l^2\dot{\theta}^2}{2g} = h.$$

Show that the energy equation is

$$l^2\dot{\theta}^2 = 2gh - 4gl \sin^2 \frac{\theta}{2}$$

and then write $y = \sin(\theta/2)$ to obtain the equation

$$\left(\frac{dy}{dt}\right)^2 = \frac{g}{l}(1 - y^2)\left(\frac{h}{2l} - y^2\right).$$

Suppose that the motion of the pendulum is oscillatory, that is $\dfrac{dy}{dt} = 0$ for some $y < 1$, whence $0 < h/2l < 1$. Write $h = 2lk^2$ and so obtain

$$\left(\frac{dy}{dt}\right)^2 = \frac{gk^2}{l}\left(1 - k^2\frac{y^2}{k^2}\right)\left(1 - \frac{y^2}{k^2}\right). \tag{1.12}$$

Replace y/k by y to obtain the Jacobi normal form (1.14), in Section 1.4, below, where the significance of this exercise will become apparent.

1.3.3 Suppose that the motion is of the circulatory type in which $h > 2l$ (so that the bob makes complete revolutions). If $2l = hk^2$ (so that the k for the oscillatory motion is replaced by $1/k$), and then, again $0 < k < 1$. Show that Equation (1.12) now reads

$$\left(\frac{dy}{dt}\right)^2 = \frac{g}{lk^2}(1 - y^2)(1 - k^2y^2).$$

1.4　The Euler and Jacobi normal equations

We have already exhibited Equation (1.10) in two different forms, and in this section we review all that and make some classical changes of variable (due originally to Euler and Jacobi) in the light of our earlier preview.

First we write (following Euler)

$$\phi = \arcsin\left(k^{-1}\sin\frac{\theta}{2}\right). \tag{1.13}$$

The map $\theta \mapsto \phi$ is a homeomorphism[3] of $[-\alpha, \alpha]$ onto $[-\pi/2, \pi/2]$ and a C^∞ – diffeomorphism of $(-\alpha, \alpha)$ onto $(-\pi/2, \pi/2)$. (Note that the requirement in the latter case, that the interval $(-\alpha, \alpha)$ be an *open* interval, is essential, since $\frac{d\phi}{d\theta}$ is meaningless when $\theta = \pm\alpha$ (and so $\phi = \pm\pi/2$)).

Now differentiate $\sin(\theta/2) = k \sin\phi$ with respect to x to obtain $\frac{1}{2} \cos\frac{\theta}{2} \cdot \frac{d\theta}{dx} = k \cos\phi \cdot \frac{d\phi}{dx}$, whence

$$\left(\frac{d\theta}{dx}\right)^2 = \frac{4k^2 \cos^2\phi}{\cos^2\frac{\theta}{2}} \left(\frac{d\phi}{dx}\right)^2 = \frac{4k^2(1 - \sin^2\phi)}{1 - \sin^2\frac{\theta}{2}} \left(\frac{d\phi}{dx}\right)^2.$$

Hence (1.10) becomes

$$\frac{4k^2(1 - \sin^2\phi)}{1 - k^2 \sin^2\phi} \left(\frac{d\phi}{dx}\right)^2 = 4k^2(1 - \sin^2\phi),$$

that is

$$\left(\frac{d\phi}{dx}\right)^2 = 1 - k^2 \sin^2\phi, \quad \left(-\frac{\pi}{2} < \phi < \frac{\pi}{2}\right). \tag{1.14}$$

Equation (1.14) is Euler's normal form; we shall see later that it remains valid when $\phi = \pm\pi/2$.

To obtain Jacobi's normal form, we can use the substitution (due to Jacobi)

$$y = \sin\phi = k^{-1} \sin\frac{\theta}{2}. \tag{1.15}$$

Then the increasing function $\theta \mapsto y$ is a C^∞ diffeomorphism of $[-\alpha, \alpha]$ onto $[-1, 1]$ and this time the end-points may be included. On differentiating (1.15) with respect to x, we obtain

$$\frac{dy}{dx} = \frac{k^{-1}}{2} \cos\frac{\theta}{2} \frac{d\theta}{dx}$$

and so

$$\left(\frac{dy}{dx}\right)^2 = \frac{k^{-2}}{4} \left(1 - \sin^2\frac{\theta}{2}\right) 4 \left(k^2 - \sin^2\frac{\theta}{2}\right)$$

$$= \left(1 - \sin^2\frac{\theta}{2}\right) \left(1 - k^{-2} \sin^2\frac{\theta}{2}\right).$$

---

[3] Recall that a *homeomorphism is a* one-to-one continuous map whose inverse exists throughout its range and a C^n-*diffeomorphism* is a bijective, n – times continuously differentiable map.

So we see that (1.10) becomes

$$\left(\frac{dy}{dx}\right)^2 = (1 - y^2)(1 - k^2 y^2), \quad -1 \le y \le 1, \tag{1.16}$$

which is *Jacobi's normal form*. (Compare with Exercise 1.3.2, and note that $k = (\sin \theta/2) y|_\alpha < 1$.)

1.5   The classical formal solutions of (1.14)

Denote by $\theta_0 = \theta_0(x|k)$ (the notation exhibits the dependence of θ_0 on k as well as on x) that solution of (1.2) such that $\theta_0(0) = 0$ and $\theta_0'(0) = 2k, 0 < k < 1$. Then in the notation of (1.8), we have $E/(mgl) = \frac{1}{2}(2k)^2 + 0 = 2k^2$ and so θ_0 satisfies (1.10) with the same k. So at time $t = 0$ the bob is in its lowest position and is moving counter-clockwise with velocity sufficient to ensure that $\theta = \alpha$ when $t = T/4$, where T is the period of the pendulum (the time required for a complete swing). All that is plausible on physical grounds; but we must give a proof.

The Euler substitution

$$\phi = \arcsin \left(k^{-1} \sin \frac{\theta_0}{2} \right)$$

yields $\phi = 0$ and $\frac{d\phi}{dx} = 1$ when $x = 0$. We shall try to solve (1.10) under those initial conditions.

In some neighbourhood of $x = 0$, we must take the positive square root in (1.14) to obtain

$$\frac{d\phi}{dx} = \sqrt{1 - k^2 \sin^2 \phi} \tag{1.17}$$

and we note that that is > 0, provided that x is sufficiently small. It follows that

$$\frac{dx}{d\phi} = \frac{1}{\sqrt{1 - k^2 \sin^2 \phi}}, \tag{1.18}$$

provided that ϕ is sufficiently small. The solution of (1.18) under the given initial conditions is

$$x = \int_0^\phi \frac{d\phi}{\sqrt{1 - k^2 \sin^2 \phi}}, \tag{1.19}$$

provided that ϕ (and so x) is sufficiently small for (1.17) and (1.18) to hold. (We have permitted ourselves a familiar 'abuse of notation' in using ϕ for the variable and for the upper limit of integration.)

Physical intuition leads us to suspect that (1.19) holds for $-K \leq x \leq K$, where

$$K = K(k) = \int_0^{\pi/2} \frac{d\phi}{\sqrt{1 - k^2 \sin^2 \phi}}, \tag{1.20}$$

and that K is the quarter period – the x-time the bob takes to go from $\phi = 0$ (at $x = 0$) to $\phi = \pi/2$. Note that $\lim_{k \to 0} K(k) = \pi/2$, the quarter period of the solution of the linearized Equation (1.3). Of course that requires proof, to which we shall return when we complete it in Section 1.6.

For the moment, we introduce some fundamental definitions.

The integral (1.19) is called an *elliptic integral of the first kind* and (1.20) is called a *complete integral of the first kind*. The number k is called the *modulus* and $k' = \sqrt{1 - k^2}$ is called the *complementary modulus*. Integrals of that kind were first studied by Euler and Legendre (further background will be given later when we look at the example of the *lemniscatic integrals*). Note that (1.19) expresses the pendulum *time* as a function of the *angle*, but we would like to have the angle as a function of the time. It was some fifty years after the time of Euler that the insight of Gauss and Abel led them to the idea of *inverting* the integral (1.19) to obtain the angle, ϕ, as a function of the time, x. (The situation is analogous to the inversion of the integral

$$x = \int_0^\phi \frac{du}{\sqrt{1 - u^2}},$$

which leads to $x = \arcsin \phi$ and thence $\phi = \sin x$. (See Bowman (1961), where the elliptic functions are introduced in outline in a manner analogous to that in which the trigonometric functions are introduced.)

Nomenclature. The adjective *elliptic* arises from the problem of the measurement of the perimeter (length of arc) of an ellipse. The coordinates of any point on the ellipse

$$\frac{x^2}{a^2} + \frac{y^2}{b^2} = 1$$

may be expressed in the form $x = a \cos t$, $y = b \sin t$, $(a \geq b)$, where t is the eccentric angle. The directed arc length, s, measured from the point where $t = \pi/2$, is given by the integral

$$s = a \int_0^u \sqrt{1 - k^2 \sin^2 u} \cdot du, \tag{1.21}$$

where $u = t - \pi/2$ and $k = (a^2 - b^2)^{1/2}/a$ is the eccentricity of the ellipse. Note that the term which was in the denominator in (1.19) is now in the numerator. Through a perversity of history, such integrals are now known as *elliptic integrals of the second kind*. There are three kinds – the *elliptic integrals of the third kind* are the integrals of the form

$$\int_0^\phi \frac{d\phi}{(1 + n\sin^2\phi)(1 - k^2\sin^2\phi)^{1/2}}, \tag{1.22}$$

which we shall encounter later; and we shall show that a very wide class of integrals involving square roots of cubic or quartic polynomials can be reduced to integrals involving those three forms.

Exercises 1.5

1.5.1 Show that the three types of integral:

$$\int_0^x \frac{dx}{\sqrt{1 - x^2}\sqrt{1 - k^2x^2}};$$

$$\int_0^x \frac{\sqrt{1 - k^2x^2}}{\sqrt{1 - x^2}}dx;$$

$$\int_0^x \frac{dx}{(1 + nx^2)\sqrt{1 - x^2}\sqrt{1 - k^2x^2}}$$

may be reduced to one of the three types above in (1.19), (1.21) and (1.22) by means of the substitution $x = \sin\phi$.

As noted at the end of Section 1.5. we shall see later that any integral of the form $\int R(x, \sqrt{X})dx$, where X is a cubic or quartic polynomial in x and where R is a rational function of x and $\sqrt{X}$, may be reduced to one of those standard forms, together with elementary integrals.

1.6 Rigorous solution of Equation (1.2)

Recall Equation (1.19):

$$x = x(\psi) = \int_0^\psi \frac{du}{\sqrt{1 - k^2\sin^2 u}}, \tag{1.23}$$

where $-\infty < \psi < \infty$, if $0 \le k < 1$, and $-\pi/2 < \psi < \pi/2$ if $k = 1$. We see that x is an odd, increasing function of ψ, with positive derivative

$$\frac{dx}{d\psi} = (1 - k^2\sin^2\psi)^{-1/2}.$$

It follows that we may invert[4] (1.19) to obtain ψ as an odd, increasing function of x, which we shall denote by $\psi = am(x), (-\infty < x < \infty)$, with positive derivative

$$\frac{d\psi}{dx} = (1 - k^2 \sin^2 \psi)^{1/2}.$$

As before, we write

$$K = K(k) = \int_0^{\frac{\pi}{2}} \frac{d\psi}{\sqrt{1 - k^2 \sin^2 \psi}}, \quad (0 \le k < 1).$$

Now write $y = \sin \psi$. Then

$$\frac{dy}{dx} = \cos \psi \cdot \frac{d\psi}{dx},$$

whence $y = 0$ and $y' = 1$, when $x = 0$. Moreover,

$$\left(\frac{dy}{dx}\right)^2 = \cos^2 \psi \cdot \left(\frac{dy}{dx}\right)^2 = (1 - \sin^2 \psi)(1 - k^2 \sin^2 \psi)$$

$$= (1 - y^2)(1 - k^2 y^2).$$

Clearly, $\psi = \psi(x) \in C^2(-\infty, \infty)$ and is not a constant and so Theorem 1.2 and the Uniqueness Theorem 1.1 imply that

$$y = \sin \phi = k^{-1} \sin \frac{\theta_0}{2}, \quad 0 < k < 1,$$

$\phi = \arcsin(\sin \psi)$, whence

$$\frac{d\phi}{dx} = \frac{\cos \psi}{\sqrt{1 - \sin^2 \psi}} \frac{d\psi}{dx} = \frac{\cos \psi}{|\cos \psi|} \sqrt{1 - k^2 \sin^2 \phi}. \tag{1.24}$$

We conclude that $\frac{d\phi}{dx}$ has a jump discontinuity at $x = \pm K$ (that is, at $\phi = \pm\pi/2$), but that implies that

$$\left(\frac{d\phi}{dx}\right)^2 = 1 - k^2 \sin^2 \phi,$$

even when $\phi = \pm\pi/2$.

In the interval $-K \le x \le K$, ψ increases from $-\pi/2$ to $\pi/2$ and so $y = \sin \psi$ increases from -1 to 1, and $\theta_0 = 2 \arcsin(ky)$ increases from $-\alpha$ to α. Moreover, $\frac{d\theta_0}{dx} > 0$ on the open interval $(-K, K)$. In the interval $K \le x \le 3K$, ψ increases from $\pi/2$ to $3\pi/2$, $y = \sin \psi$ decreases from 1 to -1, θ_0 decreases from α to $-\alpha$ and, on $(K, 3K)$, $\frac{d\theta_0}{dx} < 0$. Hence, in the interval

---

[4] See Hardy (1944), Section 110.

$-K \leq x \leq 3K$, θ_0 takes on every value $A \in [-\pi/2, \pi/2]$ twice, with opposite signs for $\frac{d\theta_0}{dx}$.

All that is physically obvious, perhaps, but it is precisely what we need to obtain Equation (1.2) completely for $0 < k < 1$ (cf. (1.10)).

Theorem 1.3 *Let $\theta = \theta(x)$ be any solution of (1.2) for $0 < k < 1$. Then there exist $n \in \mathbb{Z}$ and $a \in \mathbb{R}$ such that $\theta(x) = \theta_0(x + a) + 2\pi n$, $(-\infty < x < \infty)$.*

Proof Let $\theta(0) = A$ and $\theta'(0) = B$. Then $2k^2 = B^2/2 + (1 - \cos A)$ (see (1.19)). Without loss of generality, we may suppose that $-\pi \leq A \leq \pi$, whence $-\alpha \leq A \leq \alpha$, by Proposition 1.1. By the remarks immediately prior to the enunciation of the theorem there exists $a \in [-K, 3K]$ such that $\theta_0(a) = A$ and $B' = \theta_0'(a)$ has arbitrary sign. Then $2k^2 = B'^2/2 + (1 - \cos \theta)$, whence $B'^2 = B^2$ and we may choose $B' = B$. But then, $\theta(x)$ and $\theta_0(x + a)$ are two solutions of (1.2) satisfying the same initial conditions at $x = 0$ and so the Uniqueness Theorem, Theorem 1.1, implies that $\theta(x) = \theta_0(x + a)$, $-\infty < x < \infty$.

The case $k = 0$ is trivial: the bob stays at the lowest point. When $k > 1$, the energy of the bob is sufficient to carry it past the highest point of the circle and the pendulum rotates. When $k = 1$, the bob just reaches the top, but in infinite time.

To verify all that in the case when $k \geq 1$, set $y = \sin(\theta/2)$, $u = kx$ and transform the energy equation

$$\left(\frac{d\theta}{dx}\right)^2 = 4\left(k^2 - \sin^2\frac{\theta}{2}\right)$$

into

$$\left(\frac{dy}{du}\right)^2 = (1 - y^2)(1 - k^{-2}y^2). \tag{1.25}$$

When $k > 1$, $0 < k^{-1} < 1$ and we are back with Equation (1.16) (see Exercise 1.3.2, and Exercise 1.6.2 below). When $k = 1$, (1.19) has the solution $y = \tanh(x + a)$.

Exercises 1.6

1.6.1 Show that

$$\lim_{k \to 1^-} K(k) = \infty.$$

1.6.2 Show that the change of variable $y = \sin \psi, (-\pi/2 \leq \psi \leq \pi/2)$ transforms (1.19) and (1.20) into

$$x = \int_0^y \frac{dy}{\sqrt{1 - y^2}\sqrt{1 - k^2 y^2}} \qquad (1.26)$$

and

$$K = \int_0^1 \frac{dy}{\sqrt{(1 - y^2)}\sqrt{(1 - k^2 y^2)}}, \qquad (1.27)$$

respectively (cf. Exercise 1.5.1).

1.7 The Jacobian elliptic functions

We have seen that the function $x \mapsto \sin \psi$, introduced in Section 1.6, arises in the study of pendulum motion; so we are naturally led to the question: what can we say about $x \mapsto \cos \psi$? Again, in (1.17) to (1.19), we encountered the function $x \mapsto \sqrt{1 - k^2 \sin^2 \psi}$. We are accordingly led to the study of the three functions:

$$sn(x) = sn(x, k) = \sin \psi,$$

$$cn(x) = cn(x, k) = \cos \psi,$$

$$dn(x) = dn(x, k) = \sqrt{1 - k^2 \sin^2 \psi} = \sqrt{1 - k^2 sn^2 x}, \qquad (1.28)$$

where $0 \leq k \leq 1$ and ψ is defined in (1.23) for $-\infty < x < \infty$.

When $0 < k < 1$, the functions defined in (1.28) are the basic Jacobian elliptic functions.

For the degenerate cases $k = 0$ and $k = 1$, we recover the circular (trigonometric) functions and the hyperbolic functions, as follows.

When $k = 0$, we have $x = \psi$ and so

$$sn(x, 0) = \sin x,$$

$$cn(x, 0) = \cos x,$$

$$dn(x, 0) = 1, \quad K(0) = \frac{\pi}{2}. \qquad (1.29)$$

When $k = 1$ and $-\pi/2 < \psi < \pi/2$, we have

$$x = \int_0^\psi \sec \psi \, d\psi = \ln(\sec \psi + \tan \psi),$$

whence

$$\exp x = \sec \psi + \tan \psi = \frac{1 + \sin \psi}{\cos \psi}.$$

Moreover,

$$\exp 2x = \frac{(1 + \sin \psi)^2}{(1 - \sin^2 \psi)} = \frac{1 + \sin \psi}{1 - \sin \psi}.$$

On solving for $\sin \psi$, we obtain $\sin \psi = \tanh x$ and so, finally, since $\cos \psi = \sqrt{1 - \sin^2 \psi}$,

$$sn(x, 1) = \tanh x,$$
$$cn(x, 1) = \operatorname{sech} x,$$
$$dn(x, 1) = \operatorname{sech} x,$$
$$K(1) = \text{(definition)} \infty. \qquad (1.30)$$

Theorem 1.4 *Suppose that $0 \le k < 1$ and let the functions sn, cn, dn, be defined by (1.28). Then sn and cn have period 4K, and dn has the smaller period 2K.*

Proof We have

$$x + 2K = \int_0^\psi (1 - k^2 \sin^2 \psi)^{-1/2} \mathrm{d}\psi + \int_{-\pi/2}^{+\pi/2} (1 - k^2 \sin^2 \psi)^{-1/2} \, \mathrm{d}\psi$$

$$= \left\{ \int_0^\psi + \int_\psi^{\psi+\pi} \right\} (1 - k^2 \sin^2 \psi)^{-1/2} \mathrm{d}\psi$$

$$= \int_0^{\psi+\pi} (1 - k^2 \sin^2 \psi)^{-1/2} \, \mathrm{d}\psi,$$

using the fact that $\sin(\psi + \pi) = -\sin \psi$. Hence

$$sn(x + 2K) = \sin(\psi + \pi) = -\sin \psi = -sn(x).$$

Similarly

$$cn(x + 2K) = -cn(x).$$

So *sn* and *cn* each have period 4K. However,

$$dn(x) = (1 - k^2 sn^2 x)^{1/2}$$

has period 2K.

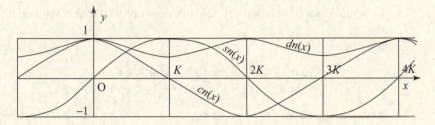

Figure 1.2 Graphs of *sn* (x), *cn* (x), *dn* (x) ($k^2 = 0.7$)

The following special values and identities are an immediate consequence of the definition (1.28).

$$cn(0) = 1, sn(0) = 0, dn(0) = 1;$$
$$cn(K) = 0, sn(K) = 1, dn(K) = \sqrt{1-k^2} = k';$$
$$cn^2 + sn^2 = 1, dn^2 + k^2 sn^2 = 1. \tag{1.31}$$

Note that *sn* is an odd function of *x* and *cn* and *dn* are even function of *x*. (Cayley (1895) described *sn* as 'a sort of sine-function' and *cn*, *dn* as 'sorts of cosine functions'.)

Since

$$\frac{d\psi}{dx} = \sqrt{1 - k^2 \sin^2 \psi} = dn(x),$$

an application of the chain rule to (1.28) yields

$$sn'(x) = cn(x)dn(x),$$
$$cn'(x) = -sn(x)dn(x),$$
$$dn'(x) = -k^2 sn(x)cn(x), \tag{1.32}$$

where the dash, ', denotes $\frac{d}{dx}$. It follows that *cn*, *sn* and *dn* all lie in $C^\infty(-\infty, \infty)$. Figure 1.2 illustrates the graphs of the three functions in the case when $k^2 = 0.7$.

Exercises 1.7

1.7.1 (The 'sticking solution'.) Write

$$y = \begin{cases} sn(x), & x \le K, \\ 1, & x > K. \end{cases}$$

Show that y is a C^1 solution of (1.16) that does *not* arise from a solution of (1.2).

1.7.2 Show that

$$(cn'x)^2 = (1 - cn^2x)(1 - k^2 + k^2cn^2x),$$
$$(dn'x)^2 = (1 - dn^2x)(k^2 - 1 + dn^2x).$$

1.7.3 From the equation $\sin(\theta/2) = k\,sn(x)$, $(-\pi < \theta < \pi)$, derive directly that $\theta''(x) + \sin\theta = 0$.

1.8 The imaginary period

The elliptic functions for $0 < k < 1$ have one very significant difference between them and the circular functions; as well as having a real period (like $4K$ in the case of sn) they also have a second, imaginary period, and so, as we shall see in Chapter 2, they are 'doubly periodic' (indeed, that is what is so special about them). To see why that is so, from the physical point of view, which has informed this chapter, we shall now consider the initial conditions

$$\theta = \alpha, \quad \frac{d\theta}{dx} = 0, \quad (x = 0).$$

We are then required to find the non-constant C^2 solutions of (1.16),

$$\left(\frac{dy}{dx}\right)^2 = (1 - y^2)(1 - k^2y^2), \quad 0 < k < 1, \tag{1.33}$$

satisfying the initial conditions $y = 1$ and $\frac{dy}{dx} = 0$, when $x = 0$. Our existence and uniqueness theorems imply that

$$y = sn(x + K, k).$$

Recall that $x = (g/l)^{1/2}t$. We now follow an ingenious trick ('a trick used twice becomes a method') due to Appell (1878) (see Whittaker (1937), Section 34, p. 47, for further details) and invert gravity (that is, replace g by $-g$). Then $x \to u = \pm ix$ and (1.16) becomes

$$\left(\frac{dy}{du}\right)^2 = (1 - y^2)(1 - k^2y^2), \tag{1.34}$$

together with the initial conditions $y = 1$ and $\frac{dy}{dx} = 0$, when $x = 0$. A *formal* solution is then given by $y = sn(\pm ix + K, k)$.

But we can also invert gravity in the real world by inverting Figure 1.1 to obtain Figure 1.3.

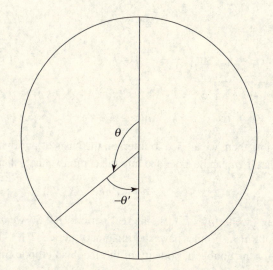

Figure 1.3 Inverting gravity – similar to Figure 1.1, but 'upside down'.

Whittaker (1937) proves the theorem that *in any dynamical system subjected to constraints independent of the time and to forces which depend only on the position of the particles, the integrals of the equations of motion are still real if t is replaced by $\sqrt{-1}t$ and the initial conditions $\beta_1, \ldots, \beta_n$, by $-\sqrt{-1}\beta_1, \ldots, -\sqrt{-1}\beta_n$, respectively; and the expressions thus obtained represent the motion which the same system would have if, with the same initial conditions, it were acted on by the same forces reversed in direction.*

When one inverts gravity, one makes the following *real* replacements: $\theta \to \theta^*$, where $\theta^* = \theta - \pi$. The initial condition $\theta = \alpha$ becomes $\theta^* = -\alpha^*$, where $\alpha^* = \pi - \alpha$ lies in $(0, \pi)$. Then $k = \sin(\alpha/2)$ becomes

$$k' = \sin \frac{\alpha^*}{2} = \sin \left(\frac{\pi}{2} - \frac{\alpha}{2} \right) = \cos \frac{\alpha}{2} = \sqrt{1 - \sin^2 \frac{\alpha}{2}} = \sqrt{1 - k^2} = k',$$

the complementary modulus.

The transformed real system accordingly satisfies the differential equation

$$\left(\frac{\mathrm{d}Y}{\mathrm{d}x} \right)^2 = (1 - Y^2)(1 - k'^2 Y^2) \tag{1.35}$$

and the initial conditions $Y = -1$ and $\frac{\mathrm{d}Y}{\mathrm{d}x} = 0$, when $x = 0$. Hence $Y = \sin \phi^* = sn(x - K', k')$, where

$$K' = K(k') = \int_0^{\pi/2} (1 - k'^2 \sin^2 \psi)^{-1/2} \, \mathrm{d}\psi.$$

But

$$y = k^{-1} \sin \frac{\theta}{2} = k^{-1} \sin \left(\frac{\theta^*}{2} + \frac{\pi}{2} \right) = k^{-1} \cos \frac{\theta^*}{2} = k^{-1} \sqrt{1 - \sin^2 \frac{\theta^*}{2}}$$

$$= k^{-1} \sqrt{1 - k'^2 \sin^2 \phi^*} = k^{-1} (1 - k'^2 sn^2 (x - K', k'))^{1/2}$$

$$= k^{-1} dn(x - K', k').$$

In conclusion, then, we are led to infer, on the basis of our dynamical considerations, that if sn can be extended into the complex plane, then

$$sn(\pm ix + K, k) = k^{-1} dn(x - K', k').$$

But, as we have seen, $dn(x, k')$ has the real period $2K'$. So we are led to the conjecture that $sn(x, k)$ must have the imaginary period $2iK'$. That, and its implications for the double periodicity of the Jacobian elliptic functions, will be the main concern of Chapter 2.

1.9 Miscellaneous exercises

1.9.1 Obtain the following expansions in ascending powers of u (the Maclaurin series):

(a) $sn(u) = u - \dfrac{(1 + k^2)u^3}{3!} + (1 + 14k^2 + k^4)\dfrac{u^5}{5!} - \cdots;$

(b) $cn(u) = 1 - \dfrac{u^2}{2!} + (1 + k^2)\dfrac{u^4}{4!} - \cdots;$

(c) $dn(u) = 1 - \dfrac{k^2 u^2}{2!} + k^2(4 + k^2)\dfrac{u^4}{4!} - \cdots.$

Use induction to try to find the order of the coefficient of u^n, for n odd and n even, when the coefficients are viewed as polynomials in k.

1.9.2 Prove that, if $k^2 < 1$, then the radius of convergence of the series in 1.9.1 is

$$K' = \int_0^1 \frac{dt}{\sqrt{(1 - t^2)(1 - k'^2 t^2)}}.$$

(Hint: use the following theorem from the theory of functions of a complex variable. The circle of convergence of the Maclaurin expansion of an analytic function passes through the singularity nearest to the origin. In the present case, that singularity is $u = iK'$, as we shall show in Chapter 2, when the definitions have been extended to complex values of u and the meromorphic character of the functions has been established.)

1.9.3 Prove that:

(i) $\dfrac{d}{du}(sn(u)cn(u)dn(u)) = 1 - 2(1+k^2)sn^2u + 3k^2sn^4u;$

(ii) $\dfrac{d}{du}\left(\dfrac{cn(u)dn(u)}{sn(u)}\right) = -\dfrac{1}{sn^2u} + k^2sn^2u;$

(iii) $\dfrac{d}{du}\dfrac{(sn(u))}{(cn(u)dn(u))} = \dfrac{1}{cn^2u} + \dfrac{1}{dn^2u} - 1;$

(iv) $\dfrac{d}{du}\left(\dfrac{cn(u)dn(u)}{1-sn(u)}\right) = k^2cn^2u + \dfrac{k'^2}{1-sn(u)};$

(v) $\dfrac{d}{du}\left(\dfrac{sn(u)cn(u)}{dn(u)-k'}\right) = cn^2u + \dfrac{k'}{dn(u)-k'}.$

1.9.4 Show that the functions $sn(u)$, $cn(u)$, $dn(u)$ satisfy the differential equations:

$$\frac{d^2y}{du^2} = -(1+k^2)y + 2k^2y^3,$$

$$\frac{d^2y}{du^2} = -(1-2k^2)y - 2k^2y^3,$$

$$\frac{d^2y}{du^2} = (2-k^2)y - 2y^3,$$

respectively. Hence obtain the first few terms of the Maclaurin expansion. (Hint: in the first equation substitute $y = u + a_3u^3 + a_5u^5 + \cdots$, with appropriate substitutions of a similar kind in the other two.)

1.9.5 Show that:

$$sn(u) = \frac{\sin\{u\sqrt{(1+k^2)}\}}{\sqrt{(1+k^2)}} + \alpha u^4;$$

$$cn(u) = \cos u + \beta u^3;$$

$$dn(u) = \cos(ku) + \gamma u^3;$$

where each of α, β, γ tends to 0 as $u \to 0$.

1.9.6 Prove that, if $0 < u < K$, then

$$\frac{1}{cn(u)} > \frac{u}{sn(u)} > 1 > dn(u) > cn(u).$$

1.9.7 (See McKean & Moll, 1997, p.57, and Chapter 2, Section 2.8 for a development of this question.) The problem of the division of the arc of a lemniscate into equal parts intrigued Gauss, Abel and Fagnano, who preceded them, and who first coined the phrase 'elliptic integral'. (See Chapter 9).

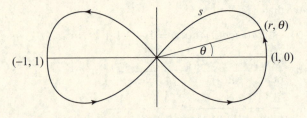

Figure 1.4 The lemniscate $r^2 = \cos 2\theta$.

Let Q, R be two fixed points in the plane and let $P(x, y)$ be a variable point such that the product of the distances $d(P, Q) \cdot d(P, R) = d$, where d is a constant. (The locus of P is called a *lemniscate; 'lemniscatus' is Latin for 'decorated with ribbons'*. See Figure 1.4.) Take $Q = \left(-1/\sqrt{2}, 0\right)$, $R = \left(+1/\sqrt{2}, 0\right)$ and $d = 1/2$. Let $r^2 = x^2 + y^2$ and so $x^2 = (r^2 + r^4)/2$, $y^2 = (r^2 - r^4)/2$. Show that the length of the arc of the lemniscate is

$$2 \int_0^1 \frac{dr}{\sqrt{1 - r^4}},$$

and verify that that is Jacobi's complete elliptic integral of the first kind, with modulus $k = \sqrt{-1}$.

1.9.8 (McKean & Moll, 1997, p. 60). Random walks in two and three dimensions.

Denote by $x_n (n = 0, 1, 2, \ldots)$ the nth position of a point in a random walk starting at the origin, $(0, 0)$ of the two dimensional lattice $\mathbb{Z}^2$. At the next step, the point is equally likely to move to any of the four neighbours of its present position. The probability that the point returns to the origin in n steps is 0 if n is odd (an even number of steps is required for a return) and if n is even ($2m$, say,) then a return requires an even number $2l(2m - 2l)$ of vertical (horizontal) steps, of which half must be up (or to the left) and half must be down (or to the right). So the probability that $x_n = 0$ is

$$P[x_n = 0] = \sum_{l=0}^m \binom{2m}{2l}\binom{2l}{l}\binom{2m - 2l}{m - l} 4^{-2m}$$

$$= 4^{-2m} \binom{2m}{m} \sum_{l=0}^m \binom{m}{l}^2 = \left[\frac{(2m - 1)(2m - 3)\cdots 3 \cdot 1}{2m \cdot (2m - 2)\cdots 4 \cdot 2}\right]^2,$$

where $\binom{n}{m}$ denotes the binomial coefficient $\frac{n!}{m!(n-m)!}$, and $\sum_{l=0}^{m} \binom{m}{l}^2 = \binom{2m}{m}$. It follows that

$$\sum_{n=0}^{\infty} P[x_n = 0]k^n = \frac{2}{\pi} \int_0^{\pi/2} (1 - k^2 \sin^2 \theta)^{-1/2} d\theta,$$

– a complete elliptical integral of the first kind, for $K(k)$.

(Hint: the formula can be checked by expanding the expression for K in a power series in the disc $|k| < 1$,

$$K(k) = \sum_{n=0}^{\infty} \binom{-1/2}{n} (-k^2)^n \int_0^{1/2} \sin^{2n} \theta d\theta,$$

and evaluating the integrals. $\binom{-1/2}{n} = (-1)^n \frac{(2n-1)(2n-3)\cdots 1}{n!}$. Recall too a formula related to Wallis's formula, $\int_0^{\pi/2} \sin^{2n} \theta \, d\theta = \frac{(2m-1)(2m-3)\cdots 1}{2m(2m-2)\cdots 2} \cdot \frac{\pi}{2}$.)

The three-dimensional random walk is similar, only now there are six neighbours for each point. If $p(<1)$ denotes the probability of a return to the origin at some time, then

$$(1-p)^{-1} = \int_0^1 \int_0^1 \int_0^1 [1 - 3[\cos 2\pi x_1 + \cos 2\pi x_2 + \cos 2\pi x_3]]^{-1} dx_1 dx_2 dx_3.$$

Polya (1921) and Watson (1939) evaluated the integral in terms of complete elliptic integrals $K(k)$ of modulus $k^2 = (2 - \sqrt{3})(\sqrt{3} - \sqrt{2})$ and found that

$$(1 - p)^{-1} = \frac{12}{\pi^2}(18 + 12\sqrt{2} - 10\sqrt{3} - \sqrt{6})K^2.$$

1.9.9 Show that the length of the arc of the curve whose equation is $y = b \sin(x/a)$, measured from the point where $x = 0$ to the point $x = a\phi$, is

$$\sqrt{a^2 + b^2} \int_0^{\phi} \sqrt{1 - k^2 \sin^2 \phi} \, d\phi$$

where $k^2 = b^2/(a^2 + b^2)$.

1.9.10 A heavy bead is projected from the lowest part of a smooth circular wire, fixed in a vertical plane; so that it describes one-third of the circumference before coming to rest. Prove that during the first half of the time it describes one-quarter of the circumference.

1.9.11 Show that the length of a pendulum which keeps seconds when swinging through an angle 2α is given by $l = g/(4k^2)$, $k = \sin(\alpha/2)$.

1.9.12 A pendulum swinging through an angle 2α makes N beats a day. If α is increased by $d\alpha$, show that the pendulum loses $(1/8)N \sin \alpha \, d\alpha$ beats a day, approximately.

2

Jacobian elliptic functions of a complex variable

2.1 Introduction: the extension problem

In this chapter we define the Jacobian elliptic functions and establish their basic properties; for the convenience of the reader a summary of those properties is included at the end of the chapter.

In the last section of Chapter 1 we sought to make clear the possibility that the Jacobian elliptic functions (arising out of our study of the simple pendulum and as defined in (1.22) of that chapter) are much more than routine generalizations of the circular functions. But to explore their suspected richness it is necessary to move off the real line into the complex plane.

Our treatment is based on Abel's original insight (1881), and is in three steps: (i) extension of the definitions, (1.28), to the imaginary axis; (ii) derivation of the addition formulae for the functions sn, cn and dn; (iii) formal extension to the complex plane by means of the addition formulae. The discussion in Section 2.6 then shows that the extended functions are doubly periodic.

What appears to be unusual (perhaps new) in our treatment is: step (iv), verification that the functions so extended are analytic except for poles (that is, meromorphic) in the finite $\mathbb{C}$ plane. (Of course, there are other ways of obtaining the functions sn, cn and dn as meromorphic functions of a complex variable, but none of them is easy. As Whittaker & Watson (1927), p. 492, write: 'unless the theory of the theta functions is assumed, it is exceedingly difficult to show that the integral formula

$$u = \int_0^y (1-t^2)^{-1/2}(1-k^2t^2)^{-1/2}\mathrm{d}t$$

defines y as a function of u, which is analytic except for simple poles. See Hancock (1958).)

2.2 The half-angle formulae

The functions defined by Abel admit of an obvious extension to the imaginary axis. However, we have chosen to work with the functions of Jacobi, one of whose methods of extending them to the imaginary axis was to start with a function, $\phi(u)$, like that in the Maclaurin expansion found in Exercises 1.9, (1); so that $sn(u) = \phi(u)$, where $\phi(u)$ is a real, odd function. But then it is plausible (by applying the formula twice) that $sn(iu) = i\phi(u)$ and one can start with that as a definition of $sn(iu)$. Another method is to use the integrals in Exercise 1.5.1 and to replace the upper limit y by iy (see Bowman (1961), Chapter IV for an outline, but note the quotation above from Whittaker and Watson – the method relies on Jacobi's 'imaginary transformation', which is treated later on and which is described in detail in Whittaker & Watson (1927), Chapter XXII). Copson (1935) treats the problem using the Weierstrass theory of σ-functions, but that is essentially the theta functions approach. Another interesting and unusual approach, based on properties of period-lattices, is to be found in Neville (1944).

None of those methods is easy to justify (and we prefer to defer our treatment of the theta functions to a later chapter); so we shall use the 'half-angle substitution' method, introduced by Weierstrass to integrate rational functions of sine and cosine.

To that end, set

$$t = \tan \frac{\psi}{2} \ (-\pi < \psi < \pi).$$

Then

$$\cos \psi = \frac{1-t^2}{1+t^2}, \quad \sin \psi = \frac{2t}{1+t^2}, \quad \frac{d\psi}{dt} = \frac{2}{1+t^2}.$$

It follows that

$$1 - k^2 \sin^2 \psi = \frac{Q(t,k)}{(1+t^2)^2},$$

where

$$Q(t,k) = 1 + 2(1 - 2k^2)t^2 + t^4. \tag{2.1}$$

Note that $Q(t,k) = [t^2 + (1 - 2k^2)]^2 + 4k^2(1 - k^2)$, whence $Q(t,k) > 0$, for $-\infty < t < +\infty$, provided that $0 \le k < 1$. If $k = 1$, then the result is true, provided that $-1 < t < 1$. Equations (1.20) and (1.19) of Chapter 1 become

$$K = K(k) = 2 \int_0^1 \frac{dt}{(Q(t,k))^{1/2}}, \tag{2.2}$$

$$x = 2 \int_0^t \frac{dt}{(Q(t, k))^{1/2}}, \quad (-2K < x < 2K), \qquad (2.3)$$

respectively, where $-\infty < t < \infty$ when $0 \le k < 1$, and where $-1 < t < 1$ when $k = 1$.

Note that

$$Q(t, 0) = (1 + t^2)^2, \quad Q(t, 1) = (1 - t^2)^2$$

and

$$Q\left(t, \frac{1}{\sqrt{2}}\right) = 1 + t^4.$$

On returning to the definitions in (1.28) in the case when u is real, we have, for $-2K < x < 2K$,

$$sn(x, k) = \sin \psi = \frac{2t}{1 + t^2},$$

$$cn(x, k) = \cos \psi = \frac{1 - t^2}{1 + t^2},$$

$$dn(x, k) = (1 - k^2 \sin^2 \psi)^{1/2} = \frac{(Q(t, k))^{1/2}}{1 + t^2}, \qquad (2.4)$$

where t is given as a function of x by inverting (2.3).

Note that the trigonometric functions are absent from our final formulae (except implicitly, in the half-angle formulae). Whether or not that is a good thing depends, to some extent, on one's familiarity with the rigorous development of the circular functions in analysis (see, for example, Apostol (1969) or, in the spirit of our present treatment, the outline in Bowman (1961). In fact, the elliptic functions may be introduced in other ways, which may also be used to introduce the circular functions; see Weil (1975).

In two earlier papers, one of us (Eberlein (1954) and (1966)) has shown how the half-angle approach to the circular functions offers an alternative to the usual treatments. We now sketch the extension of those ideas to the functions of Jacobi.

We start with (2.2) as the definition of K, taking K to be ∞ in the case when $k = 1$. Set $t = s^{-1}$ and then we see that

$$\int_1^\infty \frac{dt}{(Q(t, k))^{1/2}} = \int_0^1 \frac{ds}{(Q(s, k))^{1/2}}, \quad (0 \le k < 1).$$

Then (2.3) defines x as an odd, increasing function of t, with positive derivative

$$\frac{dx}{dt} = \frac{2}{(Q(t, k))^{1/2}}. \qquad (2.5)$$

We may then invert (2.3) to obtain t as an odd, increasing differentiable function of x, $(-2K < x < 2K)$. The formulae (2.4) then define sn, cn and dn for $-2K < x < 2K$. Note that sn is an odd function, and cn and dn are both even functions of x.

When $k = 1$, $K = \infty$ and the definition is complete.

When $k < 1$, we follow the discussion in Chapter 1, Section 1.4, and use

$$sn(\pm 2K) = \lim_{x \to \pm 2K} sn(x) = \lim_{t \to \pm \infty} \frac{2t}{1 + t^2} = 0,$$

$$cn(\pm 2K) = \lim_{x \to \pm 2K} cn(x) = \lim_{t \to \pm \infty} \frac{1 - t^2}{1 + t^2} = -1,$$

$$dn(\pm 2K) = \lim_{x \to \pm 2K} dn(x) = \lim_{t \to \pm \infty} \frac{(Q(t, k))^{\frac{1}{2}}}{(1 + t^2)} = 1. \tag{2.6}$$

Those formulae, originally defined for $-2K < x < 2K$, may be extended to all real x by periodicity, that is by the formulae:

$$sn(x + 4K) = sn\ x,$$

$$cn(x + 4K) = cn\ x,$$

$$dn(x + 4K) = dn\ x. \tag{2.7}$$

As we shall see in (2.23), below, the function dn actually has the smaller period, $2K$.

The formulae for differentiation follow from the chain rule in the range $-2K < x < 2K$; thus

$$\frac{d}{dx} sn\,(x) = \frac{\dfrac{d}{dt}\left(\dfrac{2t}{1 + t^2}\right)}{\dfrac{dx}{dt}} = \frac{1 - t^2}{1 + t^2} \frac{(Q(t, k))^{1/2}}{1 + t^2} = cn\,(x) \cdot dn\,(x). \tag{2.8}$$

From the periodicity condition (2.7), it follows that (2.8) holds for all x other than the odd multiples of $2K$, $k < 1$. The validity at those points follows from the following corollary of the mean value theorem.

Proposition 2.1 *Let $f(x)$ be continuous at $x = a$ and differentiable in a neighbourhood of a, with a deleted. Suppose that $A = \lim_{x \to a} f'(x)$ exists. Then $f'(a)$ exists and equals A.*

The treatment of cn' and dn' is similar.

Exercises 2.2

2.2.1 Prove the corollary to the mean value theorem quoted in Proposition 2.1.
2.2.2 Obtain the formulae for $cn'x$ and $dn'x$, using the method in (2.8).

2.3 Extension to the imaginary axis

We make the following changes of variable, which at this stage are purely formal (the justification for them will be given later). We write

$$x = iy, \quad t = is$$

and observe that $Q(is, k) = Q(s, k')$, where $k' = \sqrt{1 - k^2}$ is the complementary modulus. Thus (2.3) becomes (the following heuristic argument needs justification in due course)

$$y = 2 \int_0^s \frac{ds}{(Q(s, k'))^{1/2}}, \tag{2.9}$$

and, as suggested by that and by (2.2), we write

$$K' = K(k') = 2 \int_0^1 \frac{ds}{(Q(s, k'))^{1/2}}. \tag{2.10}$$

We are led accordingly to the following definitions. If $-2K' < y < 2K'$, define

$$sn(iy, k) = \frac{2t}{1 + t^2} = i\frac{2s}{1 - s^2} = i\frac{2s/(1 + s^2)}{(1 - s^2)/(1 + s^2)} = i\frac{sn(y, k')}{cn(y, k')},$$

$$cn(iy, k) = \frac{1 - t^2}{1 + t^2} = \frac{1 + s^2}{1 - s^2} = \left(\frac{1 - s^2}{1 + s^2}\right)^{-1} = \frac{1}{cn(y, k')},$$

$$dn(iy, k) = \frac{(Q(t, k))^{1/2}}{1 + t^2} = \frac{(Q(s, k'))^{1/2}}{1 - s^2}$$

$$= \frac{(Q(s, k'))^{1/2}/(1 + s^2)}{(1 - s^2)/(1 + s^2)} = \frac{dn(y, k')}{cn(y, k')}, \tag{2.11}$$

provided that $s \neq \pm 1 (y \neq \pm K')$.

The expression of the functions $sn(iy, k)$ etc. in terms of functions in which the argument is real is known as 'Jacobi's imaginary transformation'.

Now the last expression in each of the three formulae in (2.11) has a meaning for all $y \neq K' (\text{mod } 2K')$. So we can now employ them to define sn, cn and dn as functions defined on the imaginary axis and continuous, except for poles at the odd multiples of iK'. Moreover, on the imaginary axis $cn (z)$ and

dn (z) (as we may now denote them) have period $4iK'$, whilst sn (z) has period $2iK'$. That last result confirms the conjecture made at the end of Chapter 1, Section 1.8.

Now we shall show that the differentiation formulae, (1.32), remain valid for the functions in (2.11). When $t = is$, $(s \neq \pm 1)$ and $x = iy$, $(-2K' < y < 2K')$, $y \neq \pm K'$, we have

$$
\frac{d\,sn(iy, k)}{d(iy)} = \frac{\dfrac{d}{dt}\dfrac{2t}{1+t^2}}{\dfrac{dx}{dt}} = \frac{\dfrac{d}{dt}\dfrac{2t}{1+t^2}}{\dfrac{dy}{ds}}
$$

$$
= \frac{\dfrac{d}{dt}\left(\dfrac{2t}{1+t^2}\right)}{2/Q(s, k')^{1/2}} = \frac{\dfrac{d}{dt}\left(\dfrac{2t}{1+2t^2}\right)}{2/Q(t, k)^{1/2}}
$$

$$
= \frac{1-t^2}{1+t^2} \cdot \frac{Q(t, k)^{1/2}}{1+t^2}
$$

$$
= cn(iy, k)\, dn(iy, k).
$$

The validity of that formula for all $y \not\equiv K' (\mathrm{mod}\ 2K')$ then follows, using the periodicity and the corollary to the mean value theorem, Proposition 2.1. A similar argument shows that the differentiation formulae for cn and dn also remain valid in the pure imaginary case, when $x = iy$.

Exercises 2.3

2.3.1 Obtain the results for differentiating cn and dn.

2.4 The addition formulae

Euler and Legendre called the elliptic integrals 'elliptic functions' (rather like calling $\arcsin x = \sin^{-1} x$ a 'circular function'), and their addition formulae were derived by Euler in a dazzling *tour de force*. When $k = 0$ (as we have seen), that amounts to studying addition formulae for arcsin or arccos, instead of for sin and cos. (The treatment of the addition formula for arctan in Hardy (1944), pp. 435–438, is reminiscent of Euler's method.) Once one inverts the integrals, in the manner first proposed by Gauss and Abel, matters simplify.

There are many proofs of the addition formulae (see, for example, Bowman (1961), p. 12; Copson (1935), pp. 387–389; Lawden (1989), p. 33; and Whittaker & Watson (1927), pp. 494–498). We shall follow Abel's method (see Theorem 2.1, below), but first we shall try to make the forms of the addition formulae plausible in the general case, $0 \leq k \leq 1$, by interpolation between the elementary limiting cases, $k = 0$ (the circular functions) and $k = 1$ (the hyperbolic functions). In one sense, the latter afford a more reliable guide, since the imaginary poles are readily obtained.

We start on the real axis and we recall (see (1.29) and (1.30)) that

$$sn(u, 0) = \sin u, \quad sn(u, 1) = \tanh u,$$
$$cn(u, 0) = \cos u \quad cn(u, 1) = \operatorname{sech} u,$$
$$dn(u, 0) = 1, \quad dn(u, 1) = \operatorname{sech} u, \tag{2.12}$$

for $-\infty < u < \infty$.

Now

$$cn(u + v, 0) = \cos(u + v) = \cos u \cos v - \sin u \sin v$$
$$= cn(u, 0)cn(v, 0) - sn(u, 0)sn(v, 0) \tag{2.13}$$

and

$$cn(u + v, 1) = \operatorname{sech}(u + v) = \frac{1}{\cosh(u + v)}$$
$$= \frac{1}{\cosh u \cosh v + \sinh u \sinh v}.$$

To obtain only those functions appearing in (2.12), we multiply numerator and denominator in the last expression by sech u sech v to find

$$\operatorname{sech}(u + v) = \frac{\operatorname{sech} u \cdot \operatorname{sech} v}{1 + \tanh u \tanh v}. \tag{2.14}$$

Formula (2.14) is still unlike (2.13), but we can rewrite the latter as

$$\cos u \cos v - \sin u \sin v = \cos u \cos v(1 - \tan u \tan v).$$

That suggests that we multiply the numerator and denominator in (2.14) by $(1 - \tanh u \tanh v)$ and then obtain

$$\operatorname{sech}(u + v) = \frac{\operatorname{sech} u \ \operatorname{sech} v - \tanh u \tanh v \operatorname{sech} u \ \operatorname{sech} v}{1 - \tanh^2 u \tanh^2 v}. \tag{2.15}$$

Now we appeal to symmetry in u and v to write

$$cn(u + v, 1) = \frac{f(u, 1)f(v, 1) - sn(u, 1)sn(v, 1)g(u, 1)g(v, 1)}{1 - sn^2(u, 1)sn^2(v, 1)},$$

where f and g denote either cn or dn. On comparing that last equation with (2.13) we see that we should take $f = cn$ and $g = dn$ and we may then write

$$cn(u + v, 0) = cn(u, 0)cn(v, 0) - sn(u, 0)sn(v, 0)dn(u, 0)dn(v, 0)$$

$$cn(u + v, 1) = \frac{cn(u, 1)cn(v, 1) - sn(u, 1)sn(v, 1)dn(u, 1)dn(v, 1)}{1 - sn^2(u, 1)sn^2(v, 1)}$$

(2.16)

The equations in (2.16) lead us to the plausible conjecture that

$$cn(u + v) = \frac{cn(u)\, cn(v) - sn(u)\, sn(v)\, dn(u)\, dn(v)}{\Delta(u, v)}, \qquad (2.17)$$

where:

(a) $\Delta(u, v)$ is symmetric in u and v;
(b) $\Delta(u, v) = 1$ when $k = 0$;
(c) $\Delta(u, v) = 1 - sn^2u\, sn^2v$ when $k = 1$.

To obtain a further condition on Δ, put $v = -u$ in (2.17). Then $1 = cn(0) = (cn^2u + sn^2u\, dn^2u)/\Delta(u, -u)$. That is,

$$\Delta(u, -u) = cn^2u + sn^2u\, dn^2u = 1 - sn^2u + sn^2u(1 - k^2sn^2u)$$

and so finally

(d) $\Delta(u, -u) = 1 - k^2sn^4u$.

Clearly, one choice (and perhaps the simplest) for $\Delta(u, v)$ that satisfies (a), (b), (c) and (d) is

$$\Delta(u, v) = 1 - k^2\, sn^2u\, sn^2v. \qquad (2.18)$$

So our final, informed conjecture is

$$cn(u + v) = \frac{cn(u)\, cn(v) - sn(u)\, sn(v)\, dn(u)\, dn(v)}{1 - k^2sn^2u\, sn^2v} = C(u, v), \qquad (2.19)$$

say.
For brevity, put:

$$s_1 = sn\,(u),\ s_2 = sn\,(v),$$

$$c_1 = cn\,(u),\ c_2 = cn\,(v),$$

$$d_1 = dn\,(u),\ d_2 = dn\,(v),$$

$$\Delta = 1 - k^2s_1^2s_2^2.$$

We can now guess the addition formulae for sn and dn from the identities $cn^2 + sn^2 = 1$ and $dn^2 + k^2sn^2 = 1$ (see (1.31)).
We start with the conjecture in (2.19):

$$cn(u + v) = C(u, v) = (c_1c_2 - s_1s_2d_1d_2)/\Delta.$$

Then

$$sn^2(u+v) = 1 - cn^2(u+v) = 1 - C^2(u,v)$$
$$= \left[\left(1 - k^2 s_1^2 s_2^2\right)^2 - \left(c_1 s_2 - s_1 s_2 d_1 d_2\right)^2\right]/\Delta^2.$$

We use the result in Exercise 2.4.1 below to obtain (again this is at present a conjecture, a plausible inference from the cases $k = 0, k = 1$):

$$sn^2(u+v) = \left[\left(c_1^2 + s_1^2 d_2^2\right)\left(c_2^2 + s_2^2 d_1^2\right) - \left(c_1 c_2 - s_1 s_2 d_1 d_2\right)^2\right]/\Delta^2$$
$$= \left[s_1^2 c_2^2 d_2^2 + 2s_1 s_2 c_1 c_2 d_1 d_2 + s_2^2 c_1^2 d_1^2\right]/\Delta^2$$
$$= (s_1 c_2 d_2 + s_2 c_1 d_1)^2/\Delta^2.$$

So we have

$$sn(u+v) = \pm(s_1 c_2 d_2 + s_2 c_1 d_1)/\Delta.$$

We guess the sign by taking $v = 0$, and we see that we have to choose the '+'. So finally we have the conjecture

$$sn(u+v) = \frac{sn(u)\,sn(v)\,dn(v) + sn(v)\,cn(u)\,dn(u)}{\Delta(u,v)} = S(u,v). \qquad (2.20)$$

Having made the conjectures (2.19), (2.20), (2.21) (see below, Exercises 2.4.1 and 2.4.2), (in the spirit, we like to think of Polya (1954),) we shall formulate them as a theorem, which will be the central result of this chapter. Our proof follows that of Abel (see Bowman (1961), p.12).

Theorem 2.1 *Let $u, v \in (-\infty, +\infty)$. Then:*

(a) $sn(u+v)=S(u,v) \equiv [sn(u)\,cn(v)\,dn(v) + sn(v)\,cn(u)\,dn(u)]/\Delta(u,v)$,
(b) $cn(u+v)=C(u,v) \equiv [cn(u)\,cn(v) - sn(u)\,sn(v)\,dn(u)\,dn(v)]/\Delta(u,v)$,
(c) $dn(u+v)=D(u,v) \equiv [dn(u)\,dn(v) - k^2 sn(u)\,sn(v)\,cn(u)\,cn(v)]/\Delta(u,v)$,

where

$$\Delta(u,v) = 1 - k^2 sn^2 u\, sn^2 v.$$

Proof We shall use the abbreviated notation introduced above:

$$S = (s_1 c_2 d_2 + s_2 c_1 d_1)/\Delta.$$

Partial differentiation with respect to u yields

$$\Delta^2 \frac{\partial S}{\partial u} = \Delta\left[c_1 d_1 c_2 d_2 - s_1 s_2\left(d_1^2 + k^2 c_1^2\right)\right] + (s_1 c_2 d_2 + s_2 c_1 d_1)\left(2k^2 s_1 c_1 d_1 s_2^2\right)$$

$$= c_1 d_1 c_2 d_2\left(\Delta + 2k^2 s_1^2 s_2^2\right) - s_1 s_2\left[\Delta\left(d_1^2 + k^2 c_1^2\right) - 2k^2 s_2^2 c_1^2 d_1^2\right]$$

$$= c_1 d_1 c_2 d_2\left(\Delta + 2k^2 s_1^2 s_2^2\right) - s_1 s_2\left[d_1^2\left(\Delta - k^2 s_2^2 c_1^2\right) + k^2 c_1^2\left(\Delta - s_2^2 d_1^2\right)\right]$$

$$= c_1 d_1 c_2 d_2 \left(1 + k^2 s_1^2 s_2^2\right) - s_1 s_2 \left(d_1^2 d_2^2 + k^2 c_1^2 c_2^2\right)$$
$$= (c_1 c_2 - s_1 s_2 d_1 d_2)(d_1 d_2 - k^2 s_1 s_2 c_1 c_2).$$

Hence $\frac{\partial S}{\partial u} = CD$, which is symmetric in u, v. Since S is symmetric in u, v we conclude that

$$\frac{\partial S}{\partial v} = CD = \frac{\partial S}{\partial u}.$$

Let us now introduce new variables $z = u + v$, $w = u - v$ to transform that last equation into $\frac{\partial S}{\partial w} = 0$, whence $S = f(z) = f(u + v)$. Take $v = 0$. Then $f(u) = S(u, 0) = sn(u)$ and so $S = sn(u + v)$.

It is possible to use a similar argument to obtain (b) and (c) (left as an exercise!), but it is simpler to argue as follows. By Exercise 2.4.3 (see below), we have

$$cn^2(u + v) = 1 - sn^2(u + v) = 1 - S^2(u, v) = C^2(u, v).$$

Hence $cn(u + v) = \pm C(u, v)$. Now

$$CD = \frac{\partial S}{\partial u} = \frac{\partial}{\partial u} sn(u + v) = cn(u + v)dn(u + v).$$

We conclude that $dn(u + v) = \pm D(u, v)$ whenever $cn(u + v) \neq 0$. Now $dn > 0$ and $D > 0$ and so we must have $dn(u + v) = D(u, v)$ if $u + v \neq K \pmod{2K}$. On taking limits, we obtain $dn(u + v) = D(u, v)$ for all u, v. But then $CD = cn(u + v) \cdot D$, whence $C(u, v) = cn(u + v)$.

Remark It is tempting to try to determine the sign in $cn(u + v) = \pm C(u, v)$ by putting $v = 0$, whence $cn(u + 0) = C(u, 0)$. It is reasonable to guess that one should always take the positive sign, but the justification for that argument requires that the function cn is an analytic function (except for poles) and that will occupy us in the next section.

To see something of the difficulty, consider the following example. Let $f(x) = \exp(-1/x^2)$, $x \neq 0$, and $f(x) = 0$ when $x = 0$. Put $g(x) = f(x)$ when $x \leq 0$ and $g(x) = -f(x)$ when $x > 0$. Then f, $g \in C^\infty(-\infty, \infty)$, $f^2 = g^2$, $f(x) = g(x)$ for $x \leq 0$, but $f(x) \neq g(x)$ when $x > 0$.

Exercises 2.4

2.4.1 Show that

$$\Delta(u, v) = cn^2 u + sn^2 u \, dn^2 v = cn^2 v + sn^2 v \, dn^2 u$$
$$= dn^2 u + k^2 sn^2 u \, cn^2 v = dn^2 v + k^2 sn^2 v \, cn^2 u.$$

2.4.2 From the identity $dn^2u + k^2sn^2u = 1$, and by using a similar argument to that used for $sn(u + v)$, derive the conjecture

$$dn(u + v) = \frac{dn(u)\,dn(v) - k^2sn(u)\,sn(v)\,cn(u)\,cn(v)}{\Delta(u, v)} = D(u, v). \qquad (2.21)$$

Show that (given the foregoing)

$$C^2 + S^2 = 1$$

and that

$$D^2 + k^2S^2 = 1.$$

2.4.3 Show that (on the basis of the conjectures)

$$D(u, v) = \left[\left(1 - k^2s_1^2\right)^{1/2}\left(1 - k^2s_2^2\right)^{1/2} \pm k^2s_1s_2\left(1 - s_1^2\right)^{1/2}\left(1 - s_2^2\right)^{1/2}\right]/\Delta$$

and that $|s_j| \le 1, 0 \le k < 1$, and $|s_j| < 1$ when $k = 1$ ($j = 1, 2$). Conclude that $D(u, v) > 0$.

2.4.4 Obtain formulae for $sn(2u)$, $cn(2u)$, $dn(2u)$ in terms of $sn(u)$, $cn(u)$, $dn(u)$ and also sn^2u, cn^2u, dn^2u in terms of $sn(2u)$, $cn(2u)$, $dn(2u)$.

2.4.5 Show that

$$sn\frac{K}{2} = \frac{1}{\sqrt{(1 + k')}}, \quad cn\frac{K}{2} = \frac{\sqrt{k'}}{\sqrt{(1 + k')}}, \quad dn\frac{K}{2} = \sqrt{k'}.$$

2.4.6 Prove that

$$\frac{cn(u + v) - dn(u + v)}{sn(u + v)} = \frac{cn(u)dn(v) - dn(u)cn(v)}{sn(u) - sn(v)}.$$

2.4.7 Use the previous question to show that:

(i) if $u_1 + u_2 + u_3 + u_4 = 2K$, then

$$(cn(u_1)dn(u_2) - dn(u_1)cn(u_2))(cn(u_3)dn(u_4) - dn(u_3)cn(u_4))$$
$$= k'^2(sn(u_1) - sn(u_2))(sn(u_3) - sn(u_4)),$$

(ii) if $u_1 + u_2 + u_3 + u_4 = 0$, then

$$(cn(u_1)dn(u_2) - dn(u_1)cn(u_2))(sn(u_3) - sn(u_4))$$
$$+ (cn(u_3)dn(u_4) - dn(u_3)cn(u_4))(sn(u_1) - sn(u_2)) = 0.$$

2.5 Extension to the complex plane

In the course of the proof of Theorem 2.1, we proved that $\frac{\partial S}{\partial u} = CD = \frac{\partial S}{\partial v}$. Now $\frac{\partial C}{\partial u} = \frac{\partial}{\partial u}cn(u + v) = -sn(u + v)dn(u + v) = -SD$, whence, by symmetry, $\frac{\partial C}{\partial v} = -SD$. In a similar manner, we obtain $\frac{\partial D}{\partial u} = -k^2 SC = \frac{\partial D}{\partial v}$. It is convenient to summarize those results as

$$\frac{\partial S}{\partial u} = CD = \frac{\partial S}{\partial v},$$

$$\frac{\partial C}{\partial u} = -SD = \frac{\partial D}{\partial v},$$

$$\frac{\partial D}{\partial u} = -k^2 SC = \frac{\partial D}{\partial v}, \qquad (-\infty < u, v < +\infty). \qquad (2.22)$$

The first of the equations in (2.22) was derived by differentiating explicitly, and it is clear that the remaining ones can be derived in a similar manner. Since the differentiation formulae for sn, cn and dn remain valid on the imaginary axis, it follows that Equations (2.22) remain valid when u and v are pure imaginary, provided only that none of u, v and $u + v$ is of the form $(2n + 1)iK'(n \in \mathbb{Z})$ – to avoid the poles. But then the argument used in the proof of Theorem 2.1 (a) goes through again and so the addition formulae, (2.19), (2.20) and (2.21), are valid when u, v are pure imaginary, subject to the obvious restrictions, as above. Finally, it is clear that the terms in (2.22) are defined and the results remain valid when u is real and v is pure imaginary and not equal to $(2n + 1)iK'$, $n \in \mathbb{Z}$. So, following Abel, we are led to the possibility of defining $sn(x + iy)$, $cn(x + iy)$ and $dn(x + iy)$ as follows:

$$sn(x + iy) = S(x, iy) \qquad (-\infty < x, y < +\infty),$$
$$cn(x + iy) = C(x, iy),$$
$$dn(x + iy) = D(x, iy) \qquad (y \neq (2n + 1)K', n \in \mathbb{Z}). \qquad (2.23)$$

When $x = 0$ or $y = 0$, (2.23) reduces to our earlier definitions. In general, consider the possible definition

$$sn(x + iy) = S(x, iy)$$
$$= [sn(x)\, cn(iy)\, dn(iy) + sn(iy)\, cn(x)\, dn(x)]/(1 - k^2 sn^2 x\, sn^2 iy)$$
$$= \frac{\dfrac{sn(x,k)dn(y,k')}{cn^2(y,k')} + i\dfrac{sn(y,k')cn(x,k)dn(x,k)}{cn(y,k')}}{1 + k^2 sn^2(x,k)sn^2(y,k')/cn^2(y,k')}.$$
$$= \frac{sn(x,k)dn(y,k') + i\, sn(y,k')cn(y,k')cn(x,k)dn(x,k)}{cn^2(y,k') + k^2 sn^2(x,k)sn^2(y,k')}.$$

When $0 < k < 1$, that last expression is defined except when both $cn(y, k') = 0$ and $sn(x, k) = 0$. It is clear from Equations (2.4), (2.6) and (2.7) that that is equivalent to $x = 2mK$ and $y = (2n + 1)K'$. Consider then the open, connected set

$$\sum = \mathbb{C} - \{x + iy \in \mathbb{C} | x = 2mK, y = (2n + 1)K'\}. \tag{2.24}$$

Our discussion suggests that we should extend our definition of $sn(x + iy) = S(x, iy)$ to Σ in the following way. Let $0 < k < 1$ and $x + iy \in \Sigma$. Define

$$sn(x + iy) = S(x, iy)$$
$$= \frac{sn(x, k)dn(y, k') + i\, sn(y, k')cn(y, k')cn(x, k)dn(x, k)}{cn^2(y, k') + k^2sn^2(x, k)sn^2(y, k')}. \tag{2.25}$$

Clearly $f(x, y) = S(x, iy) \in C^1(\Sigma)$. By similar arguments we extend the definition of $cn(x + iy)$ and $dn(x + iy)$ to Σ as

$$cn(x + iy) = C(x, iy)$$
$$= \frac{cn(x, k)cn(y, k') - i\, sn(x, k)dn(x, k)sn(y, k')dn(y, k')}{cn^2(y, k') + k^2sn^2(x, k)sn^2(y, k')}, \tag{2.26}$$

$$dn(x + iy) = D(x, iy)$$
$$= \frac{dn(x, k)cn(y, k')dn(y, k') - ik^2sn(x, k)cn(x, k)sn(y, k')}{cn^2(y, k') + k^2sn^2(x, k)sn^2(y, k')}. \tag{2.27}$$

In (2.22) we saw that, when $y \neq (2n + 1)K'$,

$$\frac{\partial f(x, y)}{\partial x} = CD = \frac{1}{i}\frac{\partial}{\partial y}f(x, y).$$

Since all our functions are in $C^1(\Sigma)$, we may appeal to continuity to obtain

$$\frac{\partial}{\partial x}f(x, y) = CD = \frac{1}{i}\frac{\partial}{\partial y}f(x, y) \tag{2.28}$$

for $z = x + iy \in \Sigma$, where $f(x, y) = S(x, iy) = sn(x + iy)$. But (2.28) is a form of the Cauchy–Riemann equations (put $f(x, y) = u(x, y) + iv(x, y)$ and note that (2.28) implies $\frac{\partial u}{\partial x} = \frac{\partial v}{\partial y}, \frac{\partial u}{\partial y} = -\frac{\partial v}{\partial x}$) and so $sn\, z = sn(x + iy)$ is analytic in Σ. Similar arguments applied to cn and dn then yield the central result of this chapter, as follows.

Theorem 2.2 *The functions $sn\,(z)$, $cn\,(z)$, $dn\,(z)$ defined by (2.25), (2.26,) (2.27), where $z = x + iy$, are analytic in Σ and for $z \in \Sigma$ we have:*

$$\frac{d}{dz} sn\,(z) = cn\,(z)\,dn\,(z),$$

$$\frac{d}{dz} cn\,(z) = -sn\,(z)\,dn\,(z),$$

$$\frac{d}{dz} dn\,(z) = -k^2 sn\,(z)\,dn\,(z). \tag{2.29}$$

We have not yet established the nature of the exceptional points, $z = 2mK + i(2n + 1)K'$, $m, n \in \mathbb{Z}$, but we expect them to be simple poles, because of (2.11).

Our next theorem shows that the addition formulae (2.19), (2.20), (2.21) hold for all complex numbers u, v, $u + v$ in Σ. That will enable us to apply those formulae to obtain results like

$$sn(u + K + iK') = k^{-1}dn(u)/cn(u)$$

(see below), by appealing to the addition formulae.

Theorem 2.3 *The addition formulae (2.19), (2.20), (2.21) remain valid for $u, v \in \mathbb{C}$, provided that u, v and $u + v$ are in Σ.*

Proof As in the proof of Theorem 2.1, the result

$$\frac{\partial}{\partial u} S(u, v) = C(u, v)D(u, v) = \frac{\partial}{\partial v} S(u, v)$$

implies that $S(u, v) = sn(u + v)$. The results $C(u, v) = cn(u + v)$ and $D(u, v) = dn(u + v)$ follow similarly.

It is possible to prove Theorem 2.3 by appealing to a fundamental theorem of complex analysis, to which we shall frequently have recourse in the sequel and which we state here as:

Theorem 2.4 *(The Uniqueness Theorem of Complex Analysis).*

Let $f(z)$, $g(z)$ be analytic (holomorphic) in a domain G (that is, a connected open subset of $\mathbb{C}$) and suppose that $f(z) = g(z)$ in a set of points H having a limit point in G. Then $f(z) = g(z)$ identically in G.

For the proof and further background, we refer the reader to Nevanlinna & Paatero (1969), pp. 139–140.

Exercises 2.5

2.5.1 Show that:

$$sn(K + iK') = k^{-1},$$
$$cn(K + iK') = -ik'k^{-1},$$
$$dn(K + iK') = 0. \tag{2.30}$$

2.5.2 Prove Theorem 2.3 using Theorem 2.4.

2.5.3 Show that the formulae (2.11), $sn(iy, k) = i\, sn(y, k')/cn(y, k')$, etc., remain valid for $y \in \mathbb{C}$.

2.6 Periodic properties associated with K, $K + iK'$ and iK'

In this and the following sections we follow Whittaker and Watson (1927) pp. 500–505. However, it should be emphasized that those authors base their treatment on the theta functions (which we postpone to a later chapter), whereas our treatment is based essentially on that of Abel and on the addition formulae.

We recall that $sn\,(K) = 1$, $cn\,(K) = 0$ and $dn\,(K) = k'$ (see 1.37) and we take $v = K$ in the addition theorem, Theorem 2.1, as extended to $\mathbb{C}$ in Theorem 2.3, to obtain

$$sn(u + K) = cn\,(u)/dn\,(u),$$
$$cn(u + K) = -k'sn\,(u)/dn\,(u), (u \in \mathbb{C}), \tag{2.31}$$
$$dn(u + K) = k'/dn\,(u).$$

Iteration of the result in (2.31) yields

$$sn(u + 2K) = -sn\,(u)$$
$$cn(u + 2K) = -cn\,(u)$$
$$dn(u + 2K) = dn\,(u)$$

and finally

$$sn(u + 4K) = sn\,(u)$$
$$cn(u + 4K) = cn\,(u)$$
$$dn(u + 4K) = dn(u + 2K) = dn\,(u);$$

from which we see that sn and cn each have period $4K$, whereas dn has the smaller period $2K$.

Note that those last results follow from the 'Uniqueness Theorem', Theorem 2.4, and the results we already have for $u \in \mathbb{R}$ in (2.7)

We turn to periodicity in $K + iK'$. We take $v = K + iK'$, $u \in \mathbb{C}$, in Theorems 2.1 and 2.3 and use the results of (2.30) to obtain:

$$sn(u + K + iK') = k^{-1}dn\,(u)/cn\,(u),$$
$$cn(u + K + iK') = -ik'k^{-1}/cn\,(u),$$
$$dn(u + K + iK') = ik'sn\,(u)/cn\,(u). \qquad (2.32)$$

As before, iteration yields:

$$sn(u + 2K + 2iK') = -sn\,(u),$$
$$cn(u + 2K + 2iK') = cn\,(u),$$
$$dn(u + 2K + 2iK') = -dn\,(u), \qquad (2.33)$$

and, finally:

$$sn(u + 4K + 4iK') = sn\,(u),$$
$$cn(u + 4K + 4iK') = cn\,(u),$$
$$dn(u + 4K + 4iK') = dn\,(u). \qquad (2.34)$$

Hence the functions $sn\,(u)$ and $dn\,(u)$ have period $4K + 4iK'$, but $cn\,(u)$ has the smaller period $2K + 2iK'$.

We turn finally to periodicity with respect to iK'.

By the addition theorem, again, we have:

$$sn(u + iK') = sn(u - K + K + iK')$$
$$= k^{-1}dn(u - K)/cn(u - K)$$
$$= k^{-1}/sn(u), \qquad (2.35)$$

by (2.31). Similarly:

$$cn(u + iK') = -ik^{-1}dn\,(u)/sn\,(u)$$
$$dn(u + iK') = -i\,cn\,(u)/sn\,(u) \qquad (2.36)$$

and, as before, iteration yields:

$$sn(u + 2iK') = sn\,(u),$$
$$cn(u + 2iK') = -cn\,(u),$$
$$dn(n + 2iK') = -dn\,(u). \qquad (2.37)$$

Finally:

$$sn(u + 4iK') = sn(u + 2iK') = sn\,(u),$$
$$cn(u + 4iK') = cn\,(u),$$
$$dn(n + 4iK') = dn\,(u), \qquad (2.38)$$

and we see that the functions $cn\,(u)$ and $dn(u)$ have period $4\mathrm{i}K'$, while $sn\,(u)$ has the smaller period $2\mathrm{i}K'$.

Once again the Uniqueness Theorem 2.3 may be used to obtain those results.

Theorem 2.5 *The results of this section, together with (2.12), yield the identity conjectured on the basis of dynamical considerations in Section 1.8, namely:*

$$sn(\pm \mathrm{i}x + K, k) = 1/dn(x, k'),$$

and

$$k^{-1}dn(x - K', k') = 1/dn(x, k').$$

Proof We have

$$sn(\pm \mathrm{i}x + K, k) = cn(\pm \mathrm{i}x, k)/dn(\pm \mathrm{i}x, k)$$
$$= \frac{cn(\pm x, k')^{-1}cn(\pm x, k')}{dn(\pm x, k')}$$
$$= 1/dn(x, k').$$

Again

$$k^{-1}dn(x - K', k') = k^{-1}dn(-x + K', k')$$
$$= k^{-1}k/dn(-x, k')$$
$$= 1/dn(x, k').$$

The points $mK + n\mathrm{i}K'$, $m, n \in \mathbb{Z}$, form a pattern of rectangles in the complex plane, called a *lattice*. The lattice is generated as a $\mathbb{Z}$-module by the basis vectors $(K, \mathrm{i}K')$. We have seen that the sub-lattice of points of the form $z = 4mK + 4n\mathrm{i}K'$ is a period lattice for the functions sn, cn and dn and that each of the functions has a different sub-lattice of periods, given by the scheme:

$$
\begin{array}{lll}
sn\,(u) & 4K & 2\mathrm{i}K' \\
cn\,(u) & 4K & 2K + 2\mathrm{i}K' \\
dn\,(u) & 2K & 4\mathrm{i}K' \qquad\qquad\qquad (2.39)
\end{array}
$$

The fundamental parallelogram of each of those sub-lattices has area a half that of the lattice with basis $4K, 4\mathrm{i}K'$.

The function $cn\,(u)$ for example, takes the same value, $cn\,(u_0)$, at all points $u \in \Gamma$ congruent, modulo its period lattice, to the point u_0 interior to the parallelogram generated by $(4K, 2K + 2\mathrm{i}K')$. A similar result holds for each of the functions sn and dn, modulo its period parallelogram.

We have not yet described the behaviour of the Jacobian elliptic functions at points congruent to iK' (though it will come as no surprise to learn that they have poles there), but we shall see later that the properties of the Jacobian elliptic functions derived here are characteristic of elliptic functions in general and that the notion of a period lattice plays a fundamental role in their definition and the derivation of their properties.

In his book *The White Knight*[1], A. L. Taylor suggests that, in *Through the Looking Glass*, Lewis Carroll had the double periodicity of the Jacobian elliptic functions in mind when he pictured the chess-board world that Alice found there – a sort of two-dimensional time represented in terms of a (square, $K = K'$) period lattice, in which space and time have been interchanged.

Exercises 2.6

2.6.1 Prove the periodicity properties with respect to K and iK', using Theorem 2.4.

2.6.2 Obtain the formulae:
$$sn(u + 2mK + 2niK') = (-1)^m sn(u)$$
$$cn(u + 2mK + 2niK') = (-1)^{m+n} cn(u)$$
$$dn(u + 2mK + 2niK') = (-1)^n dn(u)$$

2.6.3 Prove that:
$$sn\frac{K}{2} = \frac{1}{\sqrt{1+k'}}, \quad sn\frac{iK'}{2} = \frac{i}{\sqrt{k}},$$
$$cn\frac{K}{2} = \frac{\sqrt{k'}}{\sqrt{(1+k')}}, \quad cn\frac{iK'}{2} = \frac{\sqrt{1+k}}{\sqrt{k}},$$
$$dn\frac{K}{2} = \sqrt{k'}, \quad dn\frac{iK'}{2} = \sqrt{(1+k)},$$

and that:
$$sn\frac{K+iK'}{2} = (\sqrt{1+k} + i\sqrt{1-k})/\sqrt{(2k)},$$
$$cn\frac{(K+iK')}{2} = (1-i)\sqrt{k'}/\sqrt{(2k)},$$
$$dn\frac{K+iK'}{2} = \sqrt{(kk')}\{\sqrt{1+k'} - i\sqrt{1-k'}\}/\sqrt{2k}.$$

[1] Edinburgh, Oliver and Boyd, 1952, pp. 89–93.

2.7 The behaviour of the Jacobian elliptic functions near the origin and near iK'

We have, for $u \in \Sigma$,

$$\frac{d}{du} sn(u) = cn(u)\, dn(u)$$

and

$$\frac{d^3}{du^3} sn(u) = 4k^2 sn^2(u)\, cn(u)\, dn(u) - cn(u)\, dn(u)(dn^2(u) + k^2 cn^2(u)).$$

Since $sn(u)$ is an odd function of u, Maclaurin's Theorem (see Exercises 1.9, for the case when $u \in \mathbb{R}$) yields, for small values[2] of $|u|$,

$$sn(u) = u - \frac{1}{6}(1 + k^2)u^3 + O(u^5). \tag{2.40}$$

In like manner (recalling that cn, dn are even functions)

$$cn(u) = 1 - \frac{1}{2}u^2 + O(u^4), \tag{2.41}$$

$$dn(u) = 1 - \frac{1}{2}k^2 u^2 + O(u^4). \tag{2.42}$$

The identification of *all* the zeros of sn and of those of cn and dn is given in Exercise 2.7.1.

It follows, using (2.36), that:

$$sn(u + iK') = k^{-1}/sn(u)$$

$$= \frac{1}{ku}\left\{1 - \frac{1}{6}(1 + k^2)u^2 + O(u^4)\right\}^{-1}$$

$$= \frac{1}{ku} + \frac{1 + k^2}{6k}u + O(u^3). \tag{2.43}$$

A similar argument yields:

$$cn(u + iK') = +\frac{i}{ku} + \frac{2k^2 - 1}{6k}iu + O(u^3), \tag{2.44}$$

$$dn(u + iK') = -\frac{i}{u} + \frac{2 - k^2}{6}iu + O(u^3). \tag{2.45}$$

It follows that at the point iK' the functions $sn(u), cn(u), dn(u)$ have simple poles with residues $k^{-1}, -ik^{-1}, -i$, respectively.

---

[2] The symbol O denotes 'of the order of' and $f = O(g)$ as $x \to x_0$ means that $|f(x)/g(x)| < K$ as $x \to x_0$, where K is independent of x for x close enough to x_0. Similarly $f = o(g)$ means that $\lim_{x \to u} \frac{f(x)}{g(x)} = 0$. The notation is due to Bachmann and Landau, see Whittaker and Watson (1927) p. 11.

Note that

$$sn(u + 2K + iK') = -sn(u + iK') = -\frac{1}{ku} + O(u)$$

and so, at the point $2K + iK'$, sn has a simple pole with residue $-k^{-1}$. Similarly,

$$cn(u + 2K + iK') = -cn(u + iK'),$$

whence, at $2K + iK'$, cn has a simple pole with residue ik^{-1}. Finally,

$$dn(u + 3iK') = dn(u + iK' + 2iK') = -dn(u + iK'),$$

whence, at the point $3iK'$, dn has a simple pole with residue i.

The change in sign of the residues shows that the sum of the residues at the poles in the corresponding period parallelogram is 0 and is a special case of a general theorem about the poles of doubly periodic functions in general that will be proved in Chapter 3.

Exercises 2.7

2.7.1 (*The zeros of sn, cn, dn*). Show that (2.25) implies that $sn(x + iy) = 0$ if and only if *simultaneously* $sn(x, k) = 0$ and $sn(y, k') = 0$; that is, if and only if $x + iy \equiv 0 (\mathrm{mod}\ 2K, 2iK')$. So that identifies all the zeros of sn, modulo periods.

　　(That the zeros are simple follows from

$$\frac{d}{du} sn(u)|_{u=0} = cn\,(0)\,dn\,(0) = 1 \neq 0$$

and the periodicity properties proved above.)

　　Use a similar argument to show that the zeros of cn and dn are simple and occur at the following points:

$$cn\,(z) = 0 \quad \text{if and only if } z \equiv K(\mathrm{mod}\ 2K, 2iK')$$
$$dn\,(z) = 0 \quad \text{if and only if } z \equiv K + iK'(\mathrm{mod}\ 2K, 2iK').$$

2.7.2 Define the parameter q by

$$q = e^{\pi i\tau}, \quad \tau = \frac{iK'}{K}, \quad K > 0, \quad K' > 0.$$

　　(Note that $\mathrm{Im}(\tau) > 0$.) Verify that the mappings $k \to K$ and $k \to K'$ are increasing and decreasing, respectively and hence show that the mapping $k \to q$ is a one-to-one and strictly increasing mapping of the open interval $(0, 1)$ onto $(0, 1)$. Prove that

$$\lim_{k \to 0+} q = 0, \ \lim_{k \to 1-} q = 1.$$

(The significance of the parameter q will emerge in Section 2.9.)

General description of the functions *sn, cn, dn*

The investigations of this section may be summarized, as follows (u denotes a complex variable).

1. The function *sn(u)* is a doubly periodic function of the complex variable u with periods $4K, 2iK'$. It is analytic (holomorphic) except at the points congruent to iK' and $2K + iK' \cdot (\mathrm{mod}\, 4K, 2iK')$. Those excluded points are simple poles of *sn(u)*, the residues at the first set all being k^{-1} and the residues at the second set $-k^{-1}$. The function has a simple zero at all points congruent to $0(\mathrm{mod}\, 2K, 2iK')$

   If $0 < k^2 < 1$, then K and K' are real and *sn(u)* is real for $u \in \mathbb{R}$ and is pure imaginary when u is pure imaginary.

   (We shall see later that *sn(u)* is a unique function in the sense that it is the only function satisfying the foregoing description. For if $F(u)$ were a second such function then $sn(u) - F(u)$ would be a doubly periodic function without singularities, and we shall show in Chapter 3 that such a function must be a constant, and that constant must be 0.)

2. The function *cn(u)* is a doubly periodic function of u with periods $4K$ and $2K + 2iK'$. It is analytic except at points congruent to iK' or to $2K + iK'(\mathrm{mod}\, 4K, 2K + 2iK')$, where it has simple poles with residues $-ik^{-1}$ (at the first set) and $+ik^{-1}$, (at the second set), respectively. The function *cn* has a simple zero at all points congruent to $K(\mathrm{mod}\, 2K, 2iK')$.

3. The function *dn (u)* is a doubly periodic function of u with periods $2K$ and $4iK'$. It is analytic at all points except those congruent to iK' or to $3iK' \cdot (\mathrm{mod}\, 2K, 4iK')$ where it has simple poles. The residues at the first set are $-i$ and at the second set $+i$. The function has a simple zero at all points congruent to $K + iK'(\mathrm{mod}\, 2K, 2iK')$

The functions *cn* and *dn*, like *sn*, are unique with respect to the doubly periodic functions having those properties, respectively. See Figures 2.1, 2.2 and 2.3.

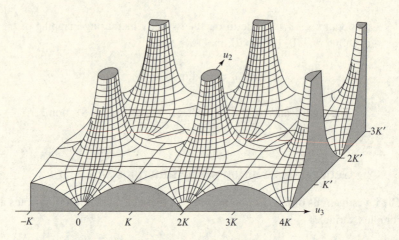

Figure 2.1 Relief of $sn\,(u)$ for $k = 0.8$ $(u = u_1 + iu_2)$

Figure 2.2 Relief of $cn\,(u)$ for $k = 0.8$ $(u = u_1 + iu_2)$

2.8 Glaisher's notation, the lemniscatic integral and the lemniscate functions

As in the case of the circular functions $(k = 0)$ it is convenient to have a notation for the reciprocals and quotients of our three functions, as will already have occurred to the reader in reading formulae like (2.32). The standard notation is due to Glaisher (1881) (who also introduced the notation s_1, c_1, d_1 etc. used above, see Glaisher, 1881). One denotes the reciprocals by reversing the order of the letters, used to express the function concerned, thus:

$$ns(u) = \frac{1}{sn(u)}, \quad nc(u) = \frac{1}{cn(u)}, \quad nd(u) = \frac{1}{dn(u)}.$$

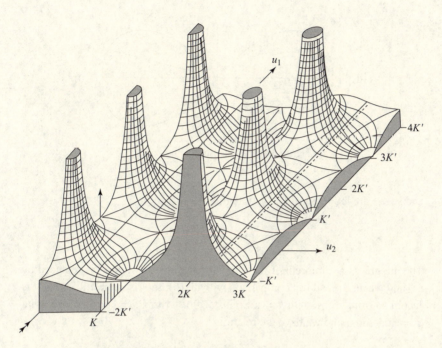

Figure 2.3 Relief of $dn\,(u)$ for $k = 0.8$ ($u = u_1 + iu_2$)

The quotients are denoted by writing, in order, the first letter of the function in the numerator and of the function in the denominator, thus:

$$sc(u) = \frac{sn(u)}{cn(u)}, \quad sd(u) = \frac{sn(u)}{dn(u)}, \quad cd(u) = \frac{cn(u)}{dn(u)},$$

$$cs(u) = \frac{cn(u)}{sn(u)}, \quad ds(u) = \frac{dn(u)}{sn(u)}, \quad dc(u) = \frac{dn(u)}{cn(u)}.$$

We return to the problem of the rectification of the arc of the lemniscate, considered in Miscellaneous Exercises 1.9.7.

The equation of the curve introduced there may also be written in terms of polar coordinates (r, θ) in the form

$$r^2 = \cos 2\theta$$

(more generally, $r^2 = a^2 \cos 2\theta$, where the fixed points are now $Q(-a, 0)$, $R(a, 0)$; so that the distance between them is $2a$ and we take the constant represented by the product of the distance $d(P, Q)d(P, R)$ to be a).

Let s denote the arc length from $\theta = \frac{\pi}{2}$ to θ (see Figure 2.4). Then $ds^2 = dr^2 + r^2 d\theta^2$ becomes

$$ds^2 = \frac{dr^2}{(1 - r^4)},$$

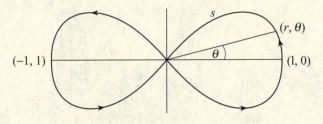

Figure 2.4 The lemniscate, $r^2 = \cos 2\theta$

and so we are led to consider the integral

$$\phi = \int_0^r (1 - r^4)^{-1/2} dr.$$

The integral is reminiscent of the integral for arcsin r (when r^4 is replaced by r^2) and we also recall (on comparison with (1.16), see Exercise 1.8.7) that it is Jacobi's normal form with $k = i$. So instead of writing $x = \arcsin \phi$, we write $x = \sin \operatorname{lemn} \phi$ and we have the relation

$$\sin \operatorname{lemn} \phi = \cos \operatorname{lemn} \left(\frac{1}{2} - \varphi \right)$$

obtained from

$$\varphi = \int_x^1 (1 - r^4)^{-1/2} dr, \qquad \frac{1}{2}\omega = \int_0^1 (1 - r^4)^{-1/2} dr. \qquad (2.46)$$

Those *lemniscate functions* were the first defined by the inversion of an integral[3] and they may be represented in terms of elliptic functions. As already noted they lead to the integral $\int_0^r (1 - r^4)^{-1/2} dr$, which is Jacobi's normal form with $k = i$, but they may also be expressed in terms of Jacobi functions with modulus $1/\sqrt{2}$, as follows.

In Exercise 2.8.1, take $k = 1/\sqrt{2}$ (also $k' = 1/\sqrt{2}$) to obtain

$$x = \int_0^{sd(x,1/\sqrt{2})} \left(1 - \frac{1}{4}t^4 \right)^{-1/2} dt$$

$$= \sqrt{2} \int_0^{sd(x,1/\sqrt{2})/\sqrt{2}} (1 - r^4)^{-1/2} dr$$

[3] Gauss (1799) *Werke* **III** p. 404. The idea of investigating the lemniscate functions by inversion of the integral occurred to Gauss on January 8th 1797.

on putting $t = r/\sqrt{2}$. Now let $x = K(1/\sqrt{2})$ and $u = r^4$ to obtain[4]:

$$K\left(\frac{1}{\sqrt{2}}\right) = \sqrt{2}\int_0^1 (1-r^4)^{-1/2}dr$$

$$= \frac{1}{\sqrt{2^3}}\int_0^1 u^{-3/4}(1-u)^{-1/2}du$$

$$= 2^{-\frac{3}{2}}\Gamma\left(\frac{1}{4}\right)\Gamma\left(\frac{1}{2}\right)\Big/\Gamma\left(\frac{3}{4}\right)$$

$$= \frac{1}{4}\pi^{-1/2}\Gamma\left(\frac{1}{4}\right)^2. \tag{2.47}$$

The formula of Exercise 2.8.1 also shows that

$$\sin \operatorname{lemn} \phi = 2^{-1/2}sd(\phi\sqrt{2}, 1/\sqrt{2})$$

and similarly

$$\cos \operatorname{lemn} \phi = cn(\phi\sqrt{2}, 1/\sqrt{2}).$$

The number $(1/2)\omega$ in (2.46) is the smallest positive value of ϕ for which

$$cn(\phi\sqrt{2}, 1/\sqrt{2}) = 0;$$

so that $\omega = \sqrt{2}K_0$, the number K_0 being the value of K associated with the modulus $k = \frac{1}{\sqrt{2}} = k'$. The result in (2.47) expresses K_0 in terms of the gamma function. Note that $K_0 = K_0'$ and so $q_0 = e^{\pi \mathrm{ii}\frac{K_0'}{K_0}} = e^{-\pi}$.

The ideas of this section played an important role in the development of the theory of elliptic functions and especially in the inversion of elliptic integrals; for further details see Siegel (1969), Chapter 1.

Exercises 2.8

2.8.1 Show that:

(a) $x = \displaystyle\int_0^{sd(x)} (1-k'^2t^2)^{-1/2}(1+k^2t^2)^{-1/2}dt;$

(Hint: how would you show that $\displaystyle\int_0^{\sin x} (1-t^2)^{-1/2}dt = x?$)

---

[4] We have used the definition of the beta function

$$B(\rho, \varepsilon) = \frac{\Gamma(\rho)\Gamma(\varepsilon)}{\Gamma(\rho+\varepsilon)} = \int_0^1 x^{\rho-1}(1-x)^{\varepsilon-1}dx$$

and the result $\Gamma(z)\Gamma(1-z) = \pi/\sin \pi z$ from the theory of the gamma functions. See Whittaker & Watson (1927) p. 253 or Bateman *et al.* (1953).

(b) $\displaystyle x = \int_0^{sc(x)} (1+t^2)^{-1/2}(1+k'^2t^2)^{-1/2}dt$

(c) $\displaystyle x = \int_{cs(x)}^{\infty} (t^2+1)^{-1/2}(t^2+k'^2)^{-1/2}dt$

(d) $\displaystyle x = \int_{cd(x)}^{1} (1-t^2)^{-1/2}(1-k^2t^2)^{-1/2}dt$

(e) $\displaystyle x = \int_{ns(x)}^{\infty} (t^2-1)^{-1/2}(t^2-k^2)^{-1/2}dt$

(f) $\displaystyle x = \int_1^{nd(x)} (t^2-1)^{-1/2}(1-k'^2t^2)^{-1/2}dt.$

2.8.2 Fagnano's[5] substitution $t \mapsto r = \left[2t^2/(1+t^4)\right]^{1/2}$ (cf. (2.6)) effects a monotone mapping of the interval $0 \le t \le 1$ on to the interval $0 \le r \le 1$. Show that

$$\frac{dr}{(1-r^4)^{1/2}} = 2^{1/2}\frac{dt}{(1+t^4)^{1/2}}.$$

Hence

$$K\left(\frac{1}{\sqrt{2}}\right) = 2\int_0^1 Q\left(t, \frac{1}{\sqrt{2}}\right)^{-1/2} dt = 2\int_0^1 (1+t^4)^{-1/2}dt$$

$$= 2^{1/2}\int_0^1 (1-r^4)^{-1/2}dr$$

(as in (2.47)).

2.8.3 Show that

$$\sin \mathrm{lemn}^2\phi = \frac{1 - \cos \mathrm{lemn}^2\phi}{1 + \cos \mathrm{lemn}^2\phi}.$$

2.8.4 Show that the length of one loop of the lemniscate $r^2 = a^2 \sin 2\theta$ is

$$\int_0^a \frac{2a^2 dr}{\sqrt{(a^4 - r^4)}} = a\sqrt{2}K\left(\frac{1}{\sqrt{2}}\right).$$

[5] Il Marchese Giulio Carlo de Toschi di Fagnano, another of the founders of the theory, who, like Maclaurin and Legendre, encountered elliptic integrals in connection with the problem of rectifying an arc of an ellipse. See also Chapter 9.

If s denotes the length of the arc measured from the pole to the point (r, θ), $0 < \theta < \frac{1}{4}\pi$, show that

$$r = a\, cn(K - s\sqrt{2}/a)\left(k = \frac{1}{\sqrt{2}}\right).$$

2.9 Fourier series and sums of two squares

Consider a function $f(z)$ of period 2π analytic in the strip $a < \mathrm{Im}(z) < b$ $(-\infty \le a < b \le +\infty)$. The mapping $z \mapsto \zeta = e^{iz}$ maps the strip in the z plane into the annulus $e^{-b} < |\zeta| < e^{-a}$ in the ζ-plane. Since $f(z)$ has period 2π, $F(\zeta) = f(z)$ is well defined and analytic in the annulus and so it possesses a Laurent expansion there (see, for example Whittaker & Watson (1927), Section 5.6, p. 100), namely:

$$F(\zeta) = \sum_{n=-\infty}^{\infty} C_n \zeta^n,$$

where

$$C_n = \frac{1}{2\pi i} \int_{|\zeta|=r} F(\zeta)\zeta^{-n-1}\mathrm{d}\zeta, \, (e^{-b} < r < e^{-a}).$$

By changing the variable back to z, the coefficients C_n give

$$f(z) = \sum_{n=-\infty}^{\infty} C_n e^{inz}, \, (a < \mathrm{Im}(z) < b) \tag{2.48}$$

where

$$C_n = \frac{1}{2\pi} \int_{z_0}^{z_0+2\pi} f(z)e^{-inz}\mathrm{d}z, \tag{2.49}$$

z_0 being any point of the strip $a < \mathrm{Im}(z) < b$ and the integration is along any path from z_0 to $z_0 + 2\pi$ that remains entirely in the strip. When $f(z)$ is an entire function (that is $a = -\infty$, $b = +\infty$) the Fourier development given by (2.48) is valid everywhere.

As functions of $z = \left(\frac{\pi}{2K}\right)u$, the functions $sn\,(u)$, $cn\,(u)$ and $dn\,(u)$ are analytic in the strip

$$|\mathrm{Im}(z)| < \left(\frac{\pi}{2K}\right)K' = \frac{\pi}{2}\mathrm{Im}(\tau),$$

where $\tau = iK'/K$, and have period π or 2π (depending on the periodicity properties of $sn\,(u)$, $cn\,(u)$ and $dn\,(u)$ with respect to K).

The function $sn\,(u)$ is an odd function of u and so has a Fourier sine series

$$sn(u) = \sum_{n=1}^{\infty} b_n \sin nx, \quad x \in \mathbb{R},$$

where

$$\pi i b_n = \int_{-\pi}^{\pi} sn(u) \cdot \exp(nix) dx.$$

Similarly, $cn(u)$ and $dn\,(u)$ are even functions and possess Fourier cosine expansions. For reasons which will become apparent and which are dictated by elegant applications to the theory of numbers, we choose to derive the series for $dn\,(u)$. (The other series are given as exercises.)

We recall that $dn\,(u)$ is an even function of $z = \left(\frac{\pi}{2K}\right)u$, with periods π and $2\pi\tau$, where $\tau = iK'/K$, and that it has a pole at $z = \pi\tau/2$ with residue $-i\pi/(2K)$. Hence we may write

$$dn(u) = \sum_{n=-\infty}^{\infty} a_n e^{2niz} = \frac{b_0}{2} + \sum_{n=1}^{\infty} b_n \cos 2nz,$$

where $a_{-n} = a_n$ and

$$b_n = 2a_n = \pi^{-1} \int_{-\pi}^{\pi} e^{-2niz} dn(u)\, dz. \tag{2.50}$$

To evaluate the integrals in (2.50) we appeal to the residue theorem and we consider the integral

$$\int_C \exp(-2niz) \cdot dn(u) \cdot dz,$$

where C denotes the boundary of the parallelogram shown in Figure 2.5.

Inside C, the integrand has simple poles at $\pi\tau/2$ and at $-\pi + \pi\tau/2$, the residue at each (in terms of the variable z) being $-i\left(\frac{\pi}{2K}\right)e^{-ni\pi\tau} = -i\left(\frac{\pi}{2K}\right)q^{-n}$ (using $q = e^{\pi i\tau}$). Hence

$$\int_C e^{-2niz} \cdot dn(u) \cdot dz = 2\pi i(\text{sum of the residues}) = \left(\frac{2\pi^2}{K}\right)q^{-n}.$$

The integrals along the left- and right-hand boundaries cancel by periodicity. For the upper boundary, write $z = t - \pi + \pi\tau$ (t goes from π to $-\pi$)

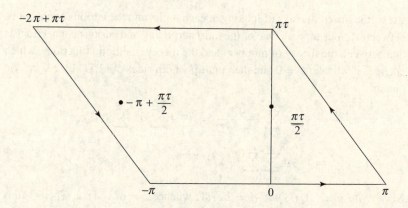

Figure 2.5 The contour C.

to obtain

$$-\int_{-2\pi+\pi\tau}^{\pi\tau} e^{-2niz} \cdot dn(u) \cdot dz = e^{-2ni\pi\tau} \int_{-\pi}^{\pi} e^{-2nit} \cdot dn\left(\frac{2K}{\pi}t\right) dt$$

$$= q^{-2n} \int_{-\pi}^{\pi} e^{-2niz} \cdot dn(u) \cdot dz.$$

Hence

$$(1+q^{-2n}) \int_{-\pi}^{\pi} e^{-2niz} dn(u) \, dz$$

$$= \int_{C} e^{-2niz} dn(u) \, dz = \left(\frac{2\pi^2}{K}\right) q^{-n}.$$

But now we have obtained

$$b_n = \frac{1}{\pi} \int_{-\pi}^{\pi} e^{-2niz} dn(u) \, dz$$

$$= \frac{2\pi}{K} \frac{q^{-n}}{1+q^{-2n}} = \frac{2\pi}{K} \frac{q^n}{1+q^{2n}}.$$

So finally we have

$$dn(u) = \frac{\pi}{2K} + \frac{2\pi}{K} \sum_{n=1}^{\infty} \frac{q^n}{1+q^{2n}} \cos 2nz, \tag{2.51}$$

in the strip $|\mathrm{Im}(z)| < \frac{\pi}{2} \mathrm{Im}(\tau)$.

In his Presidential address to the London Branch of the Mathematical Association (*Mathematical Gazette*, **XXXIV**, 1950), Davenport urged teachers to stress 'the different ways of looking at each topic, and the curious connections between various topics'. One of the most surprising and curious connections is that between the theory of numbers and the theory of elliptic functions, which emerges if we take $u = 0$ and then multiply both sides of (2.51) by $2K/\pi$. We obtain

$$\frac{2K}{\pi} = 1 + 4 \sum_{m=1}^{\infty} \frac{q^m}{1 + q^{2m}} = 1 + 4 \sum_{m=1}^{\infty} q^m \sum_{l=0}^{\infty} (-1)^l q^{2ml}$$

$$= 1 + 4 \sum_{m=1}^{\infty} \sum_{l=0}^{\infty} (-1)^l q^{(2l+1)m}.$$

Now write $n = (2l + 1)m$; $d = 2l + 1$, whence $l = (d - 1)/2$. Then d runs through the odd divisors of n and we have

$$\frac{2K}{\pi} = 1 + 4 \sum_{n=1}^{\infty} \left[\sum_{\substack{d|n \\ d \text{ odd}}} (-1)^{\frac{d-1}{2}} \right] q^n.$$

Now define

$$\chi(m) = \begin{cases} 0 & \text{if} \quad m \quad \text{is even} \\ (-1)^{(m-1)/2} & \text{if} \quad m \quad \text{is odd} \end{cases}$$

and write

$$\delta(n) = \sum_{d|n} \chi(d).$$

Clearly, $\chi(m) = 1$ if m is odd and $\equiv 1 \pmod 4$ and $\chi(m) = -1$ if m is odd and $\equiv 3 \pmod 4$. Whence

$$\delta(n) = [\text{number of divisors of } n \equiv 1 \pmod 4]$$
$$- [\text{number of divisors of } n \equiv 3 \pmod 4].$$

So finally we obtain

$$\frac{2K}{\pi} = 1 + 4 \sum_{n=1}^{\infty} \left[\sum_{d|n} \chi(d) \right] q^n$$

$$= 1 + 4 \sum_{n=1}^{\infty} \delta(n) q^n. \tag{2.52}$$

Denote by $r(n)$ the number of representations of n in the form

$$n = a^2 + b^2 \quad (a, b \in \mathbb{Z}),$$

where we count as distinct even those representations which differ only in respect of the sign or the order of the integers a, b. (Thus $r(5) = 8$, since

$5 = (\pm 1)^2 + (\pm 2)^2 = (\pm 2)^2 + (\pm 1)^2)$. In Chapter 4 we shall prove Jacobi's Theorem

$$\frac{2K}{\pi} = 1 + \sum_{n=1}^{\infty} r(n)q^n, \tag{2.53}$$

from which (together with Exercise 2.7.2 and (2.52)) we obtain:

Theorem 2.6 *Let n be an integer greater than or equal to 1, then $r(n) = 4\delta(n)$.*

Theorem 2.6 is Theorem 278 in Hardy and Wright (1979). It was first proved by Jacobi using elliptic functions, but it is equivalent to one stated by Gauss (see Hardy and Wright (1979), p. 243).

Corollary 2.1 *Every prime p of the form $4m + 1$ is representable as the sum of two squares in an essentially unique way[6].*

That a prime p of the form $4m + 3$ is never the sum of two squares also follows, but that result is trivial, since the square of an integer is congruent to 0 or 1(mod 4) and so the sum of two squares cannot be congruent to 3, mod 4.

Exercises 2.9

2.9.1 Show, using a similar argument to that used to obtain (2.51), that in the same strip

$$sn(u) = \frac{2\pi}{Kk} \sum_{n=0}^{\infty} \frac{q^{n+1/2} \sin(2n+1)z}{1 - q^{2n+1}}, \tag{2.54}$$

$$cn(u) = \frac{2\pi}{Kk} \sum_{n=0}^{\infty} \frac{q^{n+1/2} \cos(2n+1)z}{1 + q^{2n-1}}. \tag{2.55}$$

(Hint: the details of the calculation are given in pp. 510–511 of Whittaker and Watson (1927)).

2.9.2 By writing $z + \pi/2$ in place of z in the results obtained (including Exercise 2.9.1) show that, if

$$u = \frac{2Kz}{\pi}, \quad |\text{Im}(z)| < \frac{\pi}{2}\text{Im}(\tau),$$

---

[6] See the earlier remark; 'essentially unique' means that we regard representations differing only in the signs of a or b or in the order of a and b as the same.

then

$$cd(u) = \frac{2\pi}{Kk} \sum_{n=0}^{\infty} (-1)^n \frac{q^{n+1/2} \cos(2n+1)z}{1 - q^{2n+1}},$$

$$sd(u) = \frac{2\pi}{Kkk'} \sum_{n=0}^{\infty} (-1)^n \frac{q^{n+1/2} \sin(2n+1)z}{1 + q^{2n+1}},$$

$$nd(u) = \frac{\pi}{2Kk'} + \frac{2\pi}{Kk'} \sum_{n=1}^{\infty} (-1)^n \frac{q^n \cos 2nz}{1 + q^{2n}}.$$

2.9.3 Verify that $r(n) = 4\delta(n)$ for $n = 1, 2, \ldots, 10$, by explicit calculations.

2.9.4 Interchange the order of summation in the expression obtained from (2.51) to obtain

$$\frac{2K}{\pi} = 1 + 4 \sum_{l=0}^{\infty} (-1)^l \frac{q^{2l+1}}{1 - q^{2l+1}}.$$

2.10 Summary of basic properties of the Jacobian elliptic functions

The numbers in parentheses below indicate the appropriate references in the text.

Basic notation Let the modulus k, $0 < k < 1$, be given and let $\psi = 2 \arctan t$

(2.1) $Q(t, k) = 1 + 2(1 - 2k^2)t^2 + t^4 = [t^2 + (1 - 2k^2)]^2 + 4k^2(1 - k^2),$

(2.2) $K = 2 \int_0^1 \frac{dt}{Q(t, k)^{1/2}} = \int_0^{\pi/2} \frac{d\psi}{(1 - k^2 \sin^2 \psi)^{1/2}},$

(2.3) $x = 2 \int_0^t \frac{ds}{Q(s, k)^{1/2}}, \quad -2K < x < 2K.$

Complementary modulus In terms of the modulus, k, define

$$k' = (1 - k^2)^{1/2}, \quad K' = K(k') = 2 \int_0^1 \frac{dt}{Q(t, k')^{1/2}}.$$

Definition of the Jacobian elliptic functions on $(-2K, 2K)$ Let t be defined as a function of x by inverting (2.3)

$$sn(x, k) = \frac{2t}{1+t^2} = \sin \psi,$$

(2.4)
$$cn(x, k) = \frac{1-t^2}{1+t^2} = \cos \psi,$$

$$dn(x, k) = \frac{Q(t, k)^{1/2}}{1+t^2} = (1 - k^2 \sin^2 \psi)^{1/2},$$

(2.6) $$sn(\pm 2K) = 0, \quad cn(\pm 2K) = -1, \quad dn(\pm 2K) = 1.$$

Extend all the functions to $(-\infty, +\infty)$ with period $4K$. The extension of sn, cn and dn are respectively odd, even and even functions.

Further notation Let u, v be real:

$$\Delta = \Delta(u, v) = 1 - k^2 sn^2 u \, sn^2 v,$$
$$S(u, v) = [(sn(u))(cn \, (v))(dn \, (v)) + (sn \, (v))(cn \, (u))(dn(u))]/\Delta,$$
$$C(u, v) = [(cn(u))(cn \, (v)) - (sn(u))(sn \, (v))(dn(u))(dn \, (v))]/\Delta,$$
$$D(u, v) = [(dn(u))(dn \, (v)) - k^2(sn(u))(sn \, (v))(cn(u))(cn \, (v))]/\Delta.$$

Extension to the complex plane

(2.23)
$$\begin{aligned} sn(x + iy) &= S(x, iy), \\ cn(x + iy) &= C(x, iy), \quad x \text{ and } y \text{ real}, \quad y \neq (2n+1)K', \\ dn(x + iy) &= D(x, iy). \end{aligned}$$

Addition formulae (Theorem 2.1)

$$\begin{aligned} sn(u + v) &= S(u, v), \\ cn(u + v) &= C(u, v), \quad u \text{ and } v \text{ complex}, \\ dn(u + v) &= D(u, v). \end{aligned}$$

Differentiation formulae

(2.29)
$$\begin{aligned} sn'z &= cn(z) \, dn(z), \\ cn'z &= -sn(z) \, dn(z), \\ dn'z &= -k^2 sn(z) \, dn(z). \end{aligned}$$

Periodicity Let u in the formulae below be complex

$$sn(u + K) = cn(u)/dn(u),$$
$$cn(u + K) = -k'sn(u)/dn(u),$$
(2.31) $\qquad dn(u + K) = k'/dn(u), .$

$$sn(u + 2K) = -sn(u),$$
$$cn(u + 2K) = -cn(u),$$
$$dn(u + 2K) = dn(u),$$

(2.35) $\qquad sn(u + iK') = k^{-1}/sn(u),$

$$cn(u + iK') = -ik^{-1}dn(u)/sn(u),$$
(2.36) $\qquad dn(u + iK') = -icn(u)/sn (u),$

$$sn(u + 2iK') = sn(u),$$
$$cn(u + 2iK') = -cn(u),$$
(2.37) $\qquad dn(u + 2iK') = -dn(u).$

Periods

$$sn:4K, 2iK'; \quad cn:4K, 2K + 2iK', 4iK'; \quad dn:2K, 4iK'.$$

Poles and residues The functions are analytic except for simple poles at $u \equiv iK'$ mod $(2K, 2iK')$; sn, cn and dn have residues $k^{-1}, -ik^{-1}$ and $-i$ respectively at iK'.

Zeros mod$(2K, 2iK')$ (Exercise 2.7.1)

$$sn:u \equiv 0; \quad cn:u \equiv K; \quad dn:u \equiv K + iK'.$$

Formulae after change of variable

$$u = (2K/\pi)z; \quad \tau = iK'/K; \quad q = \exp(\pi i\tau) = \exp(-\pi K'/K).$$

(a) *Periods*

$$sn:2\pi, \pi\tau; \quad cn:2\pi, \pi + \pi\tau, 2\pi\tau; \quad dn: \pi, 2\pi\tau.$$

(b) *Poles* are simple and lie at $z \equiv \frac{1}{2}\pi\tau$ mod $(\pi, \pi\tau)$.
(c) *Zeros* mod$(\pi, \pi\tau)$

$$sn:z \equiv 0; \quad cn:z \equiv \frac{\pi}{2}; \quad dn:z \equiv \frac{\pi}{2} + \frac{\pi\tau}{2}.$$

2.10 Miscellaneous exercises

2.10.1 Show that, if $\Delta = 1 - k^2 s_1^2 s_2^2$, where $s_1 = sn(u)$, $s_2 = sn\ (v)$, etc., then:

(a) $sn(u + v) + sn(u - v) = 2s_1 c_2 d_2/\Delta$;

(b) $sn(u + v) - sn(u - v) = 2s_2 c_1 d_1/\Delta$;

(c) $cn(u + v) + cn(u - v) = 2c_1 c_2/\Delta$;

(d) $cn(u + v) - cn(u - v) = -2s_1 s_2 d_1 d_2/\Delta$;

(e) $dn(u + v) + dn(u - v) = 2d_1 d_2/\Delta$;

(f) $dn(u + v) - dn(u - v) = -2k^2 s_1 s_2 c_1 c_2/\Delta$.

Check that those results agree with what you would expect in the cases $k = 0$, $k = 1$.

2.10.2 In the notation of Exercise 2.10.1, prove that:

(a) $sn(u + v)sn(u - v) = (s_1^2 - s_2^2)/\Delta$;

(b) $sn(u + v)cn(u - v) = (1 - s_1^2 - s_2^2 + k^2 s_1^2 s_2^2)/\Delta$;

(c) $dn(u + v)dn(u - v) = (1 - k^2 s_1^2 - k^2 s_2^2 + k^2 s_1^2 s_2^2)/\Delta$.

2.10.3 Let $cn(u + v) = C_1$, $cn(u - v) = C_2$, $dn(u + v) = D_1$ and $dn(u - v) = D_2$. Prove that

$$sn(u)\ sn\ (v) = \frac{C_2 - C_1}{D_2 + D_1} = \frac{1}{k^2}\frac{D_2 - D_1}{C_2 + C_1}$$

and hence

$$1 - k^2 sn^2 u\ sn^2 v = \frac{2(C_1 D_2 + C_2 D_1)}{(C_1 + C_2)(D_1 + D_2)},$$

$$cn(u)\ cn\ (v) = \frac{C_1 D_2 + C_2 D_1}{D_1 + D_2},$$

$$dn(u)\ dn\ (v) = \frac{C_1 D_2 + C_2 D_1}{C_1 + C_2}.$$

By putting $u = \alpha + \beta$ and $v = \alpha - \beta$, deduce that

$$sn(\alpha + \beta)sn(\alpha - \beta) = \frac{cn(2\beta) - cn(2\alpha)}{dn(2\beta) + dn(2\alpha)},$$

$$cn(\alpha + \beta)cn(\alpha - \beta) = \frac{cn(2\alpha)\ dn(2\beta) + dn(2\alpha)\ cn(2\beta)}{dn(2\alpha) + dn(2\beta)},$$

$$dn(\alpha + \beta)dn(\alpha - \beta) = \frac{cn(2\alpha)\ dn(2\beta) + dn(2\alpha)\ cn(2\beta)}{cn(2\alpha) + cn(2\beta)}.$$

2.10.4 Prove the following:

(a) $sn\frac{1}{2}iK' = \frac{i}{\sqrt{k}}, \quad cn\frac{1}{2}iK' = \frac{\sqrt{(1+k)}}{\sqrt{k}}, \quad dn\frac{1}{2}iK' = \sqrt{(1+k)};$

(b) $sn\frac{1}{2}(K+iK') = (\sqrt{1+k} + i\sqrt{1-k})/\sqrt{2k};$

$cn\frac{1}{2}(K+iK') = (1-i)\sqrt{k'}/\sqrt{2k};$

$dn\frac{1}{2}(K+iK') = \sqrt{kk'}(\sqrt{1+k'} - i\sqrt{1-k'})/\sqrt{2k}.$

2.10.5 For $u, v \in \mathbb{R}$, prove that:

$$\left| sn\left(u + \frac{1}{2}iK'\right) \right| = \frac{1}{\sqrt{k}} = \text{constant}$$

$$\left| \frac{sn\left(\frac{1}{2}K + iv\right)}{cn\left(\frac{1}{2}K + iv\right)} \right| = \frac{1}{\sqrt{k'}} \ ;$$

$$\left| \frac{dn\left(u + \frac{1}{2}iK'\right)}{cn\left(u + \frac{1}{2}iK'\right)} \right| = \sqrt{k}.$$

2.10.6 By considering $\int_C e^{\pi iu/K} \frac{cn(u)\, dn(u)}{sn(u)} du$ round the rectangle, C, with vertices at $\pm K, \pm K + 2iK'$, suitably indented, show that

$$\int_{-K}^{K} e^{\pi iu/K} \frac{cn(u)\, dn(u)}{sn(u)} du = \pi i \tanh\frac{\pi K'}{2K}.$$

2.10.7 Prove that

$$(1 \pm cn(u+v))(1 \pm cn(u-v))$$
$$= \frac{(cn\, u \pm cn\, v)^2}{(1 - k^2 sn^2 u\, sn^2 v)}.$$

2.10.8 Prove that

$$1 + cn(u+v)cn(u-v) = \frac{cn^2 u + cn^2 v}{1 - k^2 sn^2 u\, sn^2 v}.$$

2.10.9 Let $am(u)$ be defined by the formula

$$u = \int_0^{am(u)} (1 - k^2 \sin^2\theta)^{-1/2}\, d\theta.$$

Verify that

$$sn(u) = \sin(am(u)), \quad cn(u) = \cos(am(u)), \quad dn(u) = \frac{d}{du}(am(u)).$$

2.10.10 Obtain the following Fourier expansions, valid throughout $|\text{Im}(z)| < \pi \text{Im}(\tau)$, except at the poles of the first term on the right-hand side of the respective expansions.

$$ds(u) = \frac{\pi}{2K} \text{cosec}\, z - \frac{2\pi}{K} \sum_{n=0}^{\infty} q^{2n+1} \frac{\sin(2n+1)z}{1+q^{2n+1}},$$

$$cs(u) = \frac{\pi}{2K} \cot z - \frac{2\pi}{K} \sum_{n=1}^{\infty} q^{2n} \frac{\sin 2nz}{1+q^{2n}},$$

$$dc(u) = \frac{\pi}{2K} \sec z + \frac{2\pi}{K} \sum_{n=0}^{\infty} (-1)^n q^{2n+1}:$$

2.10.11 Show that if k is so small that k^4 and higher powers of k may be neglected, then

$$sn(u) = \sin u - \frac{1}{4} k^2 \cos u (u - \sin u \cos u),$$

for small values of u.

2.10.12 Prove that, for $0 > \text{Im}(z) > -\pi \text{Im}(\tau)$,

$$ns(u) = \frac{\pi}{2K} \text{cosec}\, z + \frac{2\pi}{K} \sum_{n=0}^{\infty} \frac{q^{2n+1} \sin(2n+1)z}{1-q^{2n+1}}.$$

3

General properties of elliptic functions

3.1 Apologia

The only unconventional thing about this chapter is its placement and we believe that Gauss, Abel, and Jacobi might even contest that. Historically, the general properties of elliptic functions (that is, doubly periodic functions meromorphic in the finite complex plane) were investigated *after* those remarkable objects were known to exist, but since the time of Weierstrass it has been conventional to discuss general properties before proving that examples existed. Since we have already constructed the Jacobian elliptic functions in Chapter 2, we can afford to return to what we believe to be the proper and historical order.

3.2 Period modules and lattices

We consider functions $f(z)$ meromorphic in the whole plane. We say that ω is a *period* of $f(z)$ if $f(z + \omega) = f(z)$, for all z; we emphasize that we are interested only in functions defined in a domain Ω that is mapped onto itself by the transformation $z \mapsto z + \omega$. Clearly every $f(z)$ possesses the trivial period, $\omega = 0$, and $f(z) = z$ has only the trivial period, whilst if $f(z)$ is a constant, then $f(z)$ has every complex number for a period. We are interested in the algebraic structure of the set[1] $M = M(f)$ of periods of f. Clearly $M(\exp z) = \{2\pi n i : n \in \mathbb{Z}\}$ and we saw in Chapter 2 that $M(sn\,(z)) \supset [4n_1 K + 2n_2 iK' : n_1, n_2 \in \mathbb{Z}]$. If ω_1, ω_2 are in M, then so is $n_1\omega_1 + n_2\omega_2$ for $n_1, n_2 \in \mathbb{Z}$; whence M is a *module over* $\mathbb{Z}$, or a *lattice*, with ordered basis (ω_1, ω_2). We temporarily exclude the cases $M(f) = \{0\}$ and $M(f) = \mathbb{C}$. We wish to show that the set of periods M must in all cases be a lattice and to that end we begin with:

---

[1] The notations Λ or Ω are also commonly used for M and we shall occasionally use those in the exercises.

Lemma 3.1 *Let M denote the set of periods. Then*

$$\inf[|\omega| : \omega \in M, \omega \neq 0] > 0.$$

Proof Suppose that the inequality were not strict. Then we could find a sequence $\{\omega_n\}$ of non-zero periods with $\omega_n \to 0$. At a regular point z_0 of $f(z)$, we would then have $f(z_0 + \omega_n) = f(z_0)$ for all $n \geq N$, for some N; so the function $f(z) - f(z_0)$ would have infinitely many zeros $z_0 + \omega_n \to z_0$ as $n \to \infty$. It follows that $f(z)$ would be identically equal to $f(z_0)$, which contradicts our assumption.

Lemma 3.2 *Let M denote the set of periods. Then M is discrete; that is, M has no accumulation points in the finite plane.*

Proof Suppose that there is a point of accumulation, say ω_0, in the finite plane. Then we can find a sequence $\{\omega_n\}$ of distinct periods ω_n such that $\omega_n \to \omega_0 \neq 0$. Since $f(z + \omega_0) = \lim_{n \to \infty} f(z + \omega_n) = f(z)$, ω_0 is itself a period of $f(z)$ and so therefore are the points $\omega_0 - \omega_n (n = 1, 2, 3, \ldots)$. But then $0 \neq \omega_0 - \omega_n \to 0$ – a contradiction to Lemma 3.1

Consider now an arbitrary discrete module $M \neq \{0\}$. Then the absolute values of the non-zero elements of M are bounded below and so M contains a non-zero element, ω_1, say, whose absolute value is a minimum. The elements $n\omega_1 (n \in \mathbb{Z})$ of M lie on the straight line $l : \arg z = \arg \omega_1$ *and no other points of* l *are in* M. For if $\omega \in M$ were a point of l lying between $n_0\omega_1$ and $(n_0 + 1)\omega_1$, then $\omega - n_0\omega_1 \in M$ and $|\omega - n_0\omega_1| < |\omega_1|$ – a contradiction to our choice of ω_1.

Let us now suppose that there is a point ω of M off the line. Among all such points there is at least one, say, ω_2, whose absolute value is the smallest.

Lemma 3.3 *Every $\omega \in M$ has a unique representation*

$$\omega = n_1\omega_1 + n_2\omega_2 \quad (n_1, n_2 \in \mathbb{Z}).$$

Proof Since ω_2/ω_1 is not real, the pair (ω_1, ω_2) forms a basis for $\mathbb{C}$ regarded as a two-dimensional vector space over $\mathbb{R}$. Hence we can write $\omega = \lambda_1\omega_1 + \lambda_2\omega_2$, $\lambda_1, \lambda_2 \in \mathbb{R}$. Now choose integers m_1, m_2 such that $|\lambda_j - m_j| \leq 1/2$, $(j = 1, 2)$, and set $\omega' = \omega - m_1\omega_1 - m_2\omega_2 = (\lambda_1 - m_1)\omega_1 + (\lambda_2 - m_2)\omega_2$. Then $\omega' \in M$ and

$$|\omega'| = |(\lambda_1 - m_1)\omega_1 + (\lambda_2 - m_2)\omega_2| \leq |\lambda_1 - m_1||\omega_1|$$
$$+ |\lambda_2 - m_2||\omega_2| \leq \frac{1}{2}|\omega_1| + \frac{1}{2}|\omega_2| \leq |\omega_2|,$$

where the first inequality is strict, since, ω_2/ω_1 is not real. Because of the way in which ω_2 was chosen, it follows that $\omega' = n\omega_1$ for some integer n and hence ω has the form stated. If there were more than one such representation, one would have $0 = a_1\omega_1 + a_2\omega_2$ for integers a_1, a_2 not both zero – say $a_2 \neq 0$. But then

$$\omega_2 = -(a_1/a_2)\omega_1 \in l$$

– a contradiction.

Accordingly we have determined all discrete modules M and we may summarize our results in:

Theorem 3.1 *There are only three possibilities for the discrete module M:*

(i) $M = \{0\}$;

(ii) $M = \{n\omega : n \in \mathbb{Z} \quad \text{and} \quad \omega \neq 0\}$;

(iii) $M = \{n_1\omega_1 + n_2\omega_2 : n_1, n_2 \in \mathbb{Z}, \omega_1, \omega_2 \neq 0, \operatorname{Im}(\omega_2/\omega_1) \neq 0\}$.

Corollary 3.1 (Jacobi, 1835). *The only meromorphic functions with three independent periods are constants.*

3.3   The unimodular group $SL(2, \mathbb{Z})$

We assume henceforth that the case (iii) of Theorem 3.1 occurs; so that every $\omega \in M$ has a unique representation of the form $\omega = n_1\omega_1 + n_2\omega_2$, $n_1, n_2 \in \mathbb{Z}$. We call any ordered pair (ω_1', ω_2') with that property a *basis for M*. By replacing ω_2' by $-\omega_2'$, if necessary, we may, and henceforth shall, assume that $\operatorname{Im}(\omega_2'/\omega_1') > 0$.

Theorem 3.2 *Let (ω_1, ω_2) be an ordered basis for M, with $\operatorname{Im}(\omega_2/\omega_1) > 0$. Then every basis (ω_1', ω_2') has the matrix representation*

$$\begin{pmatrix} \omega_2' \\ \omega_1' \end{pmatrix} = \begin{pmatrix} a & b \\ c & d \end{pmatrix} \begin{pmatrix} \omega_2 \\ \omega_1 \end{pmatrix}, \tag{3.1}$$

where $a, b, c, d \in \mathbb{Z}$ and $ad - bc = 1$.

Proof Since (ω_1, ω_2) is a basis, there exist integers a, b, c, d such that

$$\omega_2' = a\omega_2 + b\omega_1,$$
$$\omega_1' = c\omega_2 + d\omega_1,$$

which we write in the matrix form (3.1). (We write ω_2 above ω_1 because in what follows the condition $\operatorname{Im}(\omega_2/\omega_1) > 0$ will be crucial.)

Since (ω_1', ω_2') is also a basis, we can write

$$\begin{pmatrix} \omega_2 \\ \omega_1 \end{pmatrix} = \begin{pmatrix} a' & b' \\ c' & d' \end{pmatrix} \begin{pmatrix} \omega_2' \\ \omega_1' \end{pmatrix}, \tag{3.2}$$

for some integers a', b', c', d'. But then

$$\begin{pmatrix} \omega_2 \\ \omega_1 \end{pmatrix} = \begin{pmatrix} a' & b' \\ c' & d' \end{pmatrix} \begin{pmatrix} a & b \\ c & d \end{pmatrix} \begin{pmatrix} \omega_2 \\ \omega_1 \end{pmatrix} = \begin{pmatrix} A & B \\ C & D \end{pmatrix} \begin{pmatrix} \omega_2 \\ \omega_1 \end{pmatrix}$$

$$= \begin{pmatrix} A\omega_2 + B\omega_1 \\ C\omega_2 + D\omega_1 \end{pmatrix}.$$

Since ω_1, ω_2 are linearly independent over $\mathbb{Z}$, $A = 1, B = 0, C = 0, D = 1$ and so we have established

$$\begin{pmatrix} a' & b' \\ c' & d' \end{pmatrix} \begin{pmatrix} a & b \\ c & d \end{pmatrix} = \begin{pmatrix} 1 & 0 \\ 0 & 1 \end{pmatrix},$$

from which follows

$$\begin{vmatrix} a' & b' \\ c' & d' \end{vmatrix} \cdot \begin{vmatrix} a & b \\ c & d \end{vmatrix} = 1.$$

Since all the matrix elements are integers (and so the determinants) we infer

$$\begin{vmatrix} a & b \\ c & d \end{vmatrix} = \pm 1.$$

However, the normalizing condition $\mathrm{Im}(\omega_2'/\omega_1') > 0$ also determines the sign. To see that, set

$$\tau = \omega_2/\omega_1, \quad \tau' = \omega_2'/\omega_1'.$$

Then (3.1) implies

$$\tau' = \frac{a\tau + b}{c\tau + d} = \frac{ac\tau\bar{\tau} + bd + ad\tau + bc\bar{\tau}}{|c\tau + d|^2}. \tag{3.3}$$

Therefore

$$\mathrm{Im}(\tau') = (ad - bc)\,\mathrm{Im}(\tau)/|c\tau + d|^2.$$

Since both $\mathrm{Im}(\tau) > 0$ and $\mathrm{Im}(\tau') > 0$, we must have

$$ad - bc > 0.$$

Matrices of the form $\begin{pmatrix} a & b \\ c & d \end{pmatrix}$ with integers a, b, c, d and $\begin{vmatrix} a & b \\ c & d \end{vmatrix} = +1$ clearly form a group under matrix multiplication, called the *unimodular group* and denoted by $SL(2, \mathbb{Z})$.

Exercises 3.3

3.3.1 Show that if (ω_1, ω_2) is a basis and if ω_1', ω_2' are given by (3.1), then (ω_1', ω_2') is also a basis.

3.3.2 Which of the following are lattices in $\mathbb{C}$?

    (i) $\mathbb{Z}$;
    (ii) $\mathbb{Q}$;
    (iii) $\{m + n\sqrt{2} : m, n \in \mathbb{Z}\}$;
    (iv) $\{a + ib : a, b \in \mathbb{Z}, a - b \text{ is even}\}$;
    (v) $\{a + ib : a, b \in \mathbb{Z}, a - b \text{ is odd}\}$
    (vi) $\{m + n\sqrt{-2} : m, n \in \mathbb{Z}\}$
    (vii) $\{a\sqrt{2} + b\sqrt{-6} + c\sqrt{8} : a, b, c \in \mathbb{Z}\}$;
    (viii) $\{m\sqrt{2i} + n\sqrt{-2i} : m, n \in \mathbb{Z}\}$;
    (ix) $\{z : z^2 - 2bz + b^2 + 3c^2 = 0; b, c \in \mathbb{Z}\}$;
    (x) $\{z : z^2 - 2bz + c^2 = 0; b, c \in \mathbb{Z}\}$.

3.3.3 Let Γ be a sub group of $\mathbb{R}^n$ and set

$$\Gamma_R = \{x \in \Gamma : \|x\| < R\},$$

where $\|x\|$ denotes the norm of x. Prove that the following are equivalent:

    (i) Γ is discrete;
    (ii) Γ_R is finite for some $R > 0$;
    (iii) Γ_R is finite for all $R > 0$.

3.3.4 Let Λ be a lattice in $\mathbb{C}$ and let (ω_1, ω_2) be a basis for Λ. The lattice Λ is said to have *complex multiplication* if for some $\alpha \in \mathbb{C}\backslash\mathbb{Z}$, the set

$$\alpha\Lambda = \{\alpha\omega : \omega \in \Lambda\}$$

is contained in Λ. Show that Λ has complex multiplication if and only if $\tau = \omega_2/\omega_1$ satisfies a quadratic equation with integer coefficients.

3.3.5 Let $\omega_1, \omega_2, \omega_3$ be three periods of a meromorphic function. Show that there exists integers a, b and c and that

$$a\omega_1 + b\omega_2 + c\omega_3 = 0.$$

3.4 The canonical basis

Among all possible bases of M one can single out one basis, almost uniquely (see below) called the *canonical basis*. It is not always necessary – indeed it is sometimes undesirable – to use such a special basis, but we shall certainly need its existence in Chapter 6.

Theorem 3.3 *There exists a basis (ω_1, ω_2) of M such that the ratio $\tau = \omega_2/\omega_1$ satisfies the following conditions:*

(i) $\operatorname{Im} \tau > 0$;
(ii) $-1/2 < \operatorname{Re} \tau \leq 1/2$;
(iii) $|\tau| \geq 1$;
(iv) $\operatorname{Re}(\tau) \geq 0$, *if* $|\tau| = 1$.

The ratio τ is uniquely determined by those conditions and there is a choice of two, four or six associated bases.

Proof Select ω_1, ω_2, as in the proof of Theorem 3.1. Then $|\omega_1| \leq |\omega_2|$ and, since $\omega_1 \pm \omega_2 \notin l$, $|\omega_2| \leq |\omega_1 \pm \omega_2|$. In terms of τ those conditions become $1 \leq |\tau|$ and $|\operatorname{Re}(\tau)| \leq 1/2$. If $\operatorname{Im}(\tau) < 0$, replace (ω_1, ω_2) by $(-\omega_1, \omega_2)$ to make $\operatorname{Im}(\tau) > 0$, without changing the condition on $\operatorname{Re}(\tau)$. If $\operatorname{Re}(\tau) = -1/2$, change to the basis $(\omega_1, \omega_1 + \omega_2)$, and if $|\tau| = 1$ and $\operatorname{Im}(\tau) < 0$, change to the new basis $(\omega_1, -\omega_2)$. After those minor adjustments, all the conditions are satisfied.

Next, we have to show that τ is uniquely defined by the conditions (i) to (iv). Those conditions mean that the point τ lies in the set $\mathfrak{I}$ shown in Figure 3.1. It is bounded by the semi-circle consisting of the upper half of $|\tau| = 1$ and by the vertical lines $\operatorname{Re}(\tau) = \pm\dfrac{1}{2}$, but actually includes only the right-hand half of that semi-circle.

Although the set $\mathfrak{I}$ is not open, it is called the *fundamental region* of the modular group $SL(2, \mathbb{Z})$.

We have seen that every change of basis is effected by a unimodular transformation, (3.1), and that the ratio ω_2/ω_1 is replaced by $\tau' = \omega_2'/\omega_1'$, where τ' is given by (3.3). Moreover,

$$\operatorname{Im}(\tau') = \operatorname{Im}(\tau)/|c\tau + d|^2. \qquad (3.4)$$

We now observe that if τ is an arbitrary complex number, with $\operatorname{Im}(\tau) > 0$, and if we put $M = \{n_1 + n_2\tau : n_1, n_2 \in \mathbb{Z}\}$, then the results proved so far tell

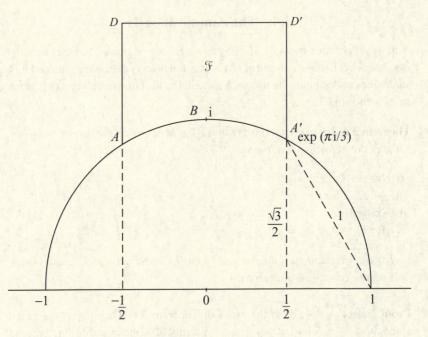

Figure 3.1 The region $\mathfrak{F}$

us that there exists a transformation $L \in SL(2, \mathbb{Z})$ such that

$$\tau' = L\tau \in \mathfrak{F}.$$

That is, every point τ in the upper half-plane is congruent under $SL(2, \mathbb{Z})$ to a point τ' in the fundamental region $\mathfrak{F}$. We now show that τ' is *unique*. Since $SL(2, \mathbb{Z})$ is a group, it suffices to establish:

Lemma 3.4 *Let τ, τ' both lie in $\mathfrak{F}$ and be connected by Equation (3.3). Then $\tau' = \tau$.*

Proof of Lemma 3.4 Since τ and τ' play symmetric roles, we may assume that $\mathrm{Im}(\tau') \geq \mathrm{Im}(\tau)$. Then (3.4) implies that $|c\tau + d| \leq 1$. Since c and d are integers, we shall see that the possibilities are restricted, as follows.

Case 1: $c = 0$;   Case 2: $c \neq 0$.

In Case 1, the determinant condition $ad - bc = 1$ reduces to $ad = 1$ whence $a = d = 1$ or $a = d = -1$. Equation (3.3) then becomes $\tau' = \tau \pm b$, whence $|b| = |\mathrm{Re}(\tau') - \mathrm{Re}(\tau)| < 1$, by condition (ii) of Theorem 3.3. Whence $b = 0$ and $\tau' = \tau$.

In Case 2, the inequality

$$\left| \tau + \frac{d}{c} \right| \le |c|^{-1}$$

implies $|c| = 1$. For if $|c| \ge 2$, the point τ would be at a distance $\le 1/2$ from the real axis, which is obviously impossible, since $\exp(\pi i/3)$, the nearest point in $\Im$, is at a distance $\sqrt{3}/2$. Thus $|\tau \pm d| \le 1$, and Figure 3.1 clearly implies that that can happen only if $d = 0$ or $d = \pm 1$.

Consider first the cases $d = \pm 1$. Since the point $\exp(2\pi i/3)$ does not lie in $\Im$, $|\tau + 1| > 1$ and $|\tau - 1| \le 1$ only when $z = \exp(\pi i/3)$. In the latter case $|c\tau + d| = 1$, whence $\mathrm{Im}(\tau') = \mathrm{Im}(\tau) = \sqrt{3}/2$, by (i). It follows from Figure 3.1 that $\tau' = \exp(\pi i/3) = \tau$.

There remains the case $d = 0$, $|c| = 1$. Then $|\tau| \le 1$, whence $|\tau| = 1$, by (iii). Moreover, $bc = -1$ implies $b/c = -1$ and $\tau' = \pm a - \tau^{-1} = \pm a - \overline{\tau}$. Hence $\mathrm{Re}(\tau + \tau') = \pm a$. But then (ii) yields $-1 < \mathrm{Re}(\tau + \tau') \le 1$, and if $a \ne 0$, then the only way equality can hold on the right is if $\tau = \exp(\pi i/3)$, $\mathrm{Re}(\tau + \tau') = 1$ and $\tau' = 1 - \tau^{-1} = \tau$. If $a = 0$, then $\tau' = -\tau^{-1}$ and there is a contradiction with (iv) unless $\tau' = i = \tau$.

That concludes the proof of Lemma 3.4.

Completion of the proof of Theorem 3.3 By Lemma 3.4, τ is unique. The canonical basis (ω_1, ω_2) can always be replaced by $(-\omega_1, \omega_2)$. There are other bases with the same τ if and only if τ is a fixed point of a unimodular transformation (3.3). It follows from the proof of Lemma 3.4 that that happens only for $\tau = i$ or $\tau = \exp(\pi i/3)$. The former is a fixed point of $\tau \mapsto -\tau^{-1}$, the latter of $\tau \mapsto -1/(\tau - 1)$ or if $\tau \mapsto (\tau - 1)/\tau$. Those are the choices given in Theorem 3.3, whose proof is now complete.

Exercises 3.4

3.4.1 Show that every lattice, Λ, satisfies $\Lambda = -\Lambda$. Show further that the only cases in which $\Lambda = k\Lambda$, $k \ne \pm 1$, are the *square lattice*, $\Lambda = i\Lambda$, and the *triangular lattice* $\Lambda = \varepsilon\Lambda$, where ε is a primitive cube root of unity.

3.4.2 Let $\omega = m\omega_1 + n\omega_2$ be an element of the lattice Λ with basis (ω_1, ω_2). Show that ω can be an element of a *basis* for Λ if and only m and n are coprime; or $m = \pm 1$, $n = 0$; or $m = 0$ and $n = \pm 1$.

3.4.3 An *automorphism* of a lattice Λ is a bijection $\alpha : \Lambda \to \Lambda$, such that $\alpha(u + v) = \alpha(u) + \alpha(v)$ for all $u, v \in \Lambda$. Prove that if $\omega \mapsto$

$\lambda \omega$ ($\omega \in \Lambda$) is an automorphism of Λ, then

$$\lambda = \pm 1, \pm i, \frac{1}{2}(1 \pm \sqrt{-3}) \text{ or } \frac{1}{2}(-1 \pm \sqrt{-3}).$$

(Hint: show that λ has modulus 1 and that it must be an eigenvalue of a 2×2 integer matrix with determinant ± 1.)

3.4.4 Use 3.4.3 to obtain the 'crystallographic restriction' (see Coxeter, 1961, Section 4.5); namely that the only possible finite orders, greater than 1, for an automorphism of a lattice are 2, 3, 4 or 6. Show that every lattice admits an automorphism of order 2, and describe those lattices that admit automorphisms of orders 3, 4 or 6.

3.5   Four basic theorems

In what follows, $f(z)$ denotes a meromorphic function with independent periods ω_1, ω_2 (and $\text{Im}(\omega_2/\omega_1) \neq 0$). We shall not assume that (ω_1, ω_2) is a canonical basis, or even a basis, for $M(f)$. By $z_1 \equiv z_2 (\text{mod } M)$, we shall mean that $z_1 - z_2 \in M$, whence $f(z_1) = f(z_2)$. It follows from the preceding remark that one can regard $f(z)$ as a function on the congruence classes, mod M. Geometrically speaking, that means that one can restrict $f(z)$ to the parallelogram or *cell* P_a with vertices $a, a + \omega_1, a + \omega_1 + \omega_2, a + \omega_2$ obtained by translating the parallelogram with vertices $0, \omega_1, \omega_1 + \omega_2, \omega_2$ (read cyclically, counter-clockwise). By including part of the boundary we may represent each congruence class by exactly one point in P_a, and then $f(z)$ is completely determined by its values in P_a. The choice of a is irrelevant, but it is convenient to choose it so that $f(z)$ has no poles in the boundary ∂P_a of P_a. We shall always do that.

Theorem 3.4 (The original form of Liouville's Theorem[2]). *An entire elliptic function is a constant.*

Proof Since $f(z)$ is continuous everywhere, it is bounded in the closure of P_a and hence in the finite plane. By what is now usually called 'Liouville's Theorem' it must reduce to a constant.

Since the poles of $f(z)$ have no accumulation point, only a finite number of them can be in P_a. By *the* poles of $f(z)$ we shall mean a full set of mutually incongruent poles; we count multiplicities in the usual manner.

---

[2] See Whittaker and Watson (1927), 5.63, p. 105 and 20.12 p. 431.

Theorem 3.5 *The sum of the residues of an elliptic function (in a period cell) is zero.*[3]

Proof The sum of the residues at the poles in P_a is equal to

$$\frac{1}{2\pi i} \int_{\partial P_a} f(z)dz,$$

where the integration round the boundary of P_a is in the counterclockwise sense. Since $f(z)$ has periods ω_1, ω_2, the integrals over opposite sides of the parallelogram P_a cancel each other and so the integral vanishes.

Corollary 3.2 *There does not exist an elliptic function with a single, simple pole.*

Theorem 3.6 *A non-constant elliptic function, $f(z)$, has exactly as many poles as zeros (counted according to multiplicities).*

Proof The poles and zeros of $f(z)$ are simple poles of the function $f'(z)/f(z)$, which is itself an elliptic function. The multiplicities are the residues of $f'(z)/f(z)$ counted positive for zeros and negative for poles. Now apply Theorem 3.5.

If c denotes a constant, then $f(z) - c$ has the same poles as $f(z)$. Therefore all values are assumed equally many times. By the *order* of an elliptic function we shall mean the number of roots of the equation $f(z) = c$, incongruent modulo the period lattice.

Theorem 3.7 *Let $a_1, \ldots, a_n$ denote the zeros and $b_1, \ldots, b_n$ the poles (counted according to multiplicities) of an elliptic function, $f(z)$. Then*

$$a_1 + \cdots + a_n \equiv b_1 + \cdots + b_n \pmod{M}. \tag{3.5}$$

Proof We may assume there are no zeros or poles on the boundary ∂P_a of P_a (by moving the parallelogram if necessary) and then consider the integral

$$\frac{1}{2\pi i} \int_{\partial P_a} \frac{zf'(z)}{f(z)} dz = (a_1 + \cdots + a_n) - (b_1 + \cdots + b_n), \tag{3.6}$$

by the *argument principle*,[4] provided that we choose the representative zeros and poles inside P_a (as we may). Consider the two opposite sides, a to $a + \omega_1$

---

[3] Compare this result with the special case of the Jacobi functions in Chapter 2, Section 2.6.
[4] See, for example, Ahlfors (1979), Chapter 4.

and $a + \omega_1 + \omega_2$ to $a + \omega_2$ (in the counterclockwise sense). The corresponding part of the integral simplifies, since $f(z) = f(z + \omega_2)$, to

$$\frac{1}{2\pi i} \left\{ \int_a^{a+\omega_1} - \int_{a+\omega_2}^{a+\omega_1+\omega_2} \right\} \frac{zf'(z)}{f(z)} dz = \frac{-\omega_2}{2\pi i} \int_a^{a+\omega_1} \frac{f'(z)}{f(z)} dz.$$

But

$$\frac{1}{2\pi i} \int_a^{a+\omega_1} \frac{f'(z)}{f(z)} dz$$

is the winding number[5] around the origin of the closed curve described by $f(z)$ as z runs from a to $a + \omega_1$; hence it is an integer.

The same argument applies to the other pair of opposite sides. We conclude that the value of (3.6) has the form $n_1\omega_1 + n_2\omega_2$.

Exercises 3.5

3.5.1 Show that a non-constant rational function cannot be periodic.
3.5.2 Show that if $f(z)$ is an elliptic function, then so is $f'(z)$. Is it true that $\int f(z)dz$ is also an elliptic function? (Illustrate your answer using simply periodic functions.)
3.5.3 If $f(z)$ has order n show that $f'(z)$ has order m, where $n + 1 \le m \le 2n$.
3.5.4 Let $f(z)$ be an *even* elliptic function with periods $2\omega_1$ and $2\omega_2$. Show that $f'(z)$ vanishes at ω_1, ω_2 and $\omega_1 + \omega_2$.
3.5.5 Show that, if $f(z)$ is an elliptic function of order m and $P(\omega)$ is a polynomial of order n, then $P(f(z))$ is an elliptic function; and find its order.

3.6   Elliptic functions of order 2

The results of (3.5) imply that the simplest non-trivial elliptic functions are of order 2 and such functions have either two simple poles with residues of opposite sign or a double pole with residue zero. The Jacobi functions of Chapter 2 are an example of the first type. The more conventional approach is to defer consideration of the Jacobi function until after the theta functions (or their equivalent, the sigma functions) have been introduced, and to begin by introducing the Weierstrass functions, $\wp(z)$, functions having a double pole with residue zero. We defer a detailed consideration of those functions to Chapter 7.

---

[5] See Ahlfors (1979), Chapter 4.

Miscellaneous exercises

1. (i) Let $N = 4k$, where k is a positive integer. Denote by Λ the set of all $x = (x_1, \dots, x_n) \in \mathbb{R}^n$ satisfying

$$2x_i \in \mathbb{Z}; \quad (x_i - x_j) \in \mathbb{Z}, \quad i, j = 1, \dots, n; \quad \frac{1}{2} \sum_{i=1}^{N} x_i \in \mathbb{Z}.$$

Show that Λ is a lattice and that $\|x\|^2$ (where $\|x\|^2 = (x_1^2 + \cdots + x_n^2)$) is always an integer if $x \in \Lambda$.

(ii) Suppose further that k *is even.* Show that $\|x\|^2$ is always even if $x \in \Lambda$ and that $x \cdot y = \sum x_i y_i$ is an integer for all $x, y \in \Lambda$.

(iii) Finally, let $k = 2$, $N = 8$. Show that there are exactly 240 elements $x \in \Lambda$ with $\|x\|^2 = 2$. Show also that the set

$$\Lambda^* = \{x \in \mathbb{R}^N : x \cdot y \text{ is an integer for every } y \in \Lambda\}$$

is precisely Λ. (Such a lattice is said to be *self-dual.*)

2. Given that the series $\sum'_{\omega \in \Lambda} |\omega|^{-\alpha}, \alpha > 2$, converges, where the summation is over all points ω of the lattice Λ, other than $\omega = 0$, prove that the function $F(z)$ defined by

$$F(z) = \sum_{m=-\infty}^{\infty} \sum_{n=-\infty}^{\infty} (z - m - ni)^{-3}$$

is an elliptic function with zeros at $z = 1/2, i/2, (1+i)/2$.

3. Let $sn(u), cn(u), dn(u)$ denote the Jacobi elliptic functions defined in Chapter 2, with respect to the lattice with basis $(4K, 4iK')$. Verify that the function

$$x\, sn(u) + y\, cn(u) + z\, dn(u) + w,$$

where x, y, z, w are complex numbers, is an elliptic function, of periods $4K, 4iK'$, of order 4.

Deduce that if u_1, u_2, u_3 and u_4 are distinct complex numbers *not* in the lattice with basis $(2K, 2iK')$ and such that $u_1 + u_2 + u_3 + u_4 = 0$, then the determinant

$$\begin{vmatrix} s_1 & c_1 & d_1 & 1 \\ s_2 & c_2 & d_2 & 1 \\ s_3 & c_3 & d_3 & 1 \\ s_4 & c_4 & d_4 & 1 \end{vmatrix} = 0,$$

where $s_k = sn(u_k), c_k = cn(u_k), d_k = dn(u_k), k = 1, 2, 3, 4$.
(Hint: use Theorem 3.7.)

4. Let $f(z)$ be an elliptic function and suppose that f has zeros of mul-
tiplicity k_i at $z = a_i, i = 1, \ldots, r$, and poles of multiplicity l_j at $z = b_j, j = 1, \ldots, s$, where no two elements of the set $\{a_i, b_j : i = 1, \ldots, r, j = 1, \ldots, s\}$ are congruent with respect to the lattice M. Prove that

$$\sum_{i=1}^{r} k_i = \sum_{j=1}^{s} l_j.$$

Show that f takes every value in $\mathbb{C}$ the same number of times, counting
multiplicities. Prove that

$$\sum_{i=1}^{r} k_i a_i - \sum_{j=1}^{s} l_j b_j = \omega$$

for some $\omega \in M$.
(Hint: use the results of Section 3.5.)

5. Let c' and c'' be two complex numbers, finite or infinite, and let a_i' and
$a_i''(i = 1, 2, \ldots, r)$ be the points where $f(z) = c'$ or $f(z) = c''$, respectively, in a period cell of an elliptic function of order r. Prove that

$$\sum_{i=1}^{r} a_i' \equiv \sum_{i=1}^{r} a_2'' \pmod{M},$$

where M denotes the period lattice with basis (ω_1, ω_2).
(Hint: the case $c'' = \infty$ is Theorem 3.7. Suppose then that c' and c'' are
both finite and unequal and use the fact that $f(z) - c'$ and $f(z) - c''$ are
elliptic functions with poles at the points $b_1, \ldots, b_r$ (say) mod M, and then
use Theorem 3.7.)

6. Prove that an elliptic function $f(z)$ of order 2 satisfies a relation $f(s - z) = f(z)$, where $s = \sum_{r=1}^{2} b_r$ and the b_r are the poles of $f(z)$ in a period cell.

7. Prove that if b_1, b_2 are two distinct poles of an elliptic function $f(z)$ of order
2 in a period cell, then $\frac{b_1+b_2}{2}, \frac{b_1+b_2}{2} + \frac{\omega_1}{2}, \frac{b_1+b_2}{2} + \frac{\omega_1+\omega_2}{2}$ and $\frac{b_1+b_2}{2} + \frac{\omega_2}{2}$ are
the zeros of $f'(z)$.

8. Prove that every half-period is a zero or a pole of an odd order for every odd
elliptic function, $f(z)$.

9. Prove that the order of the multiplicity of the pole or the zero at a half-period
of an even elliptic function is even.

4

Theta functions

In this chapter, we define and prove the basic properties of the theta functions, first studied in depth by Jacobi (1829, 1838). As in Chapter 2, we include a summary of the main results, for easy reference, at the end of the chapter.

4.1 Genesis of the theta functions

We remarked earlier that, in some respects, the limiting case $k = 1$ (the hyperbolic functions) is a better guide to the behaviour of the Jacobian elliptic functions than is the case $k = 0$ (the trigonometric functions). Recall that

$$sn(u, 1) = \tanh u = \sinh u / \cosh u$$

and

$$cn(u, 1) = dn(u, 1) = \operatorname{sech} u = 1/\cosh u.$$

Now

$$\sinh u = u \prod_{n=1}^{\infty} (1 + u^2/n^2\pi^2),$$

$$\cosh u = \prod_{n=0}^{\infty} (1 + 4u^2/((2n + 1)^2\pi^2)).$$

(4.1)

(For an elementary proof of those beautiful Eulerian formulae, which display the zeros of $\sinh u$ and $\cosh u$, see the Appendix.) It follows that

$$sn(u, 1) = \frac{u \prod_{n=1}^{\infty} (1 + u^2/n^2\pi^2)}{\prod_{n=0}^{\infty} (1 + 4u^2/((2n + 1)^2\pi^2))},$$

$$cn(u, 1) = \frac{1}{\prod_{n=0}^{\infty} (1 + 4u^2/((2n + 1)^2\pi^2))} = dn(u, 1).$$

(4.2)

75

Guided by that limiting case, let us try to write each of our three basic Jacobian elliptic functions in the form $f(u) = g(z)/h(z)$, where $z = \pi u/(2K)$ and
$g(z), h(z)$ are entire functions given by infinite products and the denominator
$h(z)$ is to be the same for each of the three functions $sn(u, k)$, $cn(u, k)$ and
$dn(u, k)$.

Since $sn(u + 4K) = sn(u)$ and $cn(u + 4K) = cn(u)$ it is reasonable to
guess that $g(z), h(z)$ should have period 2π in z, but since $dn(u + 2K) = dn(u)$
we should try to give $h(z)$ the period π. We recall (see the Summary, Section
2.10, at the end of Chapter 2) the notation

$$\tau = iK'/K, \quad q = \exp(\pi i \tau) = \exp(-\pi K'/K) \tag{4.3}$$

and that, with that notation, $sn(z), cn(z), dn(z)$ have periods $2\pi, \pi\tau; 2\pi, \pi +$
$\pi\tau, 2\pi\tau; \pi, 2\pi\tau$, respectively. So we want $h(z) = 0$ at the poles of $dn(u, k)$;
that is where $z \equiv \frac{1}{2}\pi\tau \pmod{\pi, \pi\tau}$ or $z \equiv \pm(2n - 1)\pi\tau/2 \pmod{\pi}$, $n =$
$1, 2, \ldots$ Now the function

$$1 - A_n \exp(-2iz)$$

has period π and a simple zero at $z = (2n - 1)\pi\tau/2$, if one sets $A_n =$
$\exp[(2n - 1)\pi i \tau] = q^{2n-1}$. Moreover, $1 - A_n \exp(2iz)$ then has a simple zero
when $z = -(2n - 1)\pi\tau/2$. We are accordingly led to the idea of defining $h(z)$
by

$$h(z) = G \prod_{n=1}^{\infty} (1 - q^{2n-1}e^{-2iz})(1 - q^{2n-1}e^{2iz})$$

$$= G \prod_{n=1}^{\infty} (1 - 2q^{2n-1} \cos 2z + q^{4n-2}), \tag{4.4}$$

where the constant G will be determined later.

Since $|q| < 1$ (recall that $\mathrm{Im}(\tau) > 0$), the series

$$\sum_{n=1}^{\infty} q^{2n-1}e^{\pm 2iz}$$

converges absolutely and uniformly on compact sets in the z-plane. It follows
that $h(z)$ is an entire function of period π with simple zeros at the common poles
of the three Jacobian elliptic functions. (For the properties of infinite products
appealed to here, see the Appendix and Ahlfors (1979), Chapter 5.)

Now we ask whether $h(z)$ possesses *complex* periods (as one would expect from the properties of the Jacobian elliptic functions). We find that

$$h(z + \pi\tau) = G \prod_{n=1}^{\infty} (1 - q^{-2}q^{2n-1}e^{-2iz})(1 - q^2 q^{2n-1}e^{2iz})$$

$$= \frac{1 - q^{-1}e^{-2iz}}{1 - qe^{2iz}} h(z) = -q^{-1}e^{-2iz}h(z), \qquad (4.5)$$

which looks promising – we have periodicity apart from a factor $-q^{-1}\exp(-2iz)$. We say that 1 and $-q^{-1}\exp(-2iz)$ are the *multipliers* or *periodicity factors* associated with the periods π and $\pi\tau$, respectively.

Now consider a possible representation $dn(u) = g(z)/h(z)$. Clearly we would like $g(z)$ to have period π and to vanish when $z \equiv \pi/2 + \frac{1}{2}\pi\tau(\mathrm{mod}\ \pi, \pi\tau)$. But such a function is given by

$$g(z) = h\left(z - \frac{\pi}{2}\right) = h\left(z + \frac{\pi}{2}\right)$$

$$= G \prod_{n=1}^{\infty} (1 + q^{2n-1}e^{-2iz})(1 + q^{2n-1}e^{2iz})$$

$$= G \prod_{n=1}^{\infty} (1 + 2q^{2n-1}\cos 2z + q^{4n-2}).$$

Moreover,

$$g(z + \pi\tau) = h\left(z + \frac{\pi}{2} + \pi\tau\right) = -q^{-1}\exp(-2i(z + \pi/2))h(z + \pi/2)$$

$$= q^{-1}e^{-2iz}g(z).$$

A breakthrough is imminent! Set $F(z) = g(z)/h(z)$. Then $F(z + \pi) = F(z)$ and $F(z + \pi\tau) = -F(z)$, whence $F(z + 2\pi\tau) = F(z)$. So we see that $F(z)$ is a meromorphic function with periods π and $2\pi\tau$ corresponding to the periods $2K$ and $4iK'$ of $dn(u)$ and simple zeros at the simple zeros of $dn(u)$ given by $K + iK'$ (that is at $z \equiv \pi/2 + \pi\tau/2$) and simple poles at the simple poles of $dn(u)$ (the points $z \equiv \pi\tau/2$). Hence $dn(u)/F(z)$ is a doubly periodic entire function and Liouville's Theorem implies the following:

Theorem 4.1 *Let* $u = 2Kz/\pi$. *Then*

$$dn(u) = DF(z) = Dg(z)/h(z) = Dh(z + \pi/2)/h(z)$$

for some constant D, depending on the modulus k.

4.2   The functions θ_4 and θ_3

Having met with initial success, let us look to consolidate our gains before
considering the representations of $sn\,(u)$ and $cn\,(u)$. We saw in Chapter 2,
Exercise 2.7.2 that the map $k \mapsto q = \exp(-\pi K'/K)$ is a monotonic increasing
mapping of the open interval $(0, 1)$ onto itself. Similarly, the map $k \mapsto \tau =
iK'/K$ is a one-to-one mapping of $(0, 1)$ onto the positive imaginary axis. Since
$q = \exp(\pi i \tau)$ has modulus less than one when τ lies in the upper half-plane,
we are led to the idea of writing

$$\theta_4(z) = h(z),$$

where $h(z)$ is defined in (4.4). Thus

$$\theta_4(z) = \theta_4(z|\tau) = G \prod_{n=1}^{\infty} (1 - 2q^{2n-1} \cos 2z + q^{4n-2})$$

$$= G \prod_{n=1}^{\infty} (1 - q^{2n-1} e^{-2iz})(1 - q^{2n-1} e^{2iz}) \qquad (4.6)$$

where $q = \exp(\pi i \tau)$ and $\mathrm{Im}(\tau) > 0$. The constant G, which depends on τ (but
not on z), is to be determined later. It follows from the foregoing discussion
that $\theta_4(z)$ is an entire function of period π and

$$\theta_4(z + \pi \tau) = -q^{-1} e^{-2iz} \theta_4(z). \qquad (4.7)$$

Now expand $\theta_4(z)$ in a Fourier series

$$\theta_4(z) = \sum_{n=-\infty}^{\infty} a_n \, e^{2niz},$$

with $a_{-n} = a_n$, since $\theta_4(-z) = \theta_4(z)$. The identity (4.7) becomes

$$\sum_{n=-\infty}^{\infty} a_n \, e^{2ni(z+\pi\tau)} = -q^{-1} \sum_{n=-\infty}^{\infty} a_n \, e^{2(n-1)iz};$$

that is

$$\sum_{n=-\infty}^{\infty} a_n q^{2n+1} e^{2niz} = - \sum_{n=-\infty}^{\infty} a_{n+1} e^{2niz}. \qquad (4.8)$$

The uniqueness theorem for the Fourier coefficients (which follows readily
when the series converge uniformly) then yields $a_{n+1} = -q^{2n+1} a_n$. It is then
a simple exercise in induction to show that $a_n = (-1)^n q^{n^2} a_0$, for $n \in \mathbb{Z}$. We

now *choose the multiplicative constant G in (4.6) so that $a_0 = 1$.* Then $\theta_4(z)$ is completely determined:

$$\theta_4(z) = \theta_4(z|\tau) = \sum_{n=-\infty}^{\infty} (-1)^n q^{n^2} \cos 2nz$$

$$= 1 + 2\sum_{n=1}^{\infty} (-1)^n q^{n^2} \cos(2nz). \qquad (4.9)$$

It follows from (4.6) and (4.9) that $\lim_{q\to 0} G = 1$. (We shall see in Section 4.6 that $G = \prod_{n=1}^{\infty}(1-q^{2n})$.)

Now put $\theta_3(z) = \theta_4(z + \pi/2) = \theta_4(z - \pi/2)$ to obtain

$$\theta_3(z) = \theta_3(z|\tau) = 1 + 2\sum_{n=1}^{\infty} q^{n^2} \cos 2nz$$

$$= G\prod_{n=1}^{\infty} (1 + 2q^{2n-1}\cos 2z + q^{4n-2}). \qquad (4.10)$$

Since $|q| < 1$, the series (4.9) and (4.10) converge rapidly. Theorem 4.1 now becomes $dn(u) = D\theta_3(z)/\theta_4(z)$ $(z = \pi u/2K)$ for some constant D. To evaluate D, we first set $u = 0$. Then

$$D = \theta_4(0)/\theta_3(0) = \prod_{n=1}^{\infty} (1 - q^{2n-1})^2 \bigg/ \prod_{n=1}^{\infty} (1 + q^{2n-1})^2 > 0.$$

Now take $u = K$. Then $k' = dn(K)$ and so

$$k' = dn(K) = D\theta_3(\pi/2)/\theta_4(\pi/2) = D\theta_4(0)/\theta_3(0) = D^2,$$

whence $D = k'^{1/2}$ and we have established:

Theorem 4.2 *With the foregoing notation,*

$$dn(u) = k'^{1/2}\theta_3(\pi u/2K)/\theta_4(\pi u/2K)$$

and

$$k'^{1/2} = \theta_4(0)/\theta_3(0).$$

4.3 The functions θ_1 and θ_2

It should be noted that Jacobi (1838) uses θ in place of θ_4; for a helpful survey of the different notations for the four theta functions used by different authors, see Whittaker & Watson (1927), Section 21.9, p. 487 and the summary 4.9 at the end of this chapter.

Consider the proposed representation $sn(u) = g(z)/\theta_4(z)$, where $g(z)$ now must have period 2π and vanish when $z \equiv 0 (\text{mod } \pi, \pi\tau)$. We can accomplish the second objective by setting

$$g(z) = \theta_4 \left(z + \frac{1}{2}\pi\tau \right),$$

but then $g(z)$ has the period π. Moreover,

$$g(z + \pi\tau) = \theta_4 \left(z + \frac{1}{2}\pi\tau + \pi\tau \right) = -q^{-1} e^{-2i(z+\pi\tau/2)} \theta_4 \left(z + \frac{1}{2}\pi\tau \right)$$

$$= -q^{-1} e^{-2iz} e^{-\pi i\tau} g(z),$$

a complex periodicity factor quite different from that of $\theta_4(z)$ (see (4.7)). We can remove both difficulties by writing

$$g(z) = \theta_1(z),$$

where $\theta_1(z)$ is defined by

$$\theta_1(z) = A e^{iz} \theta_4 \left(z + \frac{1}{2}\pi\tau \right), \qquad (4.11)$$

where the constant $A \neq 0$ will be chosen later. It follows from (4.11) that $\theta_1(z)$ has period 2π and

$$\theta_1(z + \pi\tau) = -q^{-1} e^{-2iz} \theta_1(z). \qquad (4.12)$$

If we now set $F(z) = \theta_1(z)/\theta_4(z)$ we find that $F(z)$ is a meromorphic function having the same periods 2π and $\pi\tau$ as $sn(u)$, simple zeros at the simple zeros of $sn(u)$ and simple poles at the simple poles of $sn(u)$. It follows as before that $sn(u)/F(z)$ is a doubly periodic entire function, whence, by Liouville's Theorem again, we have

$$sn(u) = S \frac{\theta_1(z)}{\theta_4(z)}, \qquad z = \frac{\pi u}{2K} \qquad (4.13)$$

for some constant S depending on k and A.

Finally, consider the proposed representation $cn(u) = g(z)/\theta_4(z)$, where now $g(z)$ must have period 2π and vanish when $z \equiv \pi/2 (\text{mod } \pi, \pi\tau)$. We accomplish both objectives by setting $g(z) = \theta_2(z)$, where $\theta_2(z)$ is defined by

$$\theta_2(z) = \theta_1 \left(z + \frac{\pi}{2} \right). \qquad (4.14)$$

Moreover, (4.13) gives

$$\theta_2(z + \pi\tau) = \theta_1 \left(z + \frac{\pi}{2} + \pi\tau \right) = q^{-1} e^{-2iz} \theta_2(z), \qquad (4.15)$$

by (4.12) and (4.14). Since $\theta_1(z+\pi) = -\theta_1(z)$, it follows that $\theta_2(z+\pi) = -\theta_2(z)$ and hence, if we set

$$G(z) = \frac{\theta_2(z)}{\theta_4(z)},$$

then

$$G(z+\pi) = G(z+\pi\tau) = -G(z),$$

and so $G(z)$ has periods $\pi + \pi\tau$ and 2π. Proceeding as before, we obtain

$$cn\,(u) = C\frac{\theta_2(z)}{\theta_4(z)}, \qquad z = \frac{\pi u}{2K} \tag{4.16}$$

for some constant C depending on k and A.

It remains to determine the constant A in (4.11). Clearly,

$$\theta_1(z) = Ae^{iz}\theta_4\left(z + \frac{1}{2}\pi\tau\right) = Ae^{iz}\sum_{n=-\infty}^{\infty}(-1)^n q^{n^2}e^{2ni(z+\frac{1}{2}\pi\tau)}$$

$$= Ae^{iz}\sum_{n=-\infty}^{\infty}(-1)^n q^{n^2}q^n e^{2niz}$$

$$= Aq^{-1/4}\sum_{n=-\infty}^{\infty}(-1)^n q^{(n+\frac{1}{2})^2}e^{i(2n+1)z}$$

$$= 2Aiq^{-1/4}\sum_{n=0}^{\infty}(-1)^n q^{(n+\frac{1}{2})^2}\sin(2n+1)z,$$

since $m + 1/2 = -(n + 1/2)$ when $m = -(n + 1)$. (Throughout this chapter we set $q^\lambda = \exp(\lambda\pi i\tau)$, to avoid ambiguity.) An obvious choice is to make $Aiq^{-1/4} = 1$, whence we should take

$$A = -iq^{1/4}.$$

It follows that

$$\theta_1(z) = 2\sum_{n=0}^{\infty}(-1)^n q^{(n+\frac{1}{2})^2}\sin(2n+1)z. \tag{4.17}$$

Moreover, the product formula for $\theta_4(z)$ (see (4.6)) yields

$$\theta_1(z) = -iq^{1/4}\theta_4\left(z + \frac{1}{2}\pi\tau\right)e^{iz}$$

$$= -iq^{1/4}G\prod_{n=1}^{\infty}(1 - q^{2n-2}e^{-2iz})(1 - q^{2n}e^{2iz})e^{iz}$$

$$= 2Gq^{1/4}\sin z\prod_{n=1}^{\infty}(1 - q^{2n}e^{-2iz})(1 - q^{2n}e^{2iz})$$

(on combining the first factor in the product, for $n = 1$, with the e^{iz} term). Whence

$$\theta_1(z) = 2Gq^{1/4} \sin z \prod_{n=1}^{\infty} (1 - 2q^{2n} \cos 2z + q^{4n}). \tag{4.18}$$

Since $\theta_2(z) = \theta_1(z + \pi/2)$, (4.17) and (4.18) imply

$$\theta_2(z) = 2 \sum_{n=0}^{\infty} q^{(n+1/2)^2} \cos(2n + 1)z \tag{4.19}$$

and

$$\theta_2(z) = 2Gq^{1/4} \cos z \prod_{n=1}^{\infty} (1 + q^{2n} \cos 2z + q^{4n}). \tag{4.20}$$

Now replace z by $z + \pi\tau/2$ in the definition of $\theta_1(z)$ to obtain:

$$\theta_1\left(z + \frac{1}{2}\pi\tau\right) = -iq^{3/4}e^{iz}\theta_4(z + \pi\tau) = iq^{-1/4}e^{-iz}\theta_4(z).$$

But the definition of $\theta_1(z)$ yields

$$\theta_4\left(z + \frac{1}{2}\pi\tau\right) = iq^{-1/4}e^{-iz}\theta_1(z)$$

and so

$$\frac{\theta_1\left(z + \dfrac{1}{2}\pi\tau\right)}{\theta_4\left(z + \dfrac{1}{2}\pi\tau\right)} = \frac{\theta_4(z)}{\theta_1(z)}, \tag{4.21}$$

a result we shall need in the evaluation of the constant S in (4.13), to which we now proceed.

Set $u = K$ in (4.13) to obtain

$$1 = S\frac{\theta_1\left(\dfrac{\pi}{2}\right)}{\theta_4\left(\dfrac{\pi}{2}\right)} = S\frac{\theta_2(0)}{\theta_3(0)},$$

whence

$$S = \frac{\theta_3(0)}{\theta_2(0)} = \frac{\displaystyle\prod_{n=1}^{\infty}(1 + q^{2n-1})^2}{2q^{1/4}\displaystyle\prod_{n=1}^{\infty}(1 + q^{2n})^2} > 0.$$

Now set $u = K + iK'$ in (4.13), and we find that

$$k^{-1} = sn(K + iK') = S \frac{\theta_1\left(\dfrac{\pi}{2} + \dfrac{1}{2}\pi\tau\right)}{\theta_4\left(\dfrac{\pi}{2} + \dfrac{1}{2}\pi\tau\right)} = S\frac{\theta_4\left(\dfrac{\pi}{2}\right)}{\theta_1\left(\dfrac{\pi}{2}\right)} = S\frac{\theta_3(0)}{\theta_2(0)} = S^2,$$

whence $S = k^{-1/2}$ and $k^{1/2} = S^{-1} = \theta_2(0)/\theta_3(0)$.

Finally, we set $u = 0$ in (4.16) to obtain

$$C = \frac{\theta_4(0)}{\theta_2(0)} = \frac{\theta_4(0)/\theta_3(0)}{\theta_2(0)/\theta_3(0)} = k'^{1/2}k^{-1/2}.$$

4.4 Summary

In the foregoing, for the sake of brevity, we have omitted the parameter τ and written $\theta_j(z)$ for $\theta_j(z|\tau)$. When we wish to emphasize the dependence on the parameter τ we shall use the latter expression and when we wish to emphasize dependence on q we shall write $\theta_j(z, q)$.

Note that $\theta_1(z)$ is an *odd* function and the other three theta functions are *even* functions of z. Finally, it is convenient (and a standard notation) to introduce the abbreviations $\theta_j = \theta_j(0)\,(j = 2, 3, 4)$ and

$$\theta_1' = \theta_1'(0) = \lim_{z \to 0} \frac{\theta_1(z)}{z}.$$

Our representation formulae are then given by

Theorem 4.3 *With the foregoing notation,*

$$sn\,(u) = k^{-1/2}\frac{\theta_1(z)}{\theta_4(z)}, \tag{4.22}$$

$$cn\,(u) = k'^{1/2}k^{-1/2}\frac{\theta_2(z)}{\theta_4(z)}, \tag{4.23}$$

$$dn\,(u) = k'^{1/2}\frac{\theta_3(z)}{\theta_4(z)}, \tag{4.24}$$

where

$$z = \frac{\pi u}{2K}, \quad k^{1/2} = \frac{\theta_2}{\theta_3}, \quad k'^{1/2} = \frac{\theta_4}{\theta_3}. \tag{4.25}$$

Note that

$$1 = k^2 + k'^2 = \left(\theta_2^4 + \theta_4^4\right)/\theta_3^4,$$

whence follows the 'remarkable' (to quote Jacobi (1829)) identity

$$\theta_3^4 = \theta_2^4 + \theta_4^4, \qquad (4.26)$$

or

$$16q(1 + q^{1\cdot2} + q^{2\cdot3} + q^{3\cdot4} + \cdots)^4 + (1 - 2q + 2q^4 - 2q^9 + \cdots)^4$$
$$= (1 + 2q + 2q^4 + 2q^9 + \cdots)^4.$$

Moreover, the identity

$$cn^2u + sn^2u = 1$$

translates into the theta function identity

$$\theta_4^2\theta_2^2(z) = \theta_2^2\theta_4^2(z) - \theta_3^2\theta_1^2(z), \qquad (4.27)$$

and

$$dn^2u + k^2sn^2u = 1$$

becomes

$$\theta_4^2\theta_3^2(z) = \theta_3^2\theta_4^2(z) - \theta_2^2\theta_1^2(z). \qquad (4.28)$$

On replacing z by $z + \pi/2$, we obtain two more identities:

$$\theta_4^2\theta_1^2(z) = \theta_2^2\theta_3^2(z) - \theta_3^2\theta_2^2(z), \qquad (4.29)$$

$$\theta_4^2\theta_4^2(z) = \theta_3^2\theta_3^2(z) - \theta_2^2\theta_2^2(z). \qquad (4.30)$$

(See Whittaker & Watson (1927), pp. 466–467, or Jacobi (1838), pp. 11–17.)

The proofs of those identities given in Jacobi (1838) and Whittaker & Watson (1927) are different and demand a considerable facility with the handling of complicated expressions involving the theta functions; they are valid for general values of $q = \exp(\pi i\tau)$, whereas our proofs, relying as they do on properties of the Jacobi elliptic functions obtained in Chapter 2, presuppose that q is real and $0 < q < 1$. However, since $\theta(z, q)$ is analytic in both z and q the identities for all values of q, with $|q| < 1$, follow from the Uniqueness Theorem (Theorem 2.4).

Exercises 4.4

4.4.1 Verify that

$$\theta_1' = 2q^{1/4} \sum_{n=0}^{\infty} (-1)^n (2n+1) q^{n(n+1)},$$

$$\theta_2 = 2q^{1/4} \sum_{n=0}^{\infty} q^{n(n+1)},$$

$$\theta_3 = 1 + 2 \sum_{n=1}^{\infty} q^{n^2},$$

$$\theta_4 = 1 + 2 \sum_{n=1}^{\infty} (-1)^n q^{n^2}.$$

4.4.2 Verify that the multipliers of the theta functions associated with the periods $\pi, \pi\tau$ are given by the following table (see Whittaker & Watson, (1927), p.465):

	$\theta_1(z)$	$\theta_2(z)$	$\theta_3(z)$	$\theta_4(z)$
π	-1	-1	1	1
$\pi\tau$	$-N$	N	N	$-N$

where $N = q^{-1} e^{-2iz}$.

4.4.3 Verify that

$$\theta_1' = 2q^{1/4} G \prod_{n=1}^{\infty} (1 - q^{2n})^2,$$

$$\theta_2 = 2q^{1/4} G \prod_{n=1}^{\infty} (1 + q^{2n})^2,$$

$$\theta_3 = G \prod_{n=1}^{\infty} (1 + q^{2n-1})^2,$$

$$\theta_4 = G \prod_{n=1}^{\infty} (1 - q^{2n-1})^2,$$

where G is the constant introduced above and which we shall see in Section 4.6 is given by $G = \prod_{n=1}^{\infty} (1 - q^{2n})$. Note that it follows that θ_3, θ_4 cannot vanish for $|q| < 1$.

4.4.4 From the foregoing discussion, we see that starting from any one of the four theta functions one can arrive at multiples of the other three by making the replacements $z \mapsto z + \frac{\pi}{2} \mapsto z + \frac{\pi}{2} + \frac{\pi\tau}{2} \mapsto z + \frac{\pi\tau}{2}$. Show that (see Whittaker & Watson (1927), p. 464):

$$\theta_1(z) = -\theta_2\left(z + \frac{\pi}{2}\right) = -iM\theta_3\left(z + \frac{\pi}{2} + \frac{\pi\tau}{2}\right) = -iM\theta_4\left(z + \frac{\pi\tau}{2}\right),$$

$$\theta_2(z) = M\theta_3\left(z + \frac{\pi\tau}{2}\right) = M\theta_4\left(z + \frac{\pi}{2} + \frac{\pi\tau}{2}\right) = \theta_1\left(z + \frac{\pi}{2}\right),$$

$$\theta_3(z) = \theta_4\left(z + \frac{\pi}{2}\right) = M\theta_1\left(z + \frac{\pi}{2} + \frac{\pi\tau}{2}\right) = M\theta_2\left(z + \frac{\pi\tau}{2}\right),$$

$$\theta_4(z) = -iM\theta_1\left(z + \frac{\pi\tau}{2}\right) = iM\theta_2\left(z + \frac{\pi}{2} + \frac{\pi\tau}{2}\right) = \theta_3\left(z + \frac{\pi}{2}\right),$$

where $M = q^{1/4}e^{iz}$.

4.4.5 (See Whittaker & Watson (1927), p.464.) Show that

$$\theta_3(z, q) = \theta_3(2z, q^4) + \theta_2(2z, q^4),$$
$$\theta_4(z, q) = \theta_3(2z, q^4) - \theta_2(2z, q^4).$$

4.4.6 (See Whittaker & Watson (1927), p.470.) By differentiating the series for $\theta_j(z|\tau)$ $(j = 1, 2, 3, 4)$ and taking $t = -i\tau$ $(> 0$ when $0 < k < 1)$, show that $y = \theta_j(z|\tau)$ satisfies the heat (or diffusion) equation

$$\frac{\partial y}{\partial t} = \frac{1}{4}\pi \frac{\partial^2 y}{\partial z^2}.$$

4.4.7 Verify that the zeros of $\theta_1(z), \theta_2(z), \theta_3(z), \theta_4(z)$ are, respectively, $z \equiv 0, \pi/2, \pi/2 + \pi\tau/2, \pi\tau/2 \pmod{\pi, \pi\tau}$.

4.4.8 Use the product formulae of Exercise 4.4.3 to rewrite the identity $\theta_4^4 + \theta_2^4 = \theta_3^4$ (see (4.26)) as $\prod_{n=1}^{\infty}(1 - q^{2n-1})^8 + 16q\prod_{n=1}^{\infty}(1 + q^{2n})^8 = \prod_{n=1}^{\infty}(1 + q^{2n-1})^8$.

(In his (1829) book, Jacobi describes that result as 'aequatio identica satis abstrusa'; see Whittaker & Watson (1927), p. 470.)

4.5   The pseudo-addition formulae

The formulae (4.27) to (4.30) are particular cases of identities, due to Jacobi (see (1838), Sections A, B, C, D, pp.10–17 or Whittaker & Watson (1927), p.467 and pp.487–488), which express theta functions with arguments of the form $x + y, x - y$ simultaneously in terms of those with arguments x and y separately. A prototype from the trigonometric functions is the identity

$$\cos(u + v)\cos(u - v) = \frac{1}{2}(\cos 2u + \cos 2v) = \cos^2 u - \sin^2 v.$$

That last identity generalizes to cn, as follows (the notation is that of Chapter 2, Section 2.4). We have

$$cn(u + v)cn(u - v) = (c_1c_2 - s_1s_2d_1d_2)(c_1c_2 + s_1s_2d_1d_2)/\Delta^2$$
$$= (c_1^2c_2^2 - s_1^2s_2^2d_1^2d_2^2)/\Delta^2$$
$$= [c_1^2(\Delta - s_2^2d_2^2) - s_2^2d_1^2(\Delta - c_1^2)]/\Delta^2$$
$$= (c_1^2 - s_2^2d_1^2)/\Delta;$$

that is,

$$cn(u + v)cn(u - v) = \frac{cn^2u - sn^2v\, dn^2u}{1 - k^2sn^2u\, sn^2v}. \tag{4.31}$$

Now set $x = \pi u/2K$, $y = \pi v/2K$ (cf. (4.16)) and then express the elliptic functions in (4.31) as quotients of theta functions to obtain

$$\frac{\theta_2(x + y)\theta_2(x - y)}{\theta_4(x + y)\theta_4(x - y)} = \frac{\theta_2^2(x)\theta_4^2(y) - \theta_3^2(x)\theta_1^2(y)}{\theta_4^2(x)\theta_4^2(y) - \theta_1^2(x)\theta_1^2(y)}$$

or

$$\frac{\theta_4^2(x)\theta_4^2(y) - \theta_1^2(x)\theta_1^2(y)}{\theta_4(x + y)\theta_4(x - y)} = \frac{\theta_2^2(x)\theta_4^2(y) - \theta_3^2(x)\theta_1^2(y)}{\theta_2(x + y)\theta_2(x - y)}.$$

Fix y and call the common value (of the two expressions on either side) $f(x)$. Clearly $f(x + \pi) = f(x)$ and $f(x + \pi\tau) = f(x)$. It follows that $f(x)$ is a doubly periodic function whose only possible singularities are poles at the points where

(A) $\theta_4(x + y) = 0$ or   (B) $\theta_4(x - y) = 0$

and *simultaneously*

(C) $\theta_2(x + y) = 0$ or   (D) $\theta_2(x - y) = 0$.

Now (A) and (B) cannot hold simultaneously, for then neither (C) nor (D) could hold. If $\theta_4(x + y) = 0$, then $\theta_2(x - y) = 0$ and from Exercise 4.4.7 it follows that $x \equiv \pi/4 + \pi\tau/4 \pmod{\pi, \pi\tau}$. The same conclusion for x holds if $\theta_4(x - y) = 0$. Hence $f(x)$ has at most a simple pole in any period parallelogram. Theorem 3.5 then implies that $f(x) \equiv f(0) = \theta_4^2$. It follows that

$$\theta_4^2\theta_4(x + y)\theta_4(x - y) = \theta_4^2(x)\theta_4^2(y) - \theta_1^2(x)\theta_1^2(y), \tag{4.32}$$

and

$$\theta_4^2\theta_2(x + y)\theta_2(x - y) = \theta_2^2(x)\theta_4^2(y) - \theta_3^2(x)\theta_1^2(y). \tag{4.33}$$

Our proof is valid for $0 < q < 1$, the extension to arbitrary q, $|q| < 1$, follows as usual by the Uniqueness Theorem, Theorem 2.4.

Because of the relations (4.27) to (4.30) connecting the squares of the theta functions, one can rewrite (4.32) and (4.33) in other forms; thus, for example,

$$\theta_2^2\theta_2(x+y)\theta_2(x-y) = \theta_4^{-2}\theta_2^2[\theta_2^2(x)\theta_4^2(y) - \theta_3^2(x)\theta_1^2(y)]$$

$$= \theta_4^{-2}\theta_4^2(y)[\theta_3^2\theta_3^2(x) - \theta_4^2\theta_4^2(x)]$$

$$= \theta_4^{-2}\theta_3^2(x)[\theta_3^2\theta_4^2(y) - \theta_4^2\theta_3^2(y)]$$

or

$$\theta_2^2\theta_2(x+y)\theta_2(x-y) = \theta_3^2(x)\theta_3^2(y) - \theta_4^2(x)\theta_4^2(y). \tag{4.34}$$

(Jacobi (1838), pp. 515–517, used (4.34) as the starting point for one of his proofs of the result $\theta_1' = \theta_2\theta_3\theta_4$ (see Theorem 4.5, below).

Exercises 4.5

4.5.1   Use the ideas behind (4.34) to prove that (4.32) yields

$$\theta_3^2\theta_4(x+y)\theta_4(x-y) = \theta_4^2(x)\theta_3^2(y) + \theta_2^2(x)\theta_1^2(y). \tag{4.35}$$

4.5.2   Obtain the addition formulae

$$\theta_4^2\theta_1(x+y)\theta_1(x-y) = \theta_3^2(x)\theta_2^2(y) - \theta_2^2(x)\theta_3^2(y) = \theta_1^2(x)\theta_4^2(y) - \theta_4^2(x)\theta_1^2(y),$$

$$\theta_4^2\theta_2(x+y)\theta_2(x-y) = \theta_4^2(x)\theta_2^2(y) - \theta_1^2(x)\theta_3^2(y) = \theta_2^2(x)\theta_4^2(y) - \theta_3^2(x)\theta_1^2(y),$$

$$\theta_4^2\theta_3(x+y)\theta_3(x-y) = \theta_4^2(x)\theta_3^2(y) - \theta_1^2(x)\theta_2^2(y) = \theta_3^2(x)\theta_4^2(y) - \theta_2^2(x)\theta_1^2(y),$$

$$\theta_4^2\theta_4(x+y)\theta_4(x-y) = \theta_3^2(x)\theta_3^2(y) - \theta_2^2(x)\theta_2^2(y) = \theta_4^2(x)\theta_4^2(y) - \theta_1^2(x)\theta_1^2(y).$$

4.5.3   Obtain the addition formulae

$$\theta_2^2\theta_4(x+y)\theta_4(x-y) = \theta_4^2(x)\theta_2^2(y) + \theta_3^2(x)\theta_1^2(y) = \theta_2^2(x)\theta_4^2(y) + \theta_1^2(x)\theta_3^2(y)$$

and

$$\theta_3^2\theta_2(x+y)\theta_4(x-y) = \theta_4^2(x)\theta_3^2(y) + \theta_2^2(x)\theta_1^2(y) = \theta_3^2(x)\theta_4^2(y) + \theta_1^2(x)\theta_2^2(y).$$

By replacing x by $x + \pi/2$, $x + \pi/2 + \pi\tau/2$, $x + \pi\tau/2$ in turn obtain the corresponding formulae for

$$\theta_2^2\theta_r(x+y)\theta_r(x-y), \quad \theta_3^2\theta_r(x+y)\theta_r(x-y),$$

where $r = 1, 2, 3$.

4.5.4   Obtain the duplication formulae

$$\theta_2\theta_4^2\theta_2(2x) = \theta_2^2(x)\theta_4^2(x) - \theta_1^2(x)\theta_3^2(x),$$
$$\theta_3\theta_4^2\theta_3(2x) = \theta_3^2(x)\theta_4^2(x) - \theta_1^2(x)\theta_2^2(x),$$
$$\theta_4^3\theta_4(2x) = \theta_3^4(x) - \theta_2^4(x) = \theta_4^4(x) - \theta_1^4(x).$$

(The results in Questions 1, 2, 3, 4 are all due to Jacobi; see Whittaker & Watson (1927), pp. 487–488.)

4.6   The constant G and the Jacobi identity

In Theorem 4.3, (4.22), divide both sides by u and let $u \to 0$ to obtain

$$1 = k^{-1/2}(\pi/2K)\theta_1'/\theta_4;$$

that is

$$\frac{2K}{\pi} = \frac{\theta_1'\theta_3}{\theta_2\theta_4} = \frac{\theta_1'}{\theta_2\theta_3\theta_4}\theta_3^2 = \alpha\theta_3^2, \tag{4.36}$$

where $\alpha = \alpha(q) = \theta_1'/(\theta_2\theta_3\theta_4)$. By Exercise 4.4.1, we have

$$\theta_1' = 2q^{1/4}(1 + O(q^2)),$$
$$\theta_2 = 2q^{1/4}(1 + O(q^2)),$$
$$\theta_3 = 1 + O(q),$$
$$\theta_4 = 1 + O(q),$$

as $q \to 0$ and it follows immediately that $\lim_{q \to 0} \alpha(q) = 1$. If we retain terms involving higher powers of q, we obtain

$$\theta_1' = 2q^{1/4}(1 - 3q^2 + 5q^6 + O(q^{12}))$$
$$\theta_2\theta_3\theta_4 = 2q^{1/4}(1 + q^2 + q^6 + O(q^{12}))(1 + 2(q + q^4 + q^9 + O(q^{16})))$$
$$\times (1 + 2(-q + q^4 - q^9 + O(q^{16}))),$$
$$= 2q^{1/4}(1 - 3q^2 + 5q^6 + O(q^{12})).$$

By considering the limits as $q \to 0$ we are led to the conjecture that

$$\theta_1' = \theta_2\theta_3\theta_4; \tag{4.37}$$

that is

$$\alpha(q) = 1. \tag{4.38}$$

If the conjecture were true, then we would have, from (4.36),

$$\frac{2K}{\pi} = \theta_3^2 = \theta_3^2(0|\tau). \tag{4.39}$$

But

$$\theta_3^2 = \left(\sum_{k=-\infty}^{\infty} q^{k^2}\right)\left(\sum_{l=-\infty}^{\infty} q^{l^2}\right) = \sum_{k,l=-\infty}^{\infty} q^{k^2+l^2} = \sum_{n=0}^{\infty} r(n)q^n,$$

where (as in Chapter 2, Section 2.9) $r(n)$ denotes the number of representatives of the positive integer n in the form $n = k^2 + l^2$; that is, as a sum of two squares. It follows that (4.39) now becomes

$$\frac{2K}{\pi} = 1 + \sum_{n=1}^{\infty} r(n)q^n,$$

which is (2.55), whence we obtain Theorem 2.6: *if* $n \geq 1$, *then* $r(n) = 4\delta(n)$.

Of course we must now prove (4.37) and (4.38) and we have yet to evaluate the constant G. So we turn first to the product expansions of θ_1', θ_2, θ_3 and θ_4 to derive a relation between α and the as yet unevaluated constant G.

Since every positive integer n is either *even* ($n = 2m$) or odd ($n = 2m - 1$) and since the products converge absolutely (see the Appendix) we can make the following rearrangements:

$$\theta_1' = 2q^{1/4}G \prod_{n=1}^{\infty} (1 + q^n)^2 \prod_{n=1}^{\infty} (1 - q^n)^2$$

$$= 2q^{1/4}G \prod_{m=1}^{\infty} (1 + q^{2m})^2 \prod_{m=1}^{\infty} (1 + q^{2m-1})^2 \prod_{m=1}^{\infty} (1 - q^{2m})^2$$

$$\times \prod_{m=1}^{\infty} (1 - q^{2m-1})^2 \text{ by (4.18)}$$

But, from (4.6), (4.10) and (4.20), we obtain

$$\theta_2\theta_3\theta_4 = 2q^{1/4}G^3 \prod_{m=1}^{\infty} (1 + q^{2m})^2 \prod_{m=1}^{\infty} (1 + q^{2m-1})^2 \prod_{m=1}^{\infty} (1 - q^{2m-1})^2.$$

We have accordingly proved:

Theorem 4.4 *With the foregoing notation*

$$\alpha(q) = \theta_1'/(\theta_2\theta_3\theta_4) = \prod_{n=1}^{\infty} (1 - q^{2n})^2/G^2.$$

Our discussion suggests that we should write

$$G = A(q) \prod_{n=1}^{\infty} (1 - q^{2n}) \tag{4.40}$$

and then Theorem 4.4 reads

$$\alpha(q) = 1/(A(q))^2$$

and our conjecture $\alpha(q) = 1$ is then equivalent to $A(q)^2 = 1$ or $A(q) = \pm 1$. But since $A(q)$ is continuous and $\lim_{q \to 0} A(q) = 1$ (by the remark following (4.9), the minus sign could never hold and so $\alpha(q) = 1$ is equivalent to $A(q) = 1$. Accordingly we are led to:

Theorem 4.5 *(The Jacobi identity): $A(q) = 1$, whence $\theta_1' = \theta_2\theta_3\theta_4$.*

Proof (The following device is attributed to Jacobi by Hardy and Wright (1979), p.295 and to Gauss by Rademacher (1973), p. 170.)

Put $z = \pi/4$ in the identity

$$\theta_3(z, q) = \theta_3(2z, q^4) + \theta_2(2z, q^4)$$

(Exercise 4.4.5) to obtain

$$\theta_3\left(\frac{\pi}{4}, q\right) = \theta_3\left(\frac{\pi}{2}, q^4\right).$$

Now expand each side by the product formula (4.10) and use (4.40) to obtain

$$\theta_3(z, q) = A(q)\prod_{n=1}^{\infty}(1 - q^{2n})\prod_{n=1}^{\infty}(1 + 2q^{2n-1}\cos 2z + q^{4n-2}).$$

As in the proof of Theorem 4.4, we can let the index n run separately over even integers ($n = 2m$) and over odd integers ($n = 2m - 1$). Thus

$$\theta_3\left(\frac{\pi}{4}, q\right) = A(q)\prod_{n=1}^{\infty}(1 - q^{2n})\prod_{n=1}^{\infty}(1 + q^{4n-2})$$

$$= A(q)\prod_{m=1}^{\infty}(1 - q^{4m})\prod_{m=1}^{\infty}(1 - q^{4m-2})\prod_{n=1}^{\infty}(1 + q^{4n-2})$$

$$= A(q)\prod_{l=1}^{\infty}(1 - q^{8l})\prod_{l=1}^{\infty}(1 - q^{8l-4})\prod_{n=1}^{\infty}(1 - q^{8n-4})$$

$$= A(q)\prod_{n=1}^{\infty}(1 - q^{8n})\prod_{n=1}^{\infty}(1 - q^{8n-4})^2.$$

But

$$\theta_3\left(\frac{\pi}{2}, q^4\right) = A(q^4)\prod_{n=1}^{\infty}(1 - q^{8n})\prod_{n=1}^{\infty}(1 - 2q^{8n-4} + q^{16n-8})$$

$$= A(q^4)\prod_{n=1}^{\infty}(1 - q^{8n})\prod_{n=1}^{\infty}(1 - 2q^{8n-4})^2.$$

Whence $A(q) = A(q^4)$ and iteration yields $A(q) = A(q^{4^n})$ for $n = 1, 2,$ $3, \ldots$ Since $|q| < 1, q^{4^n} \to 0$ as $n \to \infty$ and so

$$A(q) = \lim_{n \to \infty} A(q^{4^n}) = \lim_{q \to 0} A(q) = 1.$$

Corollary 4.1 *The constant G introduced in (4.4) is given by*

$$G = \prod_{n=1}^{\infty} (1 - q^{2n}).$$

Corollary 4.2 *In the notation of (2.53) and Theorem 2.6, we have*

$$\frac{2K}{\pi} = 1 + \sum_{n=1}^{\infty} r(n)q^n$$

and

$$r(n) = 4\delta(n).$$

The connection between the series and product formulae for $\theta_3(z, q)$ can now be written, on setting $w = e^{2iz}$, as

$$\prod_{n=1}^{\infty} (1 - q^{2n})(1 + q^{2n-1}w)(1 + q^{2n-1}w^{-1}) = \sum_{n=-\infty}^{\infty} q^{n^2} w^n, \quad w \neq 0.$$
$$(4.41)$$

The formula (4.41) is called 'Jacobi's triple product identity' (see Theorem 352 in Hardy & Wright, 1979) and has profound arithmetical consequences. We shall content ourselves with deriving a famous identity, due to Euler ('Euler's formula for the pentagonal numbers'; see McKean and Moll, 1997, Chapter 3, *'partitio numerorum'*, as well as Hardy & Wright, 1979).

Theorem 4.6 *(Euler's formula for the pentagonal numbers)*

For $|x| < 1$, we have

$$\prod_{n=1}^{\infty} (1 - x^n) = \sum_{n=-\infty}^{\infty} (-1)^n x^{n(3n+1)/2}.$$

Proof If we write x^k for q and $-x^l$ for w, and replace n by $n + 1$, in the left hand side of (4.41), we obtain

$$\prod_{n=0}^{\infty} (1 - x^{2kn+k-l})(1 - x^{2kn+k+l}) = \sum_{n=-\infty}^{\infty} (-1)^n x^{kn^2+ln}.$$

Now set $k = 3/2$ and $l = 1/2$ to obtain

$$\prod_{n=0}^{\infty}(1 - x^{3n+1})(1 - x^{3n+2})(1 - x^{3n+3}) = \sum_{n=-\infty}^{\infty} (-1)^n x^{n(3n+1)/2},$$

The result in Theorem 4.6 is called 'Euler's formula for the pentagonal numbers'; we explain why, primarily to offer an introduction to the theory of partitions, *'partitio numerorum'*, a beautiful branch of the theory of numbers. (See the books by Hardy & Wright, 1979 and by McKean & Moll, 1997, already referred to, and the book by Knopp, 1970, for further reading.)

The number of representations of a positive integer n as a sum

$$n = a_1 + \cdots + a_r,$$

where the a_i are positive integers, not necessarily different, and where the order is irrelevant is called the *number of 'unrestricted partitions of n'* and is denoted by $p(n)$. Thus

$$5 = 4 + 1 = 3 + 2 = 3 + 1 + 1 = 2 + 2 + 1 = 2 + 1 + 1 + 1$$
$$= 1 + 1 + 1 + 1 + 1 \quad \text{and so} \quad p(5) = 7.$$

Expand the product on the left-hand side in Theorem 4.6 to obtain

$$\prod_{n=1}^{\infty}(1 - x^n) = \sum_{d=0}^{\infty}(-1)^d \sum_{n_1,\ldots,n_d \geq 1} x^{n_1 + \cdots + n_d}$$

$$= 1 + \sum_{m=1}^{\infty} x^m \sum_{n_1 + \cdots + n_d = m} (-1)^d$$

$$= 1 + \sum_{m=1}^{\infty} x^m [p_e(m) - p_o(m)],$$

where $p_e(m)$ denotes the number of partitions of m into an even number of *unequal* parts and $p_o(m)$ denotes the number of partitions of m into an odd number of *unequal* parts. Accordingly, Theorem 4.6 asserts that

$$\sum_{n\in\mathbb{Z}}(-1)^n x^{n(3n+1)/2} = 1 + \sum_{m=1}^{\infty} x^m[p_e(m) - p_o(m)]. \qquad (4.42)$$

From that last we obtain the beautiful result, due to Euler, that $p_e(m) - p_o(m) = (-1)^n$ or 0 according as m is a pentagonal number $m = n(3n \pm 1)/2$, or not. For example, if $n = 2$ and $m = 7$, then

$$7 = 6 + 1 = 5 + 2 = 4 + 3 = 4 + 2 + 1,$$
$$p_e(7) = 3, \quad p_o(7) = 2$$

and

$$p_e(7) - p_o(7) = 1 = (-1)^2.$$

The adjective pentagonal is explained in Exercise 4.6.5.

Exercises 4.6

4.6.1 Rewrite Theorem 4.4 as the identity

$$\prod_{n=1}^{\infty}(1 + q^n) = \prod_{n=1}^{\infty}(1 - q^{2n-1})^{-1}.$$

4.6.2 Show that $\displaystyle\sum_{n=-\infty}^{\infty} (-1)^n x^{n(3n+1)/2}$ may be written as

$$1 + \sum_{n=1}^{\infty}(-1)^n \left\{ x^{n(3n-1)/2} + x^{n(3n+1)/2} \right\}$$

$$= 1 - x - x^2 + x^5 + x^7 - x^{12} - x^{15} + \cdots.$$

4.6.3 By expanding the infinite product and comparing coefficients of powers of q, show that

$$\prod_{n=1}^{\infty}(1 - q^n)^{-1} = 1 + \sum_{d=1}^{\infty}\sum_{n_1,\ldots,n_d \geq 1} q^{n_1 + \cdots + n_d} = 1 + \sum_{n=1}^{\infty} p(n)q^n.$$

4.6.4 Use the result in 4.6.3 (with x in place of q) to show that

$$(1 - x - x^2 + x^5 + x^7 - \cdots)\left\{ 1 - \sum_{n=1}^{\infty} p(n)x^n \right\}$$

$$= \frac{1 - x - x^2 + x^5 + x^7 + \cdots}{(1 - x)(1 - x^2)(1 - x^3)\cdots} = 1.$$

By equating coefficients, show that

$$p(n) - p(n - 1) - p(n - 2) + p(n - 5) + \cdots$$
$$+ (-1)^k p\left\{ n - \frac{1}{2}k(3k - 1) \right\} + (-1)^k p\left\{ n - \frac{1}{2}k(3k + 1) \right\} + \cdots = 0.$$

(In the days before electronic calculators, that last result was used by Macmahon to prove that

$$p(200) = 3\,972\,999\,029\,388.$$

(See Hardy & Wright, 1979, Chapter XIX, for further details.)

4.6.5   (Pentagonal numbers.)

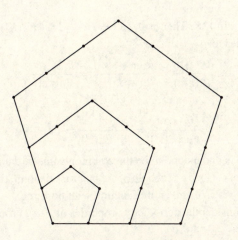

Figure 4.1 Pentagonal numbers.

Consider the diagram in Figure 4.1; it consists of pentagons with 'vertices' at the points indicated. Verify that $n(3n - 1)/2$ is the number of vertices in Figure 4.1 corresponding to the $(n - 1)$th pentagon. (Thus $n = 2$ gives the first pentagon and 5 vertices, etc.) (Hint: the increment $n(3n - 1)/2 - (n - 1)(3(n - 1) - 1)/2 = 3n - 2$ corresponds to the adjunction of three lines, with n vertices apiece, two of which are shared, whence the '-2'. See McKean & Moll, 1997, p. 144.)

4.7 Jacobi's imaginary transformation

When $0 < k < 1$ the replacement $k \to k'$ transforms $\tau = iK'/K$ into $\tau' = iK/K' = -\tau^{-1}$. We explore the relationship between $\theta_j(z|\tau)$ and $\theta_j(\tau'z|\tau')$ for $1 \le j \le 4$ and obtain the identities (4.49) to (4.52), below. The associated identities (2.11) transform into theta function identities upon expressing the elliptic functions as quotients of theta functions (cf. (2.11) and Theorem 4.3). Thus:

$$dn(iw, k) = dn(w, k')/cn(w, k'),$$
$$cn(w, k) = k'^{1/2}k^{-1/2}\theta_2(\pi w/2K|\tau)/\theta_4(\pi w/2K|\tau),$$

and

$$dn(w, k) = k'^{1/2}\theta_3(\pi w/2K|\tau)/\theta_4(\pi w/2K|\tau)$$

yield

$$k'^{1/2}\theta_3(\pi iw/2K|\tau)/\theta_4(\pi iw/2K|\tau)$$
$$= k'^{1/2}\theta_3(\pi w/2K'|\tau')/\theta_2(\pi w/2K'|\tau').$$

Now write $z = \pi i w /2K$. Then $\pi w /2K' = (\pi i w /2K)(K/iK') = -\tau' z$ and the identity simplifies to

$$\frac{\theta_3(z|\tau)}{\theta_4(z|\tau)} = \frac{\theta_3(-\tau'z|\tau')}{\theta_2(-\tau'z|\tau')} = \frac{\theta_3(\tau'z|\tau')}{\theta_2(\tau'z|\tau')}$$

or

$$\frac{\theta_3(z|\tau)}{\theta_3(\tau'z|\tau')} = \frac{\theta_4(z|\tau)}{\theta_2(\tau'z|\tau')} = F(z), \tag{4.43}$$

say, where $F(z)$ is understood to be the common value of the two expressions. Since the zeros of $\theta_2(z)$, $\theta_3(z)$ and $\theta_4(z)$ are pairwise disjoint, the identity (4.43) implies that $F(z)$ is an even, entire function with no zeros.

Let us now consider the periodicity properties of $F(z)$. From Exercise 4.5.2, we obtain

$$F(z+\pi) = \frac{\theta_3(z+\pi|\tau)}{\theta_3(\tau'z + \tau'\pi|\tau')} = e^{i\tau'(2z+\pi)} F(z) \tag{4.44}$$

and

$$F(z+\pi\tau) = \frac{\theta_3(z+\pi\tau|\tau)}{\theta_3(\tau'z - \pi|\tau')} = e^{-i(2z+\pi\tau)} F(z). \tag{4.45}$$

Now the simplest non-trivial, non-vanishing even, entire function is $f(z) = e^{az^2}$, which satisfies the functional equation

$$f(z+s) = e^{a(z+s)^2} = e^{as(2z+s)} f(z). \tag{4.46}$$

The functional equation (4.46) coincides with (4.44) when $s = \pi$ if $a\pi = i\tau'$ and with (4.45) when $s = \pi\tau$ if $a\pi\tau = -i$; that is $a\pi = -i\tau^{-1} = i\tau'$, again. Hence if we set $a = i\tau'/\pi = 1/\pi i\tau$, we have

$$f(z+\pi) = e^{i\tau'(2z+\pi)} f(z) \tag{4.47}$$

and

$$f(z+\pi\tau) = e^{-i(2z+\pi\tau)} f(z). \tag{4.48}$$

The next step is obvious! We want to relate $F(z)$ and $f(z)$ and so we define $h(z)$ by

$$h(z) = \frac{F(z)}{f(z)} = \frac{e^{-z^2/\pi i\tau}\theta_3(z|\tau)}{\theta_3(\tau'z|\tau')} = \frac{e^{-z^2/\pi i\tau}\theta_4(z|\tau)}{\theta_2(\tau'z|\tau')},$$

by (4.43). Then $h(z+\pi) = h(z)$ and $h(z+\pi\tau) = h(z)$ and, since $h(z)$ is an entire function, Liouville's Theorem, together with (4.36), says that

$$h(z) = h(0) = \frac{\theta_3(0|\tau)}{\theta_3(0|\tau')} = \frac{(2K/\pi)^{1/2}}{(2K'/\pi)^{1/2}} = \left(\frac{K'}{K}\right)^{-1/2} = (-i\tau)^{-1/2}.$$

Hence

$$\theta_3(z|\tau) = (-i\tau)^{-1/2} \exp\left(\frac{z^2}{\pi i \tau}\right) \theta_3(\tau' z|\tau') \qquad (4.49)$$

and

$$\theta_4(z|\tau) = (-i\tau)^{-1/2} \exp\left(\frac{z^2}{\pi i \tau}\right) \theta_2(\tau' z|\tau'), \qquad (4.50)$$

$(\tau' = -\tau^{-1})$ when $0 < k < 1$. Another appeal to the Uniqueness Theorem establishes (4.49) and (4.50) for arbitrary τ with $\mathrm{Im}(\tau) > 0$, provided that one adopts the convention that $|\arg(-i\tau)| < \pi/2$.

We note that (4.50) follows from (4.49), by elementary algebra, on replacing z by $z + \pi/2$ and then using Exercise 4.5.4. Again, replacing z by $z + \pi\tau/2$, (4.49) and (4.50) yield

$$\theta_2(z|\tau) = (-i\tau)^{-1/2} \exp(z^2/\pi i \tau)\theta_4(\tau' z|\tau') \qquad (4.51)$$

and

$$\theta_1(z|\tau) = -i(-i\tau)^{-1/2} \exp(z^2/\pi i \tau)\theta_1(\tau' z|\tau'). \qquad (4.52)$$

We could have started, alternatively, with the identities

$$sn(iw, k) = i\, sn(w, k')/cn(w, k')$$

or

$$cn(iw, k) = 1/cn(w, k'),$$

but the effect would have been to permute equations (4.49) to (4.52).

As an example of the numerical utility of Jacobi's imaginary transformation, set $t = -i\tau$ and $g(t) = \theta_3(0|\tau) = 1 + 2\sum_{n=1}^{\infty} e^{n^2 \pi i \tau} = 1 + 2\sum_{n=1}^{\infty} e^{-\pi n^2 t}$. Then $\tau' = -\tau^{-1} = -(it)^{-1} = it^{-1}$ and (4.49) becomes

$$g(t) = t^{-1/2}\theta_3(0|\tau') = t^{-1/2}g(t^{-1}),$$

or

$$1 + 2\sum_{n=1}^{\infty} e^{-\pi n^2 t} = g(t) = t^{-1/2}\left[1 + 2\sum_{n=1}^{\infty} e^{-\pi n^2/t}\right]. \qquad (4.53)$$

Suppose we want to evaluate the function $g(t)$ at the point $t = 0.01$ (a plausible excuse for that might be the connection with the *heat equation* – see later). Although the series on the *left* in (4.53) converges for $t > 0$, one must take

$n = 28$ to make $e^{-0.01n^2\pi} \sim 10^{-10.7}$. Consequently the first 28 terms would yield an accuracy of the order of ten significant figures. But the *right-hand* side of (4.53) converges rapidly for small t; when $n = 1$,

$$e^{-\pi n^2 \cdot (0.01)^{-1}} = 10^{-136.4}.$$

Hence one term of the series yields an accuracy of the order of one hundred significant figures.

We recall from Exercise 4.4.6 that (with the notation slightly changed) the function

$$p_t(x) = \sum_{n=-\infty}^{\infty} \exp(-2\pi^2 n^2 t) \exp(2\pi i n x)$$

is the so-called fundamental solution of the heat equation

$$\frac{\partial u}{\partial t} = \frac{1}{2} \frac{\partial^2 u}{\partial x^2}$$

which is periodic with period 1; that is, it describes heat flow on the circle $0 \le x < 1$. The temperature $u(t, x)$ at time t is determined uniquely by the distribution $u(0, x) = f(x)$ at time $t = 0$ and is given in terms of $p_t(x)$ and $f(x)$ by the convolution

$$u(t, x) = p \circ f = \int_0^1 \left\{ \sum_{n=-\infty}^{\infty} \exp(-2\pi n^2 t) \exp(2\pi i n(x - y)) \right\} f(y) dy.$$

By considering the sum

$$q = q_t(x) = (2\pi t)^{-1/2} \sum_{n=-\infty}^{\infty} \exp[-(x - n)^2/2t],$$

one can show that $q \circ f$ also solves the problem of heat flow with period 1 and initial distribution $f(x)$, and since the solution is unique one deduces that $q = p$, which is essentially Jacobi's identity for $\theta_3(x, it)$. (For further details and the appeal to 'Kelvin's method of images', which yields $q = p$, see, for example, Dym and McKean, 1972, pp.63–66. A possible relevance of our numerical example is obvious.)

Exercises 4.7

4.7.1 Show that if $0 < k < 1$, then $q = \exp(\pi^2/\log q')$, where $q' = \exp(\pi i \tau') = \exp(-\pi K'(k')/K(k')) = \exp(-\pi K(k)/K'(k))$.

4.7.2 Show that, if $\tau\tau' = -1$, then

$$\frac{\theta_4(0|\tau)}{\theta_3(0|\tau)} = \frac{\theta_2(0|\tau')}{\theta_3(0|\tau')}.$$

4.7.3 Show that

$$\frac{\theta_2(0|\tau+1)}{\theta_3(0|\tau+1)} = e^{\pi i/4}\frac{\theta_2(0|\tau)}{\theta_4(0|\tau)}.$$

4.7.4 Show that

$$\prod_{n=1}^{\infty}\frac{1-q^{2n-1}}{1+q^{2n-1}} = \pm 2^{1/2}q'^{1/8}\prod_{n=1}^{\infty}\frac{1+q'^{2n}}{1+q'^{2n-1}},$$

and check that the plus sign should be taken on the right-hand side.

4.8 The Landen transformation

In 1775, Landen discovered the formula

$$\int_0^{\phi_1}\left(1-k_1^2\sin^2\theta_1\right)^{-1/2}\mathrm{d}\theta_1 = (1+k')\int_0^{\phi}(1-k^2\sin^2\theta)^{-1/2}\mathrm{d}\theta, \qquad (4.54)$$

where $\sin\phi_1 = (1+k')\sin\phi\cos\phi(1-k^2\sin^2\phi)^{-1/2}$ and $k_1 = (1-k')/(1+k')$. His proof is to be found in the *Philosophical Transactions of the Royal Society*, **LXV**, (1775), p. 285, and a proof in the same style is to be found in the book by Osgood (1935), pp. 204–207.

The formula (4.54) may be expressed in terms of Jacobian elliptic functions, in the form

$$sn(\{1+k'\}u, k_1) = (1+k')sn(u, k)cd(u, k) \qquad (4.55)$$

on writing $\phi = am\,(u)$, $\phi_1 = am\,(u_1)$ (see (1.23) for the definition of $am\,(u)$). It can be shown (see Whittaker & Watson, 1927, pp.477–478) that the transformation (4.55) follows at once from a transformation connecting theta functions with parameters τ and 2τ, namely

$$\frac{\theta_3(z|\tau)\theta_4(z|\tau)}{\theta_4(2z|2\tau)} = \frac{\theta_3(0|\tau)\theta_4(0|\tau)}{\theta_4(0|2\tau)}, \qquad (4.56)$$

which we shall now prove.

The zeros of $\theta_3(z|\tau)\theta_4(z|\tau)$ are simple zeros at the points

$$z = \left(m+\frac{1}{2}\right)\pi + \left(n+\frac{1}{2}\right)\pi\tau$$

and

$$z = m\pi + \left(n + \frac{1}{2}\right)\pi\tau,$$

where $m, n \in \mathbb{Z}$. That is, at the points

$$2z = m\pi + \left(n + \frac{1}{2}\right)\pi(2\tau),$$

which are the zeros of $\theta_4(2z|2\tau)$. It follows that the quotient on the left in (4.56) has no zeros or poles. Moreover, associated with the periods π and $\pi\tau$, it has the multipliers 1 and $(q^{-1}e^{-2iz})(-q^{-1}e^{-2iz})/(-q^{-2}e^{-4iz}) = 1$. It is, therefore, an entire, doubly periodic function, which reduces to a constant, by Liouville's Theorem. To evaluate the constant, take $z = 0$ and then (4.56) follows.

The replacement $z \mapsto z + \pi\tau/2$ yields a similar result for the other theta functions:

$$\frac{\theta_2(z|\tau)\theta_1(z|\tau)}{\theta_1(2z|2\tau)} = \frac{\theta_3(0|\tau)\theta_4(0|\tau)}{\theta_4(0|2\tau)}. \tag{4.57}$$

Now we give another proof of the Jacobi identity (Theorem 4.5) that rests on the following two lemmas.

Lemma 4.1 *Recall that*

$$\alpha(q) = \frac{\theta_1'}{\theta_2\theta_3\theta_4}.$$

Then

$$\alpha(q) = \frac{2\theta_2(0, q^2)\theta_3(0, q^2)}{\theta_2^2(0, q)}\alpha(q^2).$$

Lemma 4.2 $\theta_2^2(0, q) = 2\theta_2(0, q^2)\theta_3(0, q^2)$.

Assuming those two lemmas, we have $\alpha(q) = \alpha(q^2)$, whence, by iteration, $\alpha(q) = \alpha(q^{2^n})$ for every positive integer n. But $q^{2^n} \to 0$ as $n \to \infty$, since $|q| < 1$. Hence

$$\alpha(q) = \lim_{n\to\infty} \alpha(q^{2^n}) = 1.$$

Proof of Lemma 4.1 Rewrite (4.57) as

$$\frac{\theta_2(z, q)\theta_1(z, q)z^{-1}}{\theta_1(2z, q^2)z^{-1}} = \frac{\theta_3(0, q)\theta_4(0, q)}{\theta_4(0, q^2)}$$

and let $z \to 0$ to obtain

$$\frac{\theta_2(0, q)\theta_1'(0, q)}{2\theta_1'(0, q^2)} = \frac{\theta_3(0, q)\theta_4(0, q)}{\theta_4(0, q^2)}.$$

Then we re-write that last identity as

$$\frac{\theta_1'(0,q)}{\theta_2(0,q)\theta_3(0,q)\theta_4(0,q)} = \frac{2\theta_2(0,q^2)\theta_3(0,q^2)}{\theta_2^2(0,q)}\frac{\theta_1'(0,q^2)}{\theta_2(0,q^2)\theta_3(0,q^2)\theta_4(0,q^2)},$$

which is the desired result.

Proof of Lemma 4.2 We have

$$\theta_2^2(0,q) - 2\theta_2(0,q^2)\theta_3(0,q^2) = 4q^{1/2}\left[\sum_{n=0}^{\infty} q^{n(n+1)}\right]^2 - 4q^{1/2}\left[\sum_{n=0}^{\infty} q^{2n(n+1)}\right]$$

$$\times\left[2\sum_{n=0}^{\infty} q^{2n^2} - 1\right].$$

$$= 4q^{1/2}\left[\sum_{m,n=0}^{\infty} q^{m(m+1)+n(n+1)} + \sum_{n=0}^{\infty} q^{2n(n+1)}\right.$$

$$\left. - 2\sum_{r,s=0}^{\infty} q^{2(r^2+s(s+1))}\right]$$

$$= 8q^{1/2}\left[\sum_{m,n=0,n\le m}^{\infty} q^{m(m+1)+n(n+1)}\right.$$

$$\left. - \sum_{r,s=0}^{\infty} q^{2(r^2+s(s+1))}\right].$$

Our desired result is then a consequence of the following two assertions:

(A) $\displaystyle\sum_{n\le m, m-n \text{ even}} q^{m(m+1)+n(n+1)} = \sum_{r\le s} q^{2(r^2+s(s+1))};$

(B) $\displaystyle\sum_{n\le m, m-n \text{ odd}} q^{m(m+1)+n(n+1)} = \sum_{r>s} q^{2(r^2+s(s+1))}.$

To establish (A), $(r \le s)$, set $m = s+r, n = s-r$. Then $0 \le n \le m, m - n = 2r$ is even and $m(m+1) + n(n+1) = 2(r^2 + s(s+1))$. The one-to-one correspondence $(r,s) \mapsto (m,n)$ accordingly rearranges the left-hand side of (A) into the right-hand side.

To establish (B), $(r > s)$, set $m = r+s, n = r-s-1$. Then $0 \le n < m$ and $m - n = 2s + 1$ is odd. Again, $m(m+1) + n(n+1) = 2(r^2 + s(s+1))$. So the one-to-one correspondence $(r,s) \mapsto (m,n)$ rearranges the left-hand side of (B) into the right-hand side.

Exercises 4.8

4.8.1 Obtain an elementary proof of Lemma 4.1, by using the method of proof
of Theorem 4.4, to show that

$$\frac{2\theta_2(0, q^2)\theta_3(0, q^2)}{\theta_2^2(0, q)} = \frac{G(q^2)^2}{G(q)^2 \displaystyle\prod_{n=1}^{\infty}(1 + q^{2n})^2}$$

and then applying Theorem 4.4.

4.8.2 In the notation of (4.55), write $\tau_1 = 2\tau$ and denote by k_1, Λ, Λ' the modulus and quarter periods formed with the parameter τ_1. Show that the
equation

$$\frac{\theta_1(z|\tau)\theta_2(z|\tau)}{\theta_3(z|\tau)\theta_4(z|\tau)} = \frac{\theta_1(2z|\tau_1)}{\theta_4(2z|\tau_1)}$$

may be written

$$k\,sn(2Kz/\pi, k)cd(2Kz/\pi, k) = k_1^{1/2}sn(4\Lambda z/\pi, k_1).$$

By taking $z = \pi/4$, show that

$$k/(1 + k') = k^{1/2}.$$

Show further that

$$\Lambda'/\Lambda = 2K'/K$$

and thence that

$$\Lambda' = (1 + k')K'.$$

4.8.3 Show that

$$cn\{(1 + k')u, k_1\} = \{1 - (1 + k')sn^2(u, k)\}nd(u, k),$$
$$dn\{(1 + k')u, k_1\} = \{k' + (1 - k')cn^2(u, k)\}nd(u, k).$$

4.8.4 Show that

$$dn(u, k) = (1 - k')cn\{(1 + k')u, k_1\} + (1 + k')dn\{(1 + k')u, k\},$$

where $k = 2k_1^{1/2}/(1 + k_1)$.

4.8.5 By considering the transformation $\tau_2 \mapsto \tau \pm 1$, show that

$$sn(k'u, k_2) = k'sd(u, k),$$

where $k_2 = \pm ik/k'$, and the upper or lower sign is taken according as
$Re(\tau) < 0$ or $Re(\tau) > 0$; and obtain formulae for $cn(k'u, k_2)$ and
$dn(k'u, k_2)$.

4.9   Summary of properties of theta functions

Numbers and other references preceding formulae in what follows indicate their location in the text.

Basic notation k is a real number such that $0 < k < 1$; $k' = (1 - k^2)^{1/2}$ is the complementary modulus.

$$K = K(k) = \int_0^{\pi/2} \frac{\mathrm{d}\psi}{(1 - k^2 \sin^2 \psi)^{1/2}}, \quad K' = K(k').$$

$$\tau = \mathrm{i}K'/K, \quad q = \exp(\pi \mathrm{i}\tau).$$

Definitions by infinite series and products   Let $G = \prod_{n=1}^{\infty} (1 - q^{2n})$ (see Corollary 4.1).

(4.9)
$$\theta_4(z) = 1 + 2 \sum_{n=1}^{\infty} (-1)^n q^{n^2} \cos(2nz);$$

(4.6)
$$\theta_4(z) = G \prod_{n=1}^{\infty} (1 - 2q^{2n-1} \cos 2z + q^{4n-2});$$

θ_4 is an even function of z;

$$\theta_3(z) = \theta_4\left(z + \frac{\pi}{2}\right) = \theta_4\left(z - \frac{\pi}{2}\right) \quad \text{(definition)};$$

(4.10)
$$\theta_3(z) = 1 + 2 \sum_{n=1}^{\infty} q^{n^2} \cos 2nz;$$

(4.10)
$$\theta_3(z) = G \prod_{n=1}^{\infty} (1 + 2q^{2n-1} \cos 2z + q^{4n-2});$$

θ_3 is an even function of z;

$$\theta_1(z) = A\mathrm{e}^{\mathrm{i}z} \theta_4\left(z + \frac{\pi \tau}{2}\right), \quad A = -\mathrm{i}q^{1/4};$$

(4.17)
$$\theta_1(z) = 2 \sum_{n=0}^{\infty} (-1)^n q^{(n+1/2)^2} \sin(2n + 1)z;$$

(4.18)
$$\theta_1(z) = 2Gq^{1/4} \sin z \prod_{n=1}^{\infty} (1 - 2q^{2n} \cos 2z + q^{4n});$$

θ_1 is an odd function of z;

$$\theta_2(z) = \theta_1\left(z + \frac{\pi}{2}\right) \quad \text{(definition)};$$

(4.19)
$$\theta_2(z) = 2 \sum_{n=0}^{\infty} q^{(n+\frac{1}{2})^2} \cos(2n + 1)z;$$

(4.20)
$$\theta_2(z) = 2Gq^{1/4} \cos z \prod_{n=1}^{\infty} (1 + 2q^{2n} \cos 2z + q^{4n});$$

θ_2 is an even function of z.

Zeros $(\mod \pi, \pi \tau)$

$$\theta_1 : z \equiv 0; \quad \theta_2 : z \equiv \frac{\pi}{2}; \quad \theta_3 : z \equiv \frac{\pi}{2} + \frac{\pi\tau}{2}; \quad \theta_4 : z \equiv \frac{\pi\tau}{2}.$$

Further notation and basic formulae

$$\theta_i = \theta_i(0), \quad 1 \le i \le 4 \quad \text{(definition);}$$

$$\theta_1' = \theta_1'(0) = \lim_{z \to 0} \frac{\theta_1(z)}{z};$$

$$\theta_1' = 2 \sum_{n=0}^{\infty} (-1)^n (2n+1) q^{(n+\frac{1}{2})^2} = 2q^{\frac{1}{4}} \prod_{n=1}^{\infty} (1 - q^{2n})^3;$$

$$\theta_1 = 0.;$$

$$\theta_2 = 2 \sum_{n=0}^{\infty} q^{(n+\frac{1}{2})^2} = 2q^{1/4} \prod_{n=1}^{\infty} (1 - q^{2n})(1 + q^{2n})^2;$$

$$\theta_3 = 1 + 2 \sum_{n=1}^{\infty} q^{n^2} = \prod_{n=1}^{\infty} (1 - q^{2n})(1 + q^{2n-1})^2;$$

$$\theta_4 = 1 + 2 \sum_{n=1}^{\infty} (-1)^n q^{n^2} = \prod_{n=1}^{\infty} (1 - q^{2n})(1 - q^{2n-1})^2;$$

$$\text{(Section 4.3)} \qquad k^{1/2} = \theta_2/\theta_3;$$

$$k'^{1/2} = \theta_4/\theta_3;$$

$$\text{(4.36) and (4.38)} \qquad \frac{2K}{\pi} = \theta_3^2;$$

$$\text{(Theorem 4.5)} \qquad \theta_1' = \theta_2\theta_3\theta_4.$$

Definition of the Jacobi elliptic functions in terms of theta functions Let $z = \pi u/2K$:

(4.22) $sn(u) = k^{-1/2} \theta_1(z)/\theta_4(z);$

(4.23) $cn(u) = (k')^{1/2} k^{-1/2} \theta_2(z)/\theta_4(z);$

(4.24) $dn(u) = (k')^{1/2} \theta_3(z)/\theta_4(z).$

Periodicity properties (See Exercise 4.4.2). Let $N = q^{-1} e^{-2iz}$,

$$\theta_1(z + \pi) = -\theta_1(z), \qquad \theta_1(z + \pi\tau) = -N\theta_1(z),$$
$$\theta_2(z + \pi) = -\theta_2(z), \qquad \theta_2(z + \pi\tau) = N\theta_2(z),$$
$$\theta_3(z + \pi) = \theta_3(z), \qquad \theta_3(z + \pi\tau) = N\theta_3(z),$$
$$\theta_4(z + \pi) = \theta_4(z), \qquad \theta_4(z + \pi\tau) = -N\theta_4(z).$$

Pseudo-addition formulae (See Section 4.5.)

(4.32) $\theta_4^2 \theta_4(x + y)\theta_4(x - y) = \theta_4^2(x)\theta_4^2(y) - \theta_1^2(x)\theta_1^2(y);$

(4.33) $\theta_4^2 \theta_2(x + y)\theta_2(x - y) = \theta_2^2(x)\theta_4^2(y) - \theta_3^2(x)\theta_1^2(y);$

(4.34) $\theta_2^2 \theta_2(x + y)\theta_2(x - y) = \theta_3^2(x)\theta_3^2(y) - \theta_4^2(x)\theta_4^2(y);$

(4.35) $\theta_3^2 \theta_4(x + y)\theta_4(x - y) - \theta_4^2(x)\theta_3^2(y) + \theta_2^2(x)\theta_1^2(y).$

Jacobi's imaginary transformation (See Section 4.7.)
The substitution $k \mapsto k' = (1 - k^2)^{1/2}$ transforms $\tau = iK'/K$ into $\tau' = iK/K' = -\tau^{-1}$.

(4.49) $\theta_3(z|\tau) = (-i\tau)^{-1/2} \exp\left(\dfrac{z^2}{\pi i\tau} \right) \theta_3(\tau'z|\tau');$

(4.50) $\theta_4(z|\tau) = (-i\tau)^{-1/2} \exp\left(\dfrac{z^2}{\pi i\tau} \right) \theta_2(\tau'z|\tau');$

(4.51) $\theta_2(z|\tau) = (-i\tau)^{-1/2} \exp\left(\dfrac{z^2}{\pi i\tau} \right) \theta_4(\tau'z|\tau');$

(4.52) $\theta_1(z|\tau) = -i(-i\tau)^{-1/2} \exp\left(\dfrac{z^2}{\pi i\tau} \right) \theta_1(\tau'z|\tau').$

Notations for the theta functions We have adopted a notation for the theta functions that combines that of Whittaker & Watson (1927) and Tannery & Molk (1893–1902) with that of Weierstrass (1883) and Briot & Bouquet (1875) and is related to that of Hermite and H. J. S. Smith (see below); more prosaically we have used the former's numbering for the subscripts and the latter's 'theta'. The notations employed by other authors are indicated in the following table, which is taken from Whittaker & Watson (1927), p.487, where there are further details.

$\vartheta_1(\pi z)$	$\vartheta_2(\pi z)$	$\vartheta_3(\pi z)$	$\vartheta(\pi z)$	Jacobi (1829)
$\vartheta_1(z)$	$\vartheta_2(z)$	$\vartheta_3(z)$	$\vartheta_4(z)$	Whittaker & Watson (1927)
				Tannery & Molk (1893–1902)
$\theta_1(\omega z)$	$\theta_2(\omega z)$	$\theta_3(\omega z)$	$\theta(\omega z)$	Briot & Bouquet (1875)
$\theta_1(z)$	$\theta_2(z)$	$\theta_3(z)$	$\theta_0(z)$	Weierstrass (1883)
$\theta(z)$	$\theta_1(z)$	$\theta_3(z)$	$\theta_2(z)$	Jordan (1893)
$\theta_1(z)$	$\theta_2(z)$	$\theta_3(z)$	$\theta_4(z)$	this book

Hermite (1861) and Smith (1965) (originally published in 1859–1865) used the
notation

$$\theta_{\mu,\nu}(x) = \sum_{n=-\infty}^{\infty} (-1)^{n\nu} q^{(2n+\mu)^2/4} e^{i\pi(2n+\mu)x/a},$$

$\mu = 0, 1; \nu = 0, 1$. In that notation, we obtain our functions in the forms

$$\theta_{0,0}(x) = \theta_3\left(\frac{\pi x}{a}\right), \quad \theta_{0,1}(x) = \theta_4\left(\frac{\pi x}{a}\right), \quad \theta_{1,0}(x) = \theta_2\left(\frac{\pi x}{a}\right),$$

$$\theta_{1,1}(x) = \theta_1\left(\frac{\pi x}{a}\right).$$

5

The Jacobian elliptic functions for complex k

5.1 Extension to arbitrary τ with $\mathrm{Im}\,(\tau) > 0$

Hitherto we have supposed (see the Summary, Section 4.9) that $0 < k < 1$, from which it follows that $\tau = it\ (t > 0)$. In that case, our basic formulae expressing the Jacobian elliptic functions in terms of the parameter τ are (see Sections 4.3 and 4.4 and Chapter 2):

$$k^{1/2} = \theta_2/\theta_3, \tag{5.1}$$

$$k'^{1/2} = \theta_4/\theta_3, \tag{5.2}$$

$$2K/\pi = \theta_3^2, \tag{5.3}$$

$$K' = -iK\tau, \tag{5.4}$$

$$sn(u) = k^{-1/2}\theta_1\left(\theta_3^{-2}u\right)/\theta_4\left(\theta_3^{-2}u\right), \tag{5.5a}$$

$$cn(u) = k'^{1/2}k^{-1/2}\theta_2\left(\theta_3^{-2}u\right)/\theta_4\left(\theta_3^{-2}u\right), \tag{5.5b}$$

$$dn(u) = k'^{1/2}\theta_3\left(\theta_3^{-2}u\right)/\theta_4\left(\theta_3^{-2}u\right), \tag{5.5c}$$

Now let τ be a complex number, not necessarily pure imaginary, but still with $\mathrm{Im}(\tau) > 0$. We take Equations (5.4) to (5.5c) to be the *definitions* of the functions. If, as before, we set $\tau' = -\tau^{-1}$, then the following formulae, valid originally for $\tau = it\ (t > 0)$ remain valid for the general values of τ under consideration, by the Uniqueness Theorem:

$$k'^{1/2} = \theta_2(0|\tau')/\theta_3(0|\tau'), \tag{5.6}$$

$$2K'/\pi = \theta_3^2(0|\tau'), \tag{5.7}$$

$$k^2 + k'^2 = 1. \tag{5.8}$$

The *definitions* (5.5a), (5.5b) and (5.5c) define doubly periodic functions whose behaviour is described in the Summary at the end of Chapter 2, again by an appeal to the Uniqueness Theorem. Similarly $y = sn(u)$ will satisfy the differential equation

$$\left(\frac{dy}{du}\right)^2 = (1 - y^2)(1 - k^2 y^2),\tag{5.9}$$

together with the initial conditions

$$y(0) = 0, \quad y'(0) > 0.$$

The addition formulae and the other identities of Chapter 2 remain valid.

Exercise 5.1

5.1.1 Obtain the following results, which are convenient for calculating k, k', K, K', when q is given:

(i) $(2kK/\pi)^{1/2} = \theta_2 = 2q^{1/4}(1 + q^2 + q^6 + q^{12} + q^{20} + \cdots)$,

(ii) $(2K/\pi)^{1/2} = \theta_3 = 1 + 2q + 2q^4 + 2q^9 + \cdots$,

(iii) $(2k'K/\pi)^{1/2} = \theta_4 = 1 - 2q + 2q^4 - 2q^9 + \cdots$,

(iv) $K' = K\pi^{-1}\log q^{-1}$.   (5.10)

5.2 The inversion problem

We have seen that there is no difficulty in extending the definitions of our elliptic functions to arbitrary values of τ in the upper half-plane, but in most applications one starts with Equation (5.9) in which the modulus, k, is given and one has no *a priori* knowledge of the value of q. To prove the existence of an analytic function $sn(u, k)$, for example, one must show that a number τ exists, with $\text{Im}(\tau) > 0$, such that the number

$$m = k^2 = \theta_2^4(0|\tau)/\theta_3^4(0|\tau).\tag{5.11}$$

Once we have succeeded in doing that, we can define $y = sn(u, k)$ by (5.5). In other words, we can invert the integral

$$u = \int_0^y [(1 - t^2)(1 - k^2 t^2)]^{-1/2}\, dt$$

to obtain $y = sn(u, k)$.

The difficulty, of course, lies in showing that Equation (5.11) has a solution – the *inversion problem*. We have already solved that problem for $0 < m < 1$, but it is helpful to look at another approach to finding a solution. Set $m' = 1 - m$. Then $0 < m' < 1$ and

$$m' = k'^2 = \theta_4^4(0|\tau)/\theta_3^4(0|\tau)$$

or

$$m' = \prod_{n=1}^{\infty} \left(\frac{1 - q^{2n-1}}{1 + q^{2n}} \right)^8.$$

As q increases from 0 to 1, the product on the right-hand side is continuous and decreases strictly from 1 to 0, whence, by the Intermediate Value Theorem, it takes the value m' exactly once.

When $0 < m < 1$, we follow (5.11) and set

$$K = K(m) = \int_0^{\pi/2} (1 - m \sin^2 \phi)^{-1/2} \mathrm{d}\phi. \tag{5.12}$$

If m denotes a complex number not on that part of the real axis from 1 to ∞ (we shall write $m \not\geq 1$), then the formula (5.12) defines an analytic function of m, provided that we integrate along the real axis and choose the principal branch of the square root – that is, the one that makes the real part positive. The integral $K'(m) = K(m') = K(1 - m)$ is similarly defined for $m \not\leq 0$ (that is for complex m not on the negative real axis). In the next section we shall prove:

Theorem 5.1 *Let $m \not\geq 1$ and $m \not\leq 0$ and set $\tau = \mathrm{i}K'(m)/K(m)$. Then $\mathrm{Im}(\tau) > 0$, $-1 < \mathrm{Re}(\tau) < 1$, τ is analytic as a function of m and τ satisfies (5.12). If $m > 1$ or $m < 0$ then one can find τ satisfying (5.11), except that now one may have $\mathrm{Re}(\tau) = \pm 1$ and there are jump discontinuities across the rays $m \geq 1$ and $m \leq 0$. However, the parameter $q = \exp(\pi \mathrm{i}\tau)$ remains analytic in m for $m \not\geq 1$.*

The question of the uniqueness of our solution to the inversion problem then arises.

Theorem 5.2 *If τ_1 is a solution of (5.12) for $m = m_0 \not\geq 1$ and $m_0 \not\leq 0$, then every solution of (5.11) for $m = m_0$ has the form (5.11)*

$$\tau_2 = \frac{a\tau_1 + b}{c\tau_1 + d},$$

where $a, b, c, d \in \mathbb{Z}$, $ad - bc = 1$, $a, d \equiv 1 \pmod 2$, and $b, c \equiv 0 \pmod 2$.

The proof of Theorem 5.2 will be deferred until Chapter 6. Assuming it, we can find all analytic solutions to (5.11), as follows

Theorem 5.3 *Let* $\tau_1 = f(m)$, $(m \not\equiv 1, m \not\equiv 0)$ *be the analytic solution of (5.11) given in Theorem 5.1. Then any other analytic solution* $\tau = g(m)$ *has a unique representation*

$$\tau_2 = \frac{af(m) + b}{cf(m) + d},$$

where a, b, c, d *satisfy the conditions of Theorem 5.2.*

Proof Choose any m_0 such that $m_0 \not\equiv 1, \not\equiv 0$ and set $\tau_0 = g(m_0)$. Then

$$\tau_0 = \frac{af(m_0) + b}{cf(m_0) + d},$$

by Theorem 5.2 and

$$\tau = h(m) = \frac{af(m) + b}{cf(m) + d}$$

is an analytic solution of (5.11) such that $h(m_0) = g(m_0)$. But then $h(m) = g(m)$, since the equation

$$m = k^2(\tau) = \theta_2^4(0|\tau)/\theta_3^4(0|\tau)$$

completely determines the derivatives $g^{(n)}(m_0)$. $(n = 1, 2, \ldots)$. That a, b, c, d are unique up to sign follows from the fact that, since $\frac{d\tau}{dm} \neq 0$, the mapping $m \mapsto f(m)$ maps a disc $|m - m_0| < \varepsilon$ conformally onto an open neighbourhood of $f(m_0)$.

We conclude this section by showing that our Jacobian elliptic functions depend only on the parameter m.

Theorem 5.4 *Let* τ_1, τ_2 *be any two solutions of (5.11) for the same m. Then:*

$$sn(u|\tau_1) = sn(u|\tau_2),$$
$$cn(u|\tau_1) = cn(u|\tau_2),$$
$$dn(u|\tau_1) = dn(u|\tau_2).$$

Proof By the Uniqueness Theorem, it is sufficient to establish the equalities in a neighbourhood of $u = 0$. Recall that $y = sn(u)$ satisfies the differential equation

$$y'^2 = (1 - y^2)(1 - my^2) = 1 - (1 + m)y^2 + my^4. \tag{5.13}$$

Differentiation of (5.13) with respect to u yields

$$2y'y'' = 2y'(2my^3 - (1 + m)y).$$

Since $y'(0) = 1 \neq 0$, we have, in some neighbourhood of $u = 0$,

$$y'' = 2my^3 - (1 + m)y.$$

But that last equation uniquely determines all the derivatives $y^{(n)}(0)(n = 0, 1, 2, \ldots)$, given that $y(0) = 0$ and $y'(0) = 1$. Hence $sn\ (u)$ really depends only on the parameter m rather than τ.

Recall that $cn\ (0) = 1 = dn\ (0)$ and the identities $cn^2 u + sn^2 u = 1, dn^2 u + m\ sn^2 u = 1$. It follows that in some neighbourhood of $u = 0$,

$$cn\ (u) = (1 - sn^2 u)^{1/2},$$
$$dn\ (u) = (1 - m\ sn^2 u)^{1/2},$$

where the principal square root is taken. Hence $cn\ (u), dn\ (u)$ also really depend only on the parameter m.

5.3 Solution of the inversion problem

Let

$$\Omega = \{z : z \nleq 0\}, \quad \Gamma = \{m : m \ngeq 1 \text{ and m} \nleq 0\} \tag{5.14}$$

We shall need the following result (whose proof is set as an exercise (see Exercise 5.3.1).

Lemma 5.1 *With the foregoing notation $m \in \Gamma$ implies $\frac{m}{1-m} \in \Omega$; that is* $\left|\arg \frac{m}{1-m}\right| < \pi$.

That $\mathrm{Re}(K(m)) > 0$ and $\mathrm{Im}(K(m)) \neq 0$ if $\mathrm{Im}(m) \neq 0$ follow immediately from:

Lemma 5.2 *(The Sharp Mean Value Theorem) Let $-\infty < a < b < \infty$ and let $f : [a, b] \to \mathbb{R}$ be continuous. Then*

$$\int_a^b f(x)\mathrm{d}x = (b - a)f(c)$$

for some $c \in (a, b)$.

Now let $m \in \Gamma$ be given. We show first that $\mathrm{Im}(\tau) > 0$, where $\tau - iK'/K, K = K(m)$ is given by (5.12) and $K'(m) = K(1 - m)$.

As ϕ runs from 0 to $\pi/2$, $1 - m \sin^2 \phi$ runs along the line segment from 1 to $1 - m$, and $(1 - m \sin^2 \phi)^{1/2}$ lies in the sector bounded by the real axis and

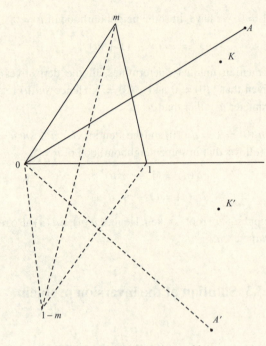

Figure 5.1 The inversion problem.

the bisector of the angle between the real axis and the line joining the origin O to $1 - m$, that is between the real axis and the line OA'. (See Figure 5.1). Again, $(1 - m \sin^2 \phi)^{-1/2}$ lies in the reflection of that sector in the real axis and the line OA. Similarly, $(1 - m' \sin^2 \phi)^{-1/2}$ lies in the sector bounded by the real axis and the reflection OA' of the bisector of the angle between the real axis and the directed line segment joining O to m.

Now if z_1, z_2 lie in an angular sector with vertex at the origin O, so also does the point $a_1 z_1 + a_2 z_2$ for arbitrary non-negative real numbers a_1, a_2. It follows that if $z_1, \ldots, z_n$ lie in the sector, so does $\sum_{1 \le j \le n} a_j z_j$ for arbitrary real $a_j \ge 0$. Since the integral defining K is the limit of such a sum, we conclude that K lies in the sector between the real axis and OA, and similarly K' lies in the sector between the real axis and OA'. It follows from the foregoing that $\mathrm{Im}(K)$ and $\mathrm{Im}(K')$ are greater than or less than 0 according as $\mathrm{Im}(m)$ is greater than or less than 0. Since the angle AOA' is half the angle subtended at O by the direct line segment joining m to $1 - m$, Lemma 5.1 implies that the angle AOA' is acute, whence $\mathrm{Re}(K'/K) > 0$ or $\mathrm{Im}(\tau) = \mathrm{Im}(iK'/K) > 0$.

Since $\mathrm{Im}(\tau) > 0$ and τ satisfies (5.11), when $0 < m < 1$, it follows from the Uniqueness Theorem, again, that τ satisfies (5.11) for all m in Γ. The remaining conditions of Theorem 5.1 require further study of the function $K(m)$, defined by (5.12). The inequality $-1 < \text{Re}(\tau) < 1$ is finally proved below, in Theorem 5.6.

Lemma 5.3 *With the foregoing notation,*

$$\lim_{m \to 1, m \not\geq 1} \text{Re} K(m) = +\infty.$$

Proof Since the integral $\int_0^{\pi/2} (1 - \sin^2 \phi)^{-1/2} d\phi = \int_0^{\pi/2} \sec \phi \, d\phi$ diverges, given $N > 0$, there exists ε $(0 < \varepsilon < \pi/2)$ such that

$$\int_0^{\pi/2-\varepsilon} (1 - \sin^2 \phi)^{-1/2} d\phi > N.$$

Hence, there exists $\delta > 0$ such that

$$\text{Re} \int_0^{\pi/2-\varepsilon} (1 - m \sin^2 \phi)^{-1/2} d\phi > N,$$

if $|m - 1| < \delta, m \not\geq 1$. Since $\text{Re}(1 - m \sin^2 \phi)^{-1/2} > 0$, it follows that $\text{Re} \int_0^{\pi/2} (1 - m \sin^2 \phi)^{-1/2} d\phi > N$, if $|m - 1| < \delta, m \not\geq 1$.

It follows from Lemma 5.3 that:

$$\lim_{\substack{m \to 0 \\ m \not\leq 0}} \text{Im}(\tau) = \lim_{\substack{m \to 0 \\ m \not\leq 0}} \text{Re}\left(\frac{K'}{K}\right) = +\infty. \tag{5.15}$$

Now assume that $|m| < 1$ and use the Binomial Theorem to expand the integrand in (5.12) to obtain

$$K = \int_0^{\pi/2} (1 - m \sin^2 \phi)^{-1/2} d\phi$$

$$= \sum_{n=0}^{\infty} \frac{1 \cdot 3 \cdots (2n-1)}{2^n n!} m^n \int_0^{\pi/2} \sin^{2n} \phi \, d\phi$$

$$= \frac{\pi}{2} \sum_{n=0}^{\infty} \frac{1^2 \cdot 3^2 \cdot 5^2 \cdots (2n-1)^2}{2^{2n} (n!)^2} m^n,$$

(cf. Miscellaneous Exercises 1.9.8, on random walks).

Now recall the definition of the *hypergeometric function* (see, for example, Whittaker & Watson (1927), Chapter XIV):

$$F(a, b; c; z) = \sum_{n=0}^{\infty} \frac{a(a+1)\cdots(a+n-1)b(b+1)\cdots(b+n-1)}{n!c(c+1)\cdots(c+n-1)} z^n, \quad |z| < 1,$$

where $c > 0$. It follows that

$$K(m) = \frac{1}{2}\pi F\left(\frac{1}{2}, \frac{1}{2}; 1; m\right), \tag{5.16}$$

(see Whittaker & Watson (1927), Chapter XXII, Section 22.3).

Since the series defining $F(a, b; c; z)$ converges for $|z| < 1$, one can differentiate term by term to show that $y = F(a, b; c; z)$ satisfies Gauss' hypergeometric differential equation:

$$z(1-z)\frac{d^2 y}{dz^2} + \{c - (a+b+1)z\}\frac{dy}{dz} - aby = 0.$$

It follows that $y = K(m)$ satisfies the *self-adjoint* differential equation:

$$m(1-m)\frac{d^2 y}{dm^2} + (1-2m)\frac{dy}{dm} - \frac{1}{4}y = 0. \tag{5.17}$$

Now Equation (5.17) is invariant under the replacement $m \mapsto 1 - m$. Hence $y = K'(m) = K(1-m)$ is also a solution, provided that we require the condition $m \not\leq 0$. Let

$$\sum = \{m : \text{Im}(m) \neq 0\} \tag{5.18}$$

and note that m, $(1-m)^{-1}$ and $(m-1)/m$ all have the same sign in the coefficient of i in the imaginary part and that m^{-1}, $1 - m$ and $m/(m-1)$ all have the opposite sign.

Lemma 5.4 *Let $y = f(m)$ be a solution of (5.15) in Σ. Then*

$$y = f(1-m), \quad (1-m)^{-1/2}f\left(\frac{m}{m-1}\right), \quad (1-m)^{-1/2}f\left(\frac{1}{1-m}\right),$$

$$m^{-1/2}f\left(\frac{m-1}{m}\right), \quad m^{-1/2}f\left(\frac{1}{m}\right)$$

are all solutions of (5.17) in Σ.

Since $0 < |\arg m| < \pi$ and $0 < |\arg(1-m)| < \pi$, we can take the principal value for the square roots in the statement of the lemma. The proof of Lemma 5.4 is straightforward; we shall indicate some typical calculations, but then leave the details of a complete proof to the reader. Let $y = f(m)$ be a solution of (5.17) and set $m_1 = m/(m-1)$. We want to show that

$$W = (1-m)^{-1/2}f\left(\frac{m}{m-1}\right) = (1-m)^{-1/2}f(m_1)$$

also satisfies (5.17). To that end, write $y_1 = f(m_1) = (1 - m)^{1/2} W$. Then

$$\frac{dy_1}{dm_1} = -(1 - m)^{3/2} \left[(1 - m) \frac{dW}{dm} - \frac{1}{2} W \right],$$

$$\frac{d^2 y}{dm_1^2} = (1 - m)^{5/2} \left[(1 - m)^2 \frac{d^2 W}{dm^2} - 3(1 - m) \frac{dW}{dm} + \frac{3}{4} W \right].$$

So

$$0 = m_1(1 - m_1) \frac{d^2 y_1}{dm_1^2} + (1 - 2m_1) \frac{dy_1}{dm_1} - \frac{1}{4} y_1$$

$$= -(1 - m)^{3/2} \left[m(1 - m) \frac{d^2 W}{dm^2} + (1 - 2m) \frac{dW}{dm} - \frac{1}{4} W \right].$$

Although such verifications are sufficient for our present purposes, we shall discuss in Chapter 6 the true origin and significance of properties of that nature.

Lemma 5.5 *With the foregoing notation,*

$$K'(m) = \log 4m^{-1/2} + o(1) \quad as \quad m \to 0, \quad |\arg m| < \pi.$$

Proof By (5.3),

$$K' = -i \frac{\pi}{2} \tau \theta_3^2 = \left(-\frac{1}{2} \log q \right) \theta_3^2 = \left(-\frac{1}{2} \log q \right) (1 + O(q)).$$

The result (5.15) now implies that $q \to 0$ as $m \to 0$, provided that $m \not\leq 0$. Now $m^{1/2}$ is analytic for $m \in \Omega$ and $m^{1/2} = k = \theta_2^2/\theta_3^2$, when $0 < m < 1$, by (5.1). Hence, for $m \in \Omega$ we have by the Uniqueness Theorem that

$$m^{1/2} = k = \frac{\theta_2^2}{\theta_3^2} = 4q^{1/2} \frac{\{1 + O(q)\}}{\{1 + O(q)\}},$$

whence $\qquad 4m^{-1/2} = q^{-1/2} \{1 + O(q)\}$

and $\qquad \log 4m^{-1/2} = -\frac{1}{2} \log q + O(q).$

The lemma now follows.

By Lemma 5.4, $K'(m) = K(1 - m)$ is also a solution to (5.17). Since $K(m) = \pi/2 + O(m)$ as $m \to \infty$, we obtain as an immediate consequence of Lemma 5.5 that *the solutions $y = K(m)$ and $y = K'(m)$ of (5.17) are linearly independent.*

It now follows from the Uniqueness Theorem for second order linear differential equations that every solution of (5.17) has the form $y = AK(m) + BK'(m)$. All that we need to prove that is contained in:

Lemma 5.6 *Consider the differential equation*

$$\frac{d^2 y}{dz^2} + p(z)\frac{dy}{dz} + q(z)y = 0, \tag{5.19}$$

where $p(z), q(z)$ are analytic in an open disc D, with centre at the point z_0. Then any two linearly independent solutions of (5.19) in D span the solution space of (5.19).

Proof (Once again we outline the proof and leave the details as an exercise.)

Any solution y of (5.19) in D can be expanded in a power series

$$y = f(z) = \sum_{n=0}^{\infty} a_n(z - z_0)^n,$$

where $a_n = f^{(n)}(z_0)/n!$ *Observe that* (5.19) *implies that* $y = f(z)$ *is uniquely determined by* $a_0 = f(z_0)$ *and* $a_1 = f'(z_0)$. Now one shows that two solutions $f(z), g(z)$ of (5.19) are linearly dependent if and only if the vectors $(f(z_0), f'(z_0))$ and $(g(z_0), g'(z_0))$ are linearly dependent in $\mathbb{R}^2$. The result follows.

Remark *We note the following:*

$$K\left(\frac{m}{m-1}\right) = K\left(1 - \frac{1}{m-1}\right) = K'\left(\frac{1}{1-m}\right),$$

$$K'\left(\frac{m}{1-m}\right) = K\left(\frac{1}{1-m}\right),$$

$$K\left(\frac{m-1}{m}\right) = K(1 - m^{-1}) = K'(m^{-1}), \tag{5.20}$$

$$K'\left(\frac{m-1}{m}\right) = K(m^{-1}).$$

Theorem 5.5 *With the foregoing notation, we have:*

$$K\left(\frac{m}{m-1}\right) = (1-m)^{1/2}K(m), \tag{5.21a}$$

$$K\left(\frac{m-1}{m}\right) = m^{1/2}K'(m), \tag{5.21b}$$

$$K'\left(\frac{m}{m-1}\right) = (1-m)^{1/2}(K'(m) \pm iK(m)), \tag{5.21c}$$

$$K'\left(\frac{m-1}{m}\right) = m^{1/2}(K(m) \mp iK'(m)), \tag{5.21d}$$

where the upper/lower sign in (5.21c), (5.21d) is to be taken according as $\text{Im}(m) > 0$ or $\text{Im}(m) < 0$, respectively.

Proof (We use Lemma 5.4 in each part of the proof)

(a) We know that $(1 - m)^{-1/2} K (m/m - 1) = AK(m) + BK'(m)$. Let $m \sim 0$. Then

$$\frac{\pi}{2} + O(m) = A\frac{\pi}{2} + B \log 4m^{-1/2} + o(1),$$

whence $A = 1$ and $B = 0$.

(b) The proof of (b) is similar on letting $m \sim 1$.
(c) We start from

$$(1 - m)^{-1/2} K' \left(\frac{m}{m - 1}\right) = AK(m) + BK'(m)$$

and then let $m \sim 0$ to obtain

$$\log 4(-m)^{-1/2} + o(1) = A\frac{\pi}{2} + B \log 4m^{-1/2}.$$

But $\log 4(-m)^{-1/2} = \log 4m^{-1/2} \pm \frac{\pi}{2}i$, according as $\text{Im}(m) > 0$ or $\text{Im}(m) < 0$. Whence $A = \pm i$, $B = 1$.

(d) In this case we start from

$$m^{-1/2} K' \left(\frac{m - 1}{m}\right) = AK(m) + BK'(m)$$

and then let $m \sim 1$ to obtain

$$\log 4(m - 1)^{-1/2} + o(1) = A \log 4(1 - m)^{-1/2} + B\frac{\pi}{2} + o(1).$$

But

$$\log 4(m - 1)^{-1/2} = \log 4(1 - m)^{-1/2} \mp \frac{\pi}{2}i,$$

according as $\text{Im}(m) > 0$ or $\text{Im}(m) < 0$. Whence $A = 1$ and $B = \mp 1$.

Theorem 5.6 *Suppose that $m \not\equiv 1$ and $m \not\equiv 0$, then $-1 < \text{Re}(\tau) < 1$.*

Proof Let $\text{Im}(m) > 0$. Then (5.21a) and (5.21c) yield

$$\frac{K' \left(\dfrac{m}{m - 1}\right)}{K \left(\dfrac{m}{m - 1}\right)} = \frac{K'(m)}{K(m)} + i.$$

Now $\text{Im}(m) > 0$ implies $\text{Im}(m/(m-1)) < 0$ and by Exercise 5.3.3 $\text{Im}(m)$ and $\text{Im}(K'/K)$ have opposite signs. Whence $\text{Im}(K'(m)/K(m)) < 0$ and $\text{Im}(K'(m/(m-1))/K(m/(m-1))) > 0$. Hence $-1 < \text{Im}(K'(m)/K(m)) < 0$ and $\text{Re}(\tau) = \text{Re}(iK'(m)/K(m)) = -\text{Im}(K'(m)/K(m))$ lies in $(0,1)$.

Similarly $\text{Im}(m) < 0$ implies $\text{Re}(\tau)$ lies in $(-1, 0)$.

We can now complete our solution of the inversion problem by letting $m \to m_1 > 1$, with $\text{Im}(m) > 0$, in the formulae (5.21):

$$K'(m) = m^{-1/2} K'(m^{-1}),$$

and (5.20), (5.21) (b and d):

$$K(m) = m^{-1/2}(K(m^{-1}) + iK'(m^{-1})).$$

Since the limits on the right-hand side exist, we are led to set

$$K_+(m_1) = m_1^{-1/2}\big(K\big(m_1^{-1}\big) + iK'\big(m_1^{-1}\big)\big),$$

$$K'_+(m_1) = m_1^{-1/2} K'\big(m_1^{-1}\big),$$

$$\tau_+(m_1) = iK'_+(m_1)/K_+(m_1). \tag{5.22}$$

Since $K(m_1^{-1})$ and $K'(m_1^{-1})$ are both positive, $\text{Im}(\tau_+(m_1)) > 0$. That $\tau_+(m_1)$ satisfies (5.11) follows by continuity.

When $m \to m_1 > 1$, with $\text{Im}(m) < 0$, one obtains, by a similar argument, the limits:

$$K_-(m_1) = m_1^{-1/2}\big(K\big(m_1^{-1}\big) - iK'\big(m_1^{-1}\big)\big) = K_+(m_1) - 2iK'_+(m_1),$$

$$K'_-(m_1) = m_1^{-1/2} K'\big(m_1^{-1}\big) = K'_+(m_1),$$

$$\tau_-(m_1) = iK'_-(m_1)/K_-(m_1). \tag{5.23}$$

The cut $\{m : m > 1\}$ accordingly becomes a line of discontinuity.

The same procedures apply to the left-hand cut $\{m : m < 0\}$. Thus $\text{Im}(m) \neq 0$, together with (5.21a) and (5.21c), implies

$$K(m) = (1-m)^{-1/2} K'\left(\frac{1}{1-m}\right),$$

$$K'(m) = (1-m)^{-1/2}\left(K\left(\frac{1}{1-m}\right) \mp iK'\left(\frac{1}{1-m}\right)\right)$$

$$\pi i\tau(m) = -\pi K'(m)/K(m)$$

$$= \text{sgn}(\text{Im}(m))\pi i - \pi K\left(\frac{1}{1-m}\right)\Big/K'\left[\frac{1}{1-m}\right]. \tag{5.24}$$

It follows that the left-hand cut is a line of discontinuity for τ, but *not* for

$$q = \exp(\pi i \tau) = -\exp\left(-\pi K\left(\frac{1}{1-m}\right)\bigg/ K'\left(\frac{1}{1-m}\right)\right). \qquad (5.25)$$

In fact, the expression (5.25) is analytic in m for $m \neq 0$ and, by Lemma 5.3, $\lim_{m\to 0, m\not\geq 0} q = 0$. Since we know already that $\lim_{m\to 0, m\not\leq 0} q = 0$, we may conclude that $q = f(m)$ is analytic in m for $m \not\geq 1$.

That concludes our discussion of the inversion problem. Note that if one wants to make a definite choice of τ on the two cuts, one should take the opposite signs in (5.23) and (5.24) in order to preserve symmetry in the point $m = 1/2$.

Exercises 5.3

5.3.1 Prove Lemma 5.1.

5.3.2 Complete the details of the poof of Lemma 5.4.

5.3.3 Prove that $\text{Im}(m)$ and $\text{Im}(K'/K)$ have opposite signs (see Figure 5.1).

5.3.4 Show that $\text{Re}(\tau) = 0$ if and only if $0 < m < 1$ (cf. Theorem 5.6).

5.3.5 Show that

$$\frac{d}{dk}\left(kk'^2 \frac{dK}{dk}\right) = kK.$$

(Hint: see Whittaker & Watson (1927), p. 499 and use the expressions for K in terms of hypergeometric functions.)

5.3.6 Show that

$$K' = \frac{1}{2}\pi F\left(\frac{1}{2}, \frac{1}{2}; 1; k'^2\right),$$

where the m'-plane is cut from 1 to $+\infty$ (that is, the m-plane is cut from 0 to $-\infty$). (See Whittaker & Watson (1927), p. 501.)

5.4 Numerical computation of elliptic functions

Equations (5.1) to (5.5) and (5.10) are rapidly convergent expressions for calculating elliptic functions, given the quantity q. But it usually happens in practice that the modulus, k, is given and one wants to calculate K, K' and $q = \exp(-\pi K'/K)$. Now

$$K = \frac{1}{2}\pi F\left(\frac{1}{2}, \frac{1}{2}; 1; k^2\right), \quad K' = \frac{1}{2}\pi F\left(\frac{1}{2}, \frac{1}{2}; 1, k'^2\right),$$

but those hypergeometric series converge slowly except when $|k|$, $|k'|$, respectively, are small. So those series are never simultaneously suitable for numerical calculations; a different procedure is necessary.

Recall that $k'^{1/2} = \theta_4(0, q)/\theta_3(0, q)$ (see (5.2)) and set

$$\varepsilon = \frac{1 - k'^{1/2}}{1 + k'^{1/2}}. \tag{5.26}$$

Since the mapping $x \mapsto (1 - x)/(1 + x)$ is a one-to-one mapping of the open interval (0,1) onto itself, we have $0 < \varepsilon < 1$, when $0 < k < 1$, which is the physically important case (see Chapter 1, Section 1.6). Moreover,

$$\varepsilon = \frac{\theta_3(0, q) - \theta_4(0, q)}{\theta_3(0, q) + \theta_4(0, q)} = \frac{\theta_2(0, q^4)}{\theta_3(0, q^4)}, \tag{5.27}$$

by Exercise 4.4.5. Thus

$$\varepsilon = \frac{2q \displaystyle\sum_{n=0}^{\infty} q^{4n(n+1)}}{1 + 2 \displaystyle\sum_{n=1}^{\infty} q^{4n^2}} = 2q \left\{ 1 + \sum_{n=1}^{\infty} a_n q^{4n} \right\}. \tag{5.28}$$

Since $\varepsilon = 0$ only when $q = 0$ and $\frac{d\varepsilon}{dq} = 2 \neq 0$ when $q = 0$, one can invert the series in (5.28) uniquely to obtain

$$q = \sum_{n=1}^{\infty} b_n \left(\frac{\varepsilon}{2} \right)^n, \tag{5.29}$$

for $|\varepsilon|$ sufficiently small. The usual methods for the reversion of power series then lead to

$$q = \frac{\varepsilon}{2} + 2 \left(\frac{\varepsilon}{2} \right)^5 + 15 \left(\frac{\varepsilon}{2} \right)^9 + 150 \left(\frac{\varepsilon}{2} \right)^{13} + 1707 \left(\frac{\varepsilon}{2} \right)^{17} + 20\,910 \left(\frac{\varepsilon}{2} \right)^{21}$$
$$+ 268\,616 \left(\frac{\varepsilon}{2} \right)^{25} + \cdots, \tag{5.30}$$

for $|\varepsilon|$ sufficiently small. (Lowan *et al.* (1942) give 14 terms of (5.30).)

Theorem 5.7 (Weierstrass, 1883) *With the foregoing notation:*

$$q = \sum_{m=0}^{\infty} c_m \left(\frac{\varepsilon}{2} \right)^{4m+1}, \tag{5.31}$$

where:

(a) *The series is convergent for* $|\varepsilon| < 1$;
(b) $c_m > 0$ *for all m.*

Corollary 5.1

$$1 = \sum_{m=0}^{\infty} c_m \left(\frac{1}{2}\right)^{4m+1}.$$

Proof of Corollary 5.1 The condition $0 \leq \varepsilon < 1$ and (b) of the theorem imply

$$\sum_{m=0}^{N} c_m \left(\frac{\varepsilon}{2}\right)^{4m+1} \leq \sum_{m=0}^{\infty} c_m \left(\frac{\varepsilon}{2}\right)^{4m+1} = q < 1.$$

Now let $\varepsilon \mapsto 1-$ to obtain

$$\sum_{m=0}^{N} c_m \left(\frac{1}{2}\right)^{4m+1} \leq 1$$

for every N. Hence $A = \sum_{m=0}^{\infty} c_m (1/2)^{4m+1}$ exists. Since $\lim_{\varepsilon \to 1-} q = 1$, Abel's Limit Theorem (see, for example, Titchmarsh, 1939, pp. 9, 10) implies $A = 1$.

Now set

$$r_n = \sum_{m=n+1}^{\infty} c_m \left(\frac{\varepsilon}{2}\right)^{4m+1}, \quad |\varepsilon| < 1.$$

Then

$$|r_n| \leq \sum_{m=n+1}^{\infty} c_m |\varepsilon|^{4m+1} \left(\frac{1}{2}\right)^{4m+1} \leq |\varepsilon|^{4n+5} \sum_{m=n+1}^{\infty} c_m \left(\frac{1}{2}\right)^{4m+1}$$

or

$$|r_n| \leq |\varepsilon|^{4n+5} \left[1 - \sum_{m=0}^{n} c_m \left(\frac{1}{2}\right)^{4m+1} \right]. \tag{5.32}$$

In particular,

$$|r_0| \leq \frac{1}{2}|\varepsilon|^5, \quad |r_1| \leq \frac{7}{16}|\varepsilon|^9, \quad |r_2| \leq \frac{209}{512}|\varepsilon|^{13},$$

$$|r_3| \leq \frac{1597}{4096}|\varepsilon|^{17}, \quad |r_4| \leq \frac{49\,397}{131\,072}|\varepsilon|^{21}.$$

The power of the method lies in the fact that when $0 \leq k^2 \leq 1/2, 0 < \varepsilon < 2/23$. When $1/2 < k^2 < 1$, one can interchange k and k' to compute q' and then $q = \exp(\pi^2/\log q')$ (see Exercise 4.7.1).

Proof of Theorem 5.7 We start with the equation

$$m = k^2 = \theta_2^4(0, q)/\theta_3^4(0, q) = 16q \left(1 + \sum_{n=1}^{\infty} \alpha_n q^n \right).$$

By the work of Section 5.3, there exists an analytic inverse

$$16q = m\left(1 + \sum_{n=1}^{\infty} \beta_n m^n\right), \quad |m| < 1.$$

Clearly,

$$\lim_{m \to 0+}(\log 16q - \log m) = \lim_{m \to 0+}\log\left(\frac{16q}{m}\right) = 0.$$

We now appeal to:

Lemma 5.7 *Let*

$$h(m) = \log 16q - \log m,$$

then $h(m)$ is analytic in the cut complex plane in which $m \not\geq 1$.

Proof of Lemma 5.7 We know already that $\log 16q - \log m$ is analytic for $m \not\leq 0$, $m \not\geq 1$. When $\mathrm{Im}(m) \neq 0$, (5.24) implies that

$$\log 16q = \log 16 + \mathrm{sgn}(\mathrm{Im}(m)) \cdot \pi i - \pi K\left(\frac{1}{1-m}\right)\bigg/K'\left(\frac{1}{1-m}\right),$$

$$\mathrm{Im}(m) \neq 0.$$

Hence

$$\log 16q - \log m = \log 16 - \pi K\left(\frac{1}{1-m}\right)\bigg/K'\left(\frac{1}{1-m}\right) - \log(-m),$$

for $\mathrm{Im}(m) \neq 0$. But that last expression is analytic for $m \not\geq 1$, except for a possible isolated singularity at $m = 0$, which is removable, since $\lim_{m \to 0} m\, h(m) = 0$, by Lemma 5.7. Clearly, we should take

$$h(0) = \lim_{m \to 0+} h(m) = 0.$$

Now return to the proof of our theorem. When $0 < k < 1$, $0 < q < 1$ and

$$\exp(h(m)) = \exp(\log 16q - \log m) = 16qm^{-1},$$

or $16q = m\exp(h(m))$. But then (5.1), (5.11) and (5.27) imply

$$\varepsilon^4 = \theta_2^4(0, q^4)/\theta_3^4(0, q^4) = k^2(0, q^4) = m(0, q^4),$$

whence

$$16q^4 = \varepsilon^4 \exp(h(\varepsilon^4)),$$

since $0 < \varepsilon < 1$ also. Moreover, $h(m) = \log 16q - \log m$ is real, whence

$$q = \left(\frac{\varepsilon}{2}\right)\exp\left(\frac{h(\varepsilon^4)}{4}\right), \tag{5.33}$$

for $0 < \varepsilon < 1$. But that last expression is analytic for $|\varepsilon| < 1$, whence, by the Uniqueness Theorem, it inverts (5.27) for all $|\varepsilon| < 1$. We have accordingly established the series structure (5.31) and part (a) of Theorem 5.7.

Part (b) of the theorem will follow from (5.33) once we have proved:

Lemma 5.8 *The coefficients in the Maclaurin expansion of $h(m)$ are non-negative.*

Proof Let W_1, W_2 be two solutions of the differential equation

$$p(z)W'' + p'(z)W' + r(z)W = 0,$$

where $p(z,)$, $r(z)$ are analytic in a domain. Then $p(z)(W_1 W_2' - W_2 W_1')$ is constant in that domain (see Exercise 5.4.1).

Now take $W_1 = K(m)$, $W_2 = K'(m)$. Then

$$h(m) = \log 16q - \log m = \log 16 + \pi i \tau - \log m$$

$$= \log 16 - \pi \frac{W_2}{W_1} - \log m, \quad \text{when } \mathrm{Im}(m) \neq 0.$$

Then

$$h'(m) = -m^{-1} - \pi \frac{W_1 W_2' - W_2 W_1'}{W_1^2} = -m^{-1} - \pi \frac{A}{m(1-m)K(m)^2},$$

for some constant A, by applying the result above to the differential equation (5.17). To evaluate A multiply by m to obtain

$$mh'(m) = -1 - \frac{\pi A}{(1-m)K(m)^2}$$

and then let $m \to 0$ to obtain $A = -\pi/4$. Whence

$$h'(m) = -m^{-1} + \frac{\pi^2}{4m(1-m)K(m)^2}. \tag{5.34}$$

Since we know that $h(0) = 0$, we need only show that the coefficients in the Maclaurin expansion of $h'(m)$ are positive.

So we turn finally to proving that property of the Maclaurin coefficients of $h'(m)$.

We begin by noting that if $z(m) = W(m)^2$, where W is a solution of the differential equation (5.17), then z satisfies the differential equation

$$z''' + 3\frac{1-2m}{m(1-m)}z'' + \frac{1-7m+7m^2}{m^2(1-m)^2}z' - \frac{1}{2}\frac{1-2m}{m^2(1-m)^2}z = 0. \tag{5.35}$$

(See Watson, 1944, pp. 145–146 and Exercise 5.4.2, below.)

If one tries to find a power series solution $z = \sum_{n=0}^{\infty} \gamma_n m^n$ of (5.35), one obtains the recurrence relation

$$(n+1)^3(\gamma_{n+1} - \gamma_n) = n^3(\gamma_n - \gamma_{n-1}) - \frac{1}{2}(2n+1)\gamma_n.$$

Now equation (5.16) implies that if we set $z = [2K(m)/\pi]^2$, then $\gamma_0 = 1, \gamma_1 = 1/2$, and $\gamma_n \geq 0$ for all n. Thus the function

$$(1-m)4K(m)^2/\pi^2 = 1 + (\gamma_1 - 1)m + \cdots + (\gamma_n - \gamma_{n-1})m^n + \cdots$$

has the form

$$(1-m)4K(m)^2/\pi^2 = 1 - \sum_{n=1}^{\infty} \delta_n m^n,$$

where $\delta_n > 0$ for all n. If we let $m \to 1-$, we see that

$$\lim_{m \to 1-} \sum_{n=1}^{\infty} \delta_n m^n = 1.$$

It follows, as in Corollary 5.1, that $\sum_{n=1}^{\infty} \delta_n = 1$, whence $\left| \sum_{n=1}^{\infty} \delta_n m^n \right| < 1$ if $|m| < 1$. Hence we can write

$$\frac{\pi^2}{4(1-m)K(m)^2} = 1 + \sum_{n=1}^{\infty} \left[\sum_{l=1}^{\infty} \delta_l m^l \right]^n = 1 + \sum_{n=1}^{\infty} \varepsilon_n m^n,$$

where $\varepsilon_n > 0$ for all n. But then, by (5.34),

$$h'(m) = -m^{-1} + \frac{\pi^2}{4m(1-m)K(m)^2} = \sum_{n=1}^{\infty} \varepsilon_n m^{n-1},$$

with $\varepsilon_n > 0$, and that completes the proof of our theorem.

Remark The proof of Theorem 5.7 (a) in Whittaker & Watson (1927), p. 486, requires an appeal to the *monodromy theorem* (see Ahlfors, 1979, pp. 295–296). Our proof is a simplification of that in Tannery & Molk (1893–1902). The idea of deriving properties of the function $K(m)$ from the differential equation (5.17) goes back to Fuchs (1870).

Exercises 5.4

5.4.1 (See the proof of Lemma 5.8). Let W_1, W_2 be two solutions of $p(z)W'' + p'(z)W' + r(z)W = 0$, where $p(z), r(z)$ are analytic in a domain, D. Show that $p(z)(W_1 W_2' - W_2 W_1')$ is a constant in D.

5.4.2 (See the proof of Theorem 5.7, following Lemma 5.8, and (5.35).) Let
$z(m) = W(m)^2$, where W is a solution of (5.17). Show that z satisfies the
differential equation

$$z''' + 3\frac{1-2m}{m(1-m)}z'' + \frac{1-7m+7m^2}{m^2(1-m)^2}z' - \frac{1}{2}\frac{1-2m}{m^2(1-m)^2}z = 0.$$

5.4.3 Show that the numbers c_m, 'the Weierstrass numbers', in Theorem 5.7,
Equation (5.31) are integers.

5.4.4 Given that $k = k' = 1/\sqrt{2}$, calculate q, K, K' (observe that $\tau = i$; so that
$q = e^{-\pi}$). (In fact $q = 0.043\,213\,9$, $K = K' = 1.854\,075$; see Whittaker
& Watson, 1927, p. 486.)

5.5 The Weierstrass ℘ function and sums of four squares

We shall suppose that $\tau = iK'/K$ is purely imaginary. Consider the functions

$$f_\alpha(z|\tau) = \frac{\theta_1'\theta_\alpha(z|\tau)}{\theta_\alpha\theta_1(z|\tau)}, \qquad \alpha = 2, 3, 4. \tag{5.36}$$

The functions f_α (as functions of z) have a pole of order one at $z = 0$, the
constant factor θ_1'/θ_α being chosen so that the residue at $z = 0$ is 1. From
Theorems 4.3 and 4.5 and from Equations 5.1, 5.2 and 5.3 we see that

$$f_2(z|\tau) = \frac{2K}{\pi}\frac{cn\dfrac{2K}{\pi}z}{sn\dfrac{2K}{\pi}z},$$

$$f_3(z|\tau) = \frac{2K}{\pi}\frac{dn\dfrac{2K}{\pi}z}{sn\dfrac{2K}{\pi}z},$$

and

$$f_4(z|\tau) = \frac{2K}{\pi}\frac{1}{sn\dfrac{2K}{\pi}z}.$$

Observe that $f_\alpha(z + \pi|\tau) = \pm f_\alpha(z|\tau)$ and $f_\alpha(z + \pi\tau|\tau) = \pm f_\alpha(z|\tau)$. Hence
$f_\alpha^2(z|\tau)$ has periods π and $\pi\tau$. Moreover, f_α^2 is an even function, whence the
residue of f_α^2 at 0 is 0; that is, f_α^2 has a pure double pole and hence

$$f_\alpha^2(z|\tau) = z^{-2} + \sum_{n=0}^{\infty} a_{\alpha_n} z^n \quad \left(|z| < \frac{1}{2}\pi|\tau|\right). \tag{5.37}$$

Now at the end of Chapter 3 we made reference to the Weierstrass $\wp$-function, having periods ω_1, ω_2 (with $\text{Im}(\omega_2/\omega_1) > 0$ and with a single double pole at $z = 0$, with residue 0, normalized so that

$$\wp(z; \omega_1, \omega_2) = z^{-2} + \sum_{n=1}^{\infty} b_n z^n. \tag{5.38}$$

(For a more complete development of the theory, see Chapter 7.)

It follows from (5.37) and (5.38) that

$$\wp(z; \pi, \pi\tau) = f_\alpha^2(z|\tau) + \text{constant}. \tag{5.39}$$

Hence

$$\wp(z; \omega_1, \omega_2) = \frac{\pi^2}{\omega_1^2} f_\alpha^2 \left(\frac{\pi z}{\omega_1} \middle| \tau \right) + \text{constant},$$

where $\tau = \omega_2/\omega_1$.

Now set $C_{\alpha\beta} = f_\alpha^2(z|\tau) - f_\beta^2(z|\tau)$. Then $C_{\alpha\beta}$ is doubly periodic with periods $\pi, \pi\tau$ and is analytic at $z = 0$ and so analytic in the cell with sides $\pi, \pi\tau$. Accordingly $C_{\alpha\beta}$ is a constant, which we shall now determine.

Since $\theta_1(z)$ is an odd function and the functions $\theta_\alpha(z)$ ($\alpha = 2, 3, 4$) are even, we have

$$f_\alpha^2(z|\tau) = \left[\frac{\theta_1' \theta_\alpha + \theta_\alpha'' z^2/2 + \cdots}{\theta_\alpha \theta_1' + \theta_1''' z^3/6 + \cdots} \right]^2 = \frac{1}{z^2} \frac{1 + \dfrac{\theta_\alpha''}{\theta_\alpha} z^2 + \cdots}{1 + \dfrac{\theta_1'''}{\theta_1'} z^2/3 + \cdots}$$

$$= z^{-2} \left[1 + \left(\frac{\theta_\alpha''}{\theta_\alpha} - \frac{\theta_1'''}{3\theta_1'} \right) z^2 + \cdots \right].$$

Thus $C_{\alpha\beta} = \theta_\alpha''/\theta_\alpha - \theta_\beta''/\theta_\beta$, where $\theta_\alpha'' = \theta_\alpha''(0|\tau)$.
Hence, in view of Exercise 4.4.6, we have

$$C_{\alpha\beta} = \frac{4\text{i}}{\pi} \left[\frac{\dfrac{\partial}{\partial \tau} \theta_\alpha}{\theta_\alpha} - \frac{\dfrac{\partial}{\partial \tau} \theta_\beta}{\theta_\beta} \right] = \frac{4\text{i}}{\pi} \frac{\partial}{\partial \tau} \log \frac{\theta_\alpha}{\theta_\beta}. \tag{5.40}$$

For an arithmetical application, we are particularly interested in

$$C_{42} = \frac{4K^2}{\pi^2} \frac{1 - cn^2 \dfrac{2K}{\pi} z}{sn^2 \dfrac{2K}{\pi} z} = \left(\frac{2K}{\pi} \right)^2 = \theta_3^4,$$

since $\theta_3 = \sum_{m=-\infty}^{\infty} q^{m^2}$ implies

$$\theta_3^4 = \sum_{m_1, m_2, m_3, m_4 \in \mathbb{Z}} q^{m_1^2 + m_2^2 + m_3^2 + m_4^2} = 1 + \sum_{n=1}^{\infty} A_4(n) q^n, \tag{5.41}$$

where $A_4(n)$ denotes the number of representations of n as a sum of four squares. Note that the integers m_1, m_2, m_3, m_4 may be positive, negative or zero.

Now $q = e^{\pi i \tau}$ implies $\dfrac{\partial}{\partial \tau} = \pi i q \dfrac{\partial}{\partial q}$, whence (5.40) becomes

$$C_{42} = 4q \frac{\partial}{\partial q} \log \frac{\theta_2(0, q)}{\theta_4(0, q)}. \tag{5.42}$$

But by Exercise 4.4.3,

$$\frac{\theta_2(0, q)}{\theta_4(0, q)} = 2q^{1/4} \frac{\displaystyle\prod_{m=1}^{\infty} (1 + q^{2m})^2}{\displaystyle\prod_{m=1}^{\infty} (1 - q^{2m-1})^2} = 2q^{1/4} \frac{\displaystyle\prod_{m=1}^{\infty} (1 - q^{4m})^2}{\displaystyle\prod_{m=1}^{\infty} (1 - q^m)^2},$$

and hence

$$C_{42} = 4q \left[\frac{1}{4q} - 8 \sum_{m=1}^{\infty} \frac{m q^{4m-1}}{1 - q^{4m}} + 2 \sum_{m=1}^{\infty} \frac{m q^{m-1}}{1 - q^m} \right]$$

$$= 1 + 8 \sum_{m=1}^{\infty} \frac{m q^m}{1 - q^m} - 8 \sum_{m=1}^{\infty} \frac{4 m q^{4m}}{1 - q^{4m}}$$

$$= 1 + 8 \sum_{m \geq 1, m \not\equiv 0 (\text{mod } 4)} \frac{m q^m}{1 - q^m}$$

$$= 1 + 8 \sum_{m=1, m \not\equiv 0 (\text{mod } 4)}^{\infty} \sum_{k=1}^{\infty} m q^{km},$$

whence

$$C_{42} = 1 + 8 \sum_{r=1}^{\infty} q^r \left(\sum_{m \mid r, 4 \nmid m} m \right). \tag{5.43}$$

On equating the two expressions, (5.41) and (5.43), for C_{42} we obtain

$$A_4(n) = 8 \sum_{d \mid n, 4 \nmid d} d. \tag{5.44}$$

When n is odd, the right-hand side is eight times the sum of the odd divisors of n. When n is even,

$$\sum_{d\mid n, 4\nmid d} d = \sum_{d\mid n, d \text{ odd}} d + 2\sum_{d\mid n, d \text{ odd}} d = 3\sum_{d\mid n, d \text{ odd}} d.$$

We summarize our results as:

Theorem 5.8 (Jacobi, 1829). *The number, $A_4(n)$, of representations of a positive integer, n, as a sum of four squares is eight times the sum of its odd divisors when n is odd, and twenty-four times the sum of its odd divisors when n is even.*

Corollary 5.2 (Lagrange) *A positive integer is the sum of four squares.*

Examples

(a) $A_4(2) = 24$. We have $2 = 1^2 + 1^2 + 0^2 + 0^2$; there are six possible positions for the pairs 1, 1 and each possibility leads to four subcases on replacing 1 by -1.

(b) $A_4(3) = 32$. Again, $3 = 1^2 + 1^2 + 1^2 + 0^2$ and now there are four possible positions for the 0. Each possibility leads to eight subcases on replacing 1 by -1.

Exercises 5.5

5.5.1 Show that every elliptic function can be expressed in terms of theta functions by elaborating the details of the argument outlined (cf. Whittaker & Watson, 1927, p. 474). Let $f(z)$ denote an elliptic function with a fundamental set of zeros $(\alpha_1, \ldots, \alpha_n)$ and poles $(\beta_1, \ldots, \beta_n)$ chosen so that $\sum_{r=1}^{n} (\alpha_r - \beta_r) = 0$ (see Theorem 3.7). Show that

$$f(z) = A_3 \prod_{r=1}^{n} \frac{[\theta_1((\pi z - \pi\alpha_r)/\omega_1 \mid \omega_2/\omega_1)]}{[\theta_1((\pi z - \pi\beta_r)/\omega_1 \mid \omega_2/\omega_1)]},$$

for some constant A_3. Now let

$$\sum_{m=1}^{m_r} A_{r,m}(z - \beta_r)^{-m}$$

be the principal part of $f(z)$ at the pole β_r and show that

$$f(z) = A_2 + \sum_{r=1}^{n}\left[\sum_{m=1}^{m_r} \frac{(-1)^{m-1}A_{r,m}}{(m-1)!} \frac{d^m}{dz^m} \log\theta_1((\pi z - \pi\beta_r)/\omega_1 \mid \omega_2/\omega_1)\right]$$

for some constant A_2. (See Whittaker & Watson, 1927, p. 474.)

5.5.2 Show that

$$\frac{\theta_3^2(z)}{\theta_1^2(z)} = -\frac{\theta_3^2}{\theta_1'^2}\frac{d}{dz}\frac{\theta_1'(z)}{\theta_1(z)} + \frac{\theta_3\theta_3''}{\theta_1'^3},$$

and deduce that

$$\int_0^{\pi/2}\frac{\theta_3^2(z)}{\theta_1^2(z)}dz = \frac{\theta_3^2}{\theta_1'^2}\frac{\theta_1'(z)}{\theta_1(z)} + \left(\frac{\pi}{2} - z\right)\frac{\theta_3\theta_3''}{\theta_1'^3}.$$

(Whittaker & Watson, 1927, p. 474.)

5.6 The eta-function of Jacobi

In Chapter 8, we shall consider the general problem of the evaluation of *elliptic integrals*, that is of integrals of the form $\int R(w, x)dx$, where R denotes a rational function of w and x and where w^2 is a quartic or cubic polynomial in x, without repeated roots. We shall prove that any such elliptic integral can be evaluated in terms of the three basic kinds of integral, one of which is the integral

$$E(k, \phi) = E(u) = \int_0^u dn^2 u\, du \qquad (5.45)$$

(see Section 1.7 and Section 8.2). Since the residues of $dn^2 u$ are zero, the integral defining $E(u)$ is independent of the path chosen, given that the path does not pass through a pole of $dn\,(u)$.

To evaluate $E(u)$ we return to a notation employed by Jacobi in his *Fundamenta Nova* (1829) and subsequently discarded. The basic function in that notation is the *theta function*

$$\Theta(u) = \theta_4\big(\theta_3^{-2}u\,|\,\tau\big); \qquad (5.46)$$

so that the periods associated with $\Theta(u)$ are $2K$ and $2iK'$. The function

$$\Theta(u + K) = \theta_3\big(\theta_3^{-2}u\big)$$

then replaces $\theta_3(z)$; $\theta_1(z)$ is replaced by the *eta function*

$$H(u) = -iq^{i/4}e^{\pi iu/2K}\Theta(u + iK') = \theta_1\big(\theta_3^{-2}u\,|\,\tau\big), \qquad (5.47)$$

and $\theta_2(z)$ is replaced by $H(u + K)$.

We recall that the integral $E(K)$ obtained from (5.45), namely

$$E = E(K) = \int_0^K dn^2 u\,du = \int_0^{\pi/2}(1 - k^2\sin^2\phi)^{1/2}d\phi,$$

is called the *complete elliptic integral of the second kind*. We have:

Theorem 5.9 *In the notation introduced above,*

$$E(u) = \Theta'(u)/\Theta(u) + uE/K.$$

Proof The function $\frac{d}{du}[\Theta'(u)/\Theta(u)]$ is a doubly periodic function with double poles at the zeros of $\Theta(u)$; that is at the poles of $dn(u)$. Hence, for some constant A,

$$dn^2u - A\frac{d}{du}[\Theta'(u)/\Theta(u)]$$

is a doubly periodic function of u, with periods $2K$, $2iK'$, having at most a simple pole in any cell. By Corollary 3.1 of Chapter 3, it is therefore a constant which we write as B/K. To determine the constant A, observe that the principal part of dn^2u at iK' is $-(u - iK')^{-2}$, by Chapter 2, Section 2.7. Since the residue of $\Theta'(u)/\Theta(u)$ at iK' is 1, the principal part of $\frac{d}{du}[\Theta'(u)/\Theta(u)]$ is also $-(u - iK)^{-2}$. Hence $A = 1$, whence

$$dn^2u = \frac{d}{du}[\Theta'(u)/\Theta(u)] + B/K.$$

Note that $\theta'_4(0) = 0$, since $\theta_4(z)$ is an even function. Hence $\Theta'(0) = 0$ and integrating yields

$$E(u) = \Theta'(u)/\Theta(u) + uB/K.$$

Now set $u = K$ and observe that $\Theta'(K) = 0$ (since $\theta_3(z)$ is even) to obtain $B = E$.

The function $E(u)$ does not have period $2K$ or $2iK'$, but

$$E(u + 2K) = \int_0^u dn^2u\,du + \int_u^{u+2K} dn^2u\,du$$

$$= E(u) + \int_{-K}^K dn^2u\,du$$

$$= E(u) + 2E.$$

Hence, if we set

$$Z(u) = E(u) - uE/K = \Theta'(u)/\Theta(u) \qquad (5.48)$$

we find that $Z(u + 2K) = Z(u)$. Moreover, the differential equation

$$\Theta'(u) = Z(u)\Theta(u)$$

implies that

$$\Theta(u) = \Theta(0)\exp\left[\int_0^u Z(u)du\right]. \qquad (5.49)$$

The function $Z(u)$ is called *Jacobi's zeta function*, but we have come full circle historically: Jacobi *defined* $\Theta(u)$, up to a multiplicative constant, by (5.49) (see Jacobi, 1829). To make the proper choice of multiplicative constant, he considered an imaginary transformation of $Z(u)$ as given by:

Theorem 5.10 *With the foregoing notation,*

$$Z(iu, k) = i\,dn(u, k')sc(u, k') - iZ(u, k') - \pi iu/(2KK').\qquad(5.50)$$

Proof Translate the formula (see 4.50)

$$\theta_2(ix|\tau) = (-i\tau)^{-1/2} \exp(-i\tau'x^2/\pi) \cdot \theta_4(ix\tau'|\tau')$$

into Jacobi's earlier notation (using (4.50) and the substitutions $z = ix\tau'$, $x = \pi u/2K$) to obtain

$$H(iu + K, k) = (-i\tau)^{-1/2} \exp(\pi u^2/4KK')\Theta(u, k'),\qquad(5.51)$$

whence (by (5.5b))

$$cn(iu, k) = (-i\tau)^{-1/2} \exp(\pi u^2/4KK')\frac{\theta_4(0|\tau)}{\theta_2(0|\tau)}\frac{\Theta(u, k')}{\Theta(iu, k)}.$$

The desired result now follows on taking the logarithmic derivative of each side and making use of Equations (2.11).

The imaginary transformation of $E(u)$ follows directly from the definition in terms of an integral:

$$E(iu, k) = \int_0^{iu} dn^2(t, k)dt = \int_0^u dn^2(it', k)i\,dt'$$

$$= i\int_0^u dc^2(t', k')dt',$$

on setting $t = it'$ and making use of Equation (5.5c) again. But then Exercise 5.6.2 yields

$$E(iu, k) = i\left[u + dn(u, k')sc(u, k') - \int_0^u dn^2(t', k')dt'\right]$$

or

$$E(iu, k) = iu + i\,dn(u, k')sc(u, k') - iE(u, k').\qquad(5.52)$$

The imaginary transformations above of $E(u)$ and $Z(u)$ now lead to a relation between the two kinds of complete elliptic integrals

Theorem 5.11 (Legendre) *We have:*

$$EK' + E'K - KK' = \frac{1}{2}\pi.\qquad(5.53)$$

Proof The imaginary transformations as given above yield

$$E(iu, k) - Z(iu, k) = iu - i[E(u, k') - Z(u, k')] + \frac{\pi iu}{2KK'}.$$

Since $E(u) - Z(u) = uE/K$, we obtain

$$iu\, E/K = iu - iu E'/K' + \pi iu/2KK'$$

and since we may take $u \neq 0$, (5.53) follows.

Theorem 5.12 (*The addition theorem for E and for Z*) *With the foregoing notation, we have:*

$$E(u + v) = E(u) + E(v) - k^2 sn(u)sn(v)sn(u + v), \qquad (5.54)$$

$$Z(u + v) = Z(u) + Z(v) - k^2 sn(u)sn(v)sn(u + v). \qquad (5.55)$$

Proof Recall that

$$Z(u) = \Theta'(u)/\Theta(u),$$

$$Z(u) = E(u) - uE/K$$

and so

$$\Theta(u) = \Theta(0)\exp\left\{\int_0^u Z(t)dt\right\}.$$

(The exponential is one-valued, even though $\Theta(u)$ is not; for $Z(t)$ has residue 1 at its poles, but the difference in the values of the integrals taken along two paths is $2n\pi i$, where n is the number of poles enclosed, whence the exponential is single-valued. See Whittaker & Watson, 1927, Section 22.731.)

Now consider the expression

$$\frac{\Theta'(u + v)}{\Theta(u + v)} - \frac{\Theta'(u)}{\Theta(u)} - \frac{\Theta'(v)}{\Theta(v)} + k^2 sn(u)sn(v)sn(u + v), \qquad (5.56)$$

as a function of the complex variable u. It is doubly periodic, with periods $2K$ and $2iK'$, and has simple poles, congruent to iK' and to $iK' - v$. The residue of the first two terms in (5.56) at iK' is -1 and the residue of $sn(u)sn(v)sn(u + k)$ there is $k^{-1}sn(v)sn(iK' + v) = k^{-2}$.

It follows that the function in (5.56) is doubly periodic and has no poles at the points congruent to iK', or, by a similar argument, at the points congruent to $iK' - v$. By Liouville's Theorem it is a constant (as a function of u) and by taking $u = 0$ we find that that constant is zero.

Whence

$$Z(u) + Z(v) - Z(u + v) = k^2 sn(u)sn(v)sn(u + v),$$
$$E(u) + E(v) - E(u + v) = k^2 sn(u)sn(v)sn(u + v),$$

as required.

We may now pick up the thread of our discussion of the history of Jacobi's approach to theta functions. Jacobi derived (5.34) from the imaginary transformation in much the same way as we have done, and then proved (5.50) directly from Legendre's relation (5.53). He then showed that

$$\frac{\Theta(2Kx/\pi)}{\Theta(0)} = \frac{\prod_{n=1}^{\infty}(1 - 2q^{2n-1}\cos 2x + q^{4n-2})}{\prod_{n=1}^{\infty}(1 - q^{2n-1})^2},$$

the expression on the left being $\theta_4(x|\tau)/\theta_4(0|\tau)$ in our notation. He was then able to show that

$$\frac{\Theta(2Kx/\pi)}{\Theta(0)} = (2k'K/\pi)^{-1/2} \sum_{n=-\infty}^{\infty} (-1)^n q^{n^2} \cos 2nx. \qquad (5.57)$$

Then he set $\Theta(0) = (2k'K/\pi)^{1/2}$ to obtain

$$\Theta(2Kx/\pi) = \sum_{n=-\infty}^{\infty} (-1)^n q^{n^2} \cos 2nx,$$

which in our notation is Equation (4.9). Accordingly, the evaluation of $\Theta(0)$ resulting from Equation (5.57) is equivalent to the evaluation of the constant G in our notation. (See Corollary 4.1.)

Exercises 5.6

5.6.1 (See Whittaker & Watson, 1927, p. 480.) Show that the only singularities of $\Theta'(u)/\Theta(u)$ are simple poles at $u \equiv iK'(\mathrm{mod}\, 2K, 2iK')$, with residue 1.

5.6.2 (See Whittaker & Watson, 1927, p. 516 and also Chapter 8, Exercise 8.2.) Prove by differentiation that

$$\int dc^2 u\, du = u + dn(u)su(u) - \int dn^2\, du.$$

5.6.3 (See Whittaker & Watson, 1927, p. 516.) Show by differentiation that:

(a) $\int sn(u)du = \frac{1}{2}k^{-1}\log\frac{1-kcd(u)}{1+kcd(u)}$;

(b) $\int cn(u)du = k^{-1}\arctan(ksd(u))$;

(c) $\int dn(u)du = am(u)$;

(d) $\int sc(u)du = \frac{1}{2}k'^{-1}\log\frac{dn(u)+k'}{dn(u)-k'}$;

(e) $\int ds(u)du = \frac{1}{2}\log\frac{1-cn(u)}{1+cn(u)}$;

(f) $\int dc(u)du = \frac{1}{2}\log\frac{1+sn(u)}{1-sn(u)}$.

Obtain six similar formulae by replacing u by $u + K$.

5.6.4 (See Whittaker & Watson, 1927, p. 480.) Show that $H'(0) = \frac{1}{2}\pi K^{-1}H(K)\Theta(0)\Theta(K)$.

5.6.5 (See Whittaker & Watson, 1927, p. 518.) Show that $E = \frac{1}{2}\pi F\left(-1/2, 1/2; 1; k^2\right)$, where F denotes the hypergeometric function $F(a, b; c; z)$.

5.6.6 Show that[1] $E(u + K) - E(u) = E - k^2 sn(u)cd(u)$.

5.6.7 Show that $E(2u + 2iK') = E(2u) + 2i(K' - E')$.

5.6.8 Use Exercise 5.6.6 to show that

$$E(u + iK') = \frac{1}{2}E(2u + 2iK') + \frac{1}{2}k^2 sn^2(u + iK')sn(2u - 2iK')$$
$$= E(u) + cn(u)ds(u) + i(K' - E').$$

5.6.9 Show that

$$E(u + K + iK') = E(u) - sn(u)dc(u) + E + i(K' - E').$$

5.6.10 Prove the addition theorem for $E(u)$ (Theorem 5.12) by carrying out the details of the following outline proof.

Show that

$$dn^2(x + y) - dn^2(x - y) = -\frac{4k^2 sn(x)cn(x)dn(x)sn(y)cn(y)dn(y)}{(1 - k^2 sn^2 x sn^2 y)^2}$$

and then integrate that expression *with respect to* y to obtain

$$E(x + y) + E(x - y) = C - \frac{2sn(x)cn(x)dn(x)}{sn^2 x(1 - k^2 sn^2 x\, sn^2 y)},$$

---

[1] Questions 5.6.4 to 5.6.8 are taken from Whittaker & Watson (1927), p. 518, where further background details may be found. For Question 5.6.10, see Bowman (1961), p. 22.

where C may depend on x, but not on y. Deduce that

$$E(2x) = C - \frac{2sn(x)cn(x)dn(x)}{sn^2x(1 - k^2sn^4x)}$$

and then

$$E(x + y) + E(x - y) - E(2x) = \frac{2k^2 sn\,(x)\,cn\,(x)\,dn\,(x)}{1 - k^2sn^4x}$$
$$\cdot \frac{sn^2x - sn^2y}{(1 - k^2sn^2x\,sn^2y)}.$$

Show that the right-hand side is $k^2sn(2x)sn(x + y)sn(x - y)$ and then take $u = x + y, v = x - y$ to obtain the result.

6

Introduction to transformation theory

6.1 The elliptic modular function

In Chapter 5, we considered Equation (5.11)

$$m \equiv k^2 = \theta_2^4(0|\tau)/\theta_3^4(0|\tau).$$ (6.1)

Then we formulated the problem of determining all the solutions τ of (6.1) for a given m, and in Theorem 5.2 we stated that if τ_1 is one such solution then every solution is of the form

$$\tau_2 = \frac{a\tau_1 + b}{c\tau_1 + d},$$

where $a, b, c, d \in \mathbb{Z}, ad - bc = 1, a, d \equiv 1 \pmod 2$ and $b, c \equiv 0 \pmod 2$. The proof of Theorem 5.2 was deferred to this chapter and it is to that theme that we now return. We begin with:

Definition 6.1 *In the notation of* (6.1) *we define*

$$\lambda(\tau) = \theta_2^4(0|\tau)/\theta_3^4(0|\tau), \quad \mathrm{Im}(\tau) > 0.$$

Clearly, in the notation of (6.1), $\lambda(\tau) = m = k^2$ and we are concerned with the equation

$$\lambda(\tau_1) = \lambda(\tau_2).$$ (6.2)

Theorem 5.2 asserts a necessary and sufficient condition for (6.2) to hold; we begin by obtaining a necessary condition.

Theorem 6.1 *A necessary condition for* (6.2) *to hold is that*

$$\tau_2 = \frac{a\tau_1 + b}{c\tau_1 + d},$$

where $a, b, c, d \in \mathbb{Z}, ad - bc = 1$ *and*

$$\begin{pmatrix} a & b \\ c & d \end{pmatrix} \equiv \begin{pmatrix} 1 & 0 \\ 0 & 1 \end{pmatrix} (\text{mod } 2).$$

Proof We saw before (in Theorem 5.4) that (6.2) implies $sn(u|\tau_1) = sn(u|\tau_2)$. Now the first expression has the module of zeros

$$[2k_1 K(\tau_1) + 2l_1 i K'(\tau_1) : k_1, l_1 \in \mathbb{Z}]$$

and the second the module

$$[2k_2 K(\tau_2) + 2l_2 i K'(\tau_2) : k_2, l_2 \in \mathbb{Z}].$$

Since $\tau_1 = iK'(\tau_1)/K(\tau_1)$ and $\tau_2 = iK'(\tau_2)/K(\tau_2)$ both have positive imaginary part, we can apply Theorem 3.2 to obtain

$$\begin{pmatrix} 2iK'(\tau_2) \\ 2K(\tau_2) \end{pmatrix} = \begin{pmatrix} a & b \\ c & d \end{pmatrix} \begin{pmatrix} 2iK'(\tau_1) \\ 2K(\tau_1) \end{pmatrix}$$

or

$$\begin{aligned} \text{(a)} \quad & iK'(\tau_2) = aiK'(\tau_1) + bK(\tau_1), \\ \text{(b)} \quad & K(\tau_2) = ciK'(\tau_1) + dK(\tau_1). \end{aligned} \Bigg\} \tag{6.3}$$

In (6.3a), $iK'(\tau_2)$ is a pole of sn, by (2.35), and that implies $a \equiv 1(\text{mod } 2)$ and $b \equiv 0(\text{mod } 2)$ by (2.31) and (2.32). We may rewrite (b) of (6.3) as

$$K(\tau_2) - dK(\tau_1) = ciK'(\tau_1).$$

The left-hand side is a non-singular point of sn, whence $c \equiv 0(\text{mod } 2)$, for otherwise we would have a pole at $K(\tau_2)$, by (2.31) and (2,32), in the case $d \equiv 0(\text{mod } 2)$, or an incompatible value at $K(\tau_2)$, by (2.32), in the case $d \equiv 1(\text{mod } 2)$. If we now divide (a) by (b) we obtain the theorem.

We observe that, since $c \equiv 0(\text{mod } 2)$, (6.3b) implies $1 = sn(K(\tau_2)) = sn(dK(\tau_1))$, by (2.37), whence we must have $d \equiv 1(\text{mod } 4)$, by (2.37). But then $ad - bc = 1$ implies $a \equiv 1(\text{mod } 4)$. Since

$$\begin{pmatrix} a & b \\ c & d \end{pmatrix}, \begin{pmatrix} -a & -b \\ -c & -d \end{pmatrix}$$

define the same transformation of τ, $a, d \equiv 1(4)$ implies $-a, -d \equiv 3(\text{mod } 4)$ and this apparent sharpening of Theorem 6.1 is illusory. There is some confusion over that point in the literature.

The function $\lambda(\tau)$ is called the *elliptic modular function*. In the next section we shall show that the *necessary* condition of Theorem 6.1 is also *sufficient* – that is, $\lambda(\tau)$ is invariant under unimodular transformations

$$\begin{pmatrix} a & b \\ c & d \end{pmatrix} \equiv \begin{pmatrix} 1 & 0 \\ 0 & 1 \end{pmatrix} \pmod{2}.$$

6.2   Return to the Weierstrass $\wp$-function

We investigate the relationship between $\lambda(\tau)$ and $\lambda(\tau')$, where $\tau' = f(\tau)$ and $f \in SL(2, \mathbb{Z})$ is arbitrary. We saw in Section 5.5 of Chapter 5, Equation (5.39) that

$$\wp(z; \omega_1, \omega_2) = A \frac{\theta_4^2(\pi z/\omega_1 | \tau)}{\theta_1^2(\pi z/\omega_1 | \tau)} + B, \qquad (6.4)$$

where B and $A \neq 0$ are constants and $\tau = \omega_2/\omega_1$.

We introduce the values of $\wp(z)$ at the half-periods $\omega_1/2, \omega_2/2$ and[1] $\omega_3/2 = \omega_1/2 + \omega_2/2$ (see Chapter 7 for further details), as follows.

Definition 6.2 *Let $\omega_1/2$, $\omega_2/2$ and $\omega_3/2 = \omega_1/2 + \omega_2/2$ denote the half-periods of the Weierstrass $\wp$-function, then we define*

$$e_1 = \wp(\omega_1/2),$$

$$e_2 = \wp(\omega_2/2),$$

$$e_3 = \wp(\omega_1/2 + \omega_2/2).$$

Theorem 6.2 *With the foregoing notation,*

$$\lambda(\tau) = \frac{e_3 - e_2}{e_1 - e_2}.$$

Proof Equation (6.4) and Exercises 4.4.1 and 4.4.2 imply that

$$e_1 = A \frac{\theta_4^2 \left(\dfrac{\pi}{2} \right)}{\theta_1^2 \left(\dfrac{\pi}{2} \right)} + B = A \frac{\theta_3^2}{\theta_2^2} + B,$$

---

[1] Some authors take $\omega_1 + \omega_2 + \omega_3 = 0$ and so $\omega_3/2 = -\omega_1/2 - \omega_2/2$. Since $\wp$ is an even function, the value of e_3 is the same in both conventions.

$$e_2 = A\frac{\theta_4^2\left(\frac{1}{2}\pi\tau\right)}{\theta_1^2\left(\frac{1}{2}\pi\tau\right)} + B = B,$$

$$e_3 = A\frac{\theta_4^2\left(\frac{1}{2}\pi + \frac{1}{2}\pi\tau\right)}{\theta_1^2\left(\frac{1}{2}\pi + \frac{1}{2}\pi\tau\right)} + B = A\frac{\theta_2^2}{\theta_3^2} + B;$$

whence

$$\frac{e_3 - e_2}{e_1 - e_2} = \frac{\theta_2^2/\theta_3^2}{\theta_3^2/\theta_2^2} = \frac{\theta_2^4}{\theta_3^4} = \lambda(\tau).$$

Theorem 6.3 *A sufficient condition that $\lambda(\tau) = \lambda(\tau')$ is that*

$$\tau' = \frac{a\tau + b}{c\tau + d},$$

where $ad - bc = 1$ and

$$\begin{pmatrix} a & b \\ c & d \end{pmatrix} \equiv \begin{pmatrix} 1 & 0 \\ 0 & 1 \end{pmatrix} \pmod{2}.$$

Proof Let a, b, c, d be integers satisfying the hypothesis of the theorem. Consider the $\wp$ function with period basis (ω_1, ω_2) and let (ω_1', ω_2') be the basis defined by:

$$\omega_2' = a\omega_2 + b\omega_1,$$
$$\omega_1' = c\omega_2 + d\omega_1.$$

Because of its axiomatic characterization in terms of a lattice, the $\wp$ function does not change. Clearly $\omega_2'/\omega_1' = \tau'$ and $e_1' = \wp\left(c\frac{1}{2}\omega_2 + d\frac{1}{2}\omega_1\right) = \wp\left(\frac{1}{2}\omega_1\right) = e_1$, by periodicity, since $c \equiv 0\pmod{2}$ and $d \equiv 1\pmod{2}$. Similarly,

$$e_2' = \wp\left(a\frac{1}{2}\omega_2 + b\frac{1}{2}\omega_1\right) = \wp\left(\frac{1}{2}\omega_2\right) = e_2$$

and

$$e_3' = \wp\left((a+c)\frac{1}{2}\omega_2 + (b+d)\frac{1}{2}\omega_1\right) = \wp\left(a\frac{1}{2}\omega_2 + d\frac{1}{2}\omega_1\right)$$

$$= \wp\left(\frac{1}{2}\omega_2 + \frac{1}{2}\omega_1\right) = e_3.$$

Hence

$$\lambda(\tau') = \frac{e_3' - e_2'}{e_1' - e_2'} = \frac{e_3 - e_2}{e_1 - e_2} = \lambda(\tau).$$

Further progress requires the evaluation of the constants A and B in (6.4). We know A from Section 5.5:

$$\wp(z, \omega_1, \omega_2) = \left(\frac{\pi^2}{\omega_1^2}\right) f_4^2 \left(\frac{\pi z}{\omega_1}\middle| \tau\right) + B,$$

where

$$f_4(z|\tau) = \frac{\theta_1' \theta_4(z|\tau)}{\theta_4 \theta_1(z|\tau)} = \frac{2K}{\pi} \frac{1}{sn(2Kz|\tau)}$$

$$= \frac{\alpha \theta_3^2}{sn(\alpha \theta_3^2 z)},$$

as follows from the discussion in Section 4.6, and

$$\alpha = \frac{\theta_1'}{\theta_2 \theta_3 \theta_4}.$$

Now Theorem 4.5 tells us that $\alpha = 1$, but for structural reasons let us appeal to that deep theorem only when absolutely necessary. Now, by (2.40),

$$sn(u) = u - \frac{1}{6}(1 + k^2)u^3 + O(u^5) \ (u \to 0)$$

and so

$$f_4(z) = \frac{\alpha \theta_3^2}{\alpha \theta_3^2 z - \frac{1}{6}(1 + \theta_2^4/\theta_3^4)\alpha^3 \theta_3^6 z^3 + O(z^3)}$$

$$= z^{-1}\left[1 - \frac{1}{6}(\theta_3^4 + \theta_2^3)\alpha^2 z^2 + O(z^4)\right]^{-1}$$

$$= z^{-1}\left[1 + \frac{1}{6}(\theta_3^4 + \theta_2^4)\alpha^2 z^2 + O(z^4)\right].$$

Hence

$$(f_4(z))^2 = z^{-2} + \frac{1}{3}(\theta_3^4 + \theta_2^4)\alpha^2 + O(z^2), \quad (z \to 0).$$

It follows that, for small $|z|$, we have

$$\wp(z, \omega_1, \omega_2) = z^{-2} + \frac{\pi^2/\omega_1^2}{3}(\theta_3^4 + \theta_2^4)\alpha^2 + O(z^2) + B.$$

By definition, the constant term in the Laurent expansion of $\wp$ about 0 vanishes; so we must have

$$B = -\frac{\pi^2/\omega_1^2}{3}(\theta_3^4 + \theta_2^4)\alpha^2.$$

Our next theorem concerns the values of the Weierstrass $\wp$ function at the half-periods $\omega_1/2, \omega_2/2$ and $\omega_3/2$; we shall give another proof, relating, as does Theorem 6.3, to the differential equation satisfied by $\wp$, in Chapter 7. But in Chapter 7 we *derive* the relation $e_1 + e_2 + e_3 = 0$ from the differential equation; here we use it to *obtain* the differential equation.

Theorem 6.4 *Let e_1, e_2, e_3 denote the values of $\wp$ at the half-periods $\omega_1/2, \omega_2/2, \omega_3/2$, respectively. Then $e_1 + e_2 + e_3 = 0$.*

Proof We know that

$$\wp(z; \omega_1, \omega_2) = \frac{\pi^2}{\omega_1^2}\left[\frac{\theta_1'^2}{\theta_4^2}\frac{\theta_4^2(\pi z/\omega_1)}{\theta_1^2(\pi z/\omega_1)} - \frac{1}{3}(\theta_3^4 + \theta_2^4)\alpha^2\right]$$

$$= \frac{\pi^2}{\omega_1^2}\left[\alpha^2\theta_2^2\theta_3^2\frac{\theta_4^2(\pi z/\omega_1)}{\theta_1^2(\pi z/\omega_1)} - \frac{1}{3}(\theta_3^4 + \theta_2^4)\alpha^2\right].$$

We can now return to our calculations in Theorem 6.2 and introduce the constants appearing in (6.4) to obtain:

$$A = \frac{\pi^2}{\omega_1^2}(\alpha^2\theta_2^2\theta_3^2)$$

$$B = \frac{\pi^2}{\omega_1^2}\left[-\frac{1}{3}(\theta_3^4 + \theta_2^4)\alpha^2\right],$$

whence

$$e_1 = \frac{\pi^2}{\omega_1^2}\left[(\alpha^2\theta_3^4) - \frac{1}{3}(\theta_3^4 + \theta_2^4)\alpha^2\right],$$

$$e_2 = \frac{\pi^2}{\omega_1^2}\left[-\frac{1}{3}(\theta_3^4 + \theta_2^4)\alpha^2\right],$$

$$e_3 = \frac{\pi^2}{\omega_1^2}\left[(\alpha^2\theta_2^4) - \frac{1}{3}(\theta_3^4 + \theta_2^4)\alpha^2\right].$$

Evidently $e_1 + e_2 + e_3 = 0$.

We can now obtain the differential equation satisfied by $\wp$ once we have established:

Lemma 6.1 *Let $\wp'$ denote the derivative of $\wp$, then $\wp'(\omega_1/2), \wp'(\omega_2/2)$ and $\wp'((\omega_1 + \omega_2/2))$ are all zero.*

Proof The symmetry and periodicity of $\wp$ imply $\wp(\omega_1 - z) = \wp(z - \omega_1) = \wp(z)$, whence $\wp'(z) = -\wp'(\omega_1 - z)$. Put $z = \omega_1/2$ to obtain $\wp'(\omega_1/2) = -\wp'(\omega_1/2)$, whence $\wp'(\omega_1/2) = 0$. Similarly $\wp'(\omega_2/2) = 0 = \wp'((\omega_1 + \omega_2)/2)$.

Now observe that $\wp'(z)$ has a triple pole at the origin and so has exactly three zeros in any period parallelogram, by Theorem 3.6. Since the numbers $\omega_1/2$, $\omega_2/2$ and $((\omega_1 + \omega_2)/2)$ are mutually incongruent modulo (ω_1, ω_2), they must be three fundamental zeros of $\wp'$ and all those zeros are simple.

Theorem 6.5 *With the notation introduced above,*

$$\wp'(z)^2 = 4(\wp(z) - e_1)(\wp(z) - e_2)(\wp(z) - e_3).$$

Proof We know that

$$\wp'(z) = -2z^{-3} + O(z) \quad (z \to 0)$$

and so

$$\wp'(z)^2 = 4z^{-6} + O(z^{-2}).$$

Now let

$$F(z) = \frac{1}{4}\wp'(z)^2 \prod_{j=1}^{3} (\wp(z) - e_j).$$

Then $F(z)$ has periods ω_1 and ω_2. At $\omega_1/2$, $\omega_2/2$ and $\omega_3/2 = (\omega_1 + \omega_2)/2$, numerator and denominator have double zeros which cancel. At $z = 0$ the sixth order poles in the numerator and denominator cancel. Moreover, $\lim_{z \to 0} F(z) = 1$, so $F(z)$ is identically equal to 1, by Liouville's Theorem.

If we multiply out the brackets on the right-hand side in Theorem 6.5, we obtain, since $e_1 + e_2 + e_3 = 0$ (by Theorem 6.4),

$$\wp'(z)^2 = 4\wp(z)^3 - g_2\wp(z) - g_3,$$

where $-g_2 = 4(e_1e_2 + e_2e_3 + e_1e_3)$ and $g_3 = 4e_1e_2e_3$. That g_2 and g_3 are invariants under the full unimodular group $SL(2, \mathbb{Z})$ is now immediate. If one makes a change of period basis, $\wp'(z)$ and $\wp(z)$ remain invariant, whence

$$g_2 = \lim_{z \to 0} z^2(g_2\wp(z) + g_3)$$

is also invariant and thence so is g_3.

It follows that under a general unimodular transformation the roots e_1, e_2, e_3 of $4\omega^3 - g_2\omega - g_3$ can only be permuted. Here are two fundamental examples:

recall from Section 2.3 that if

$$\begin{pmatrix} \omega_2' \\ \omega_1' \end{pmatrix} = f \begin{pmatrix} \omega_2 \\ \omega_1 \end{pmatrix},$$

for $f \in SL(2, \mathbb{Z})$, then $\tau' = f(\tau)$, where $\tau = \omega_2/\omega_1$ and $\tau' = \omega_2'/\omega_1'$. Let f be given by either

$$f(\tau) = S\tau \quad \text{or} \quad f(\tau) = T\tau,$$

where T, S are given, respectively, by[2]

$$T = \begin{pmatrix} 1 & 1 \\ 0 & 1 \end{pmatrix}, \quad S = \begin{pmatrix} 0 & -1 \\ 1 & 0 \end{pmatrix};$$

so that $\tau' = f(\tau)$ is given by $\tau' = \tau + 1$, $\tau' = -\tau^{-1}$, respectively. Under T we obtain $\omega_1' = \omega_1 + \omega_2$, $\omega_1' = \omega$, whence $e_1' = e_1$, $e_2' = e_3$ and $e_3' = e_2$. Accordingly,

$$\lambda(\tau') = \frac{e_3' - e_2'}{e_1' - e_2'} = \frac{e_2 - e_3}{e_1 - e_3} = \frac{\lambda(\tau)}{\lambda(\tau) - 1}.$$

Under S, one obtains $\omega_2' = -\omega_1$, $\omega_1' = \omega_2$. Thus $e_1' = e_2$, $e_2' = e_1$, $e_3' = e_3$; whence

$$\lambda(\tau') = \frac{e_3' - e_2'}{e_1' - e_2'} = \frac{e_3 - e_1}{e_2 - e_1} = 1 - \lambda(\tau).$$

We summarize our results in

$$\lambda(\tau + 1) = \frac{\lambda(\tau)}{\lambda(\tau) - 1}, \quad \lambda(-\tau^{-1}) = 1 - \lambda(\tau). \tag{6.5}$$

6.3 Generators of $SL(2, \mathbb{Z})$

The choice of S, T in the previous section was no accident. We had previously encountered S as 'Jacobi's imaginary transformation' in Section 4.7.

Theorem 6.6 *The matrices S, T generate $SL(2, \mathbb{Z})$.* (We recall that that means that every element of $SL(2, \mathbb{Z})$ can be written as a product of expressions of the form $T^n (n \in \mathbb{Z})$ and $S^m (m = 0, 1, 2, 3)$. Note that if we are interested only in the τ transformations, $S^2 = -\begin{pmatrix} 1 & 0 \\ 0 & 1 \end{pmatrix}$ implies that we need to let m take on only the values 0 and 1.)

---

[2] We follow the notation of Serre (1970) and Prasolov & Solovyev (1997); some authors interchange S and T.

Proof of Theorem 6.6 We need to show that every unimodular matrix $\left(\begin{smallmatrix} a & b \\ c & d \end{smallmatrix}\right)$ can be expressed as a product of powers of $\left(\begin{smallmatrix} 1 & 1 \\ 0 & 1 \end{smallmatrix}\right) = T$ and $\left(\begin{smallmatrix} 0 & -1 \\ 1 & 0 \end{smallmatrix}\right) = S$.

If $|a| < |c|$, we have

$$
\begin{pmatrix} a_1 & b_1 \\ c_1 & d_1 \end{pmatrix} = \begin{pmatrix} 0 & -1 \\ 1 & 0 \end{pmatrix} \begin{pmatrix} a & b \\ c & d \end{pmatrix} = \begin{pmatrix} -c & -d \\ a & b \end{pmatrix},
$$

whence $|c_1| = |a| < |c|$ – a reduction in the value of $|c|$. If $|a| \geq |c| \geq 1$, one can choose $m \neq 0$ so that $|a + mc| < |c|$.
Then

$$
\begin{pmatrix} a_1 & b_1 \\ c_1 & d_1 \end{pmatrix} = \begin{pmatrix} 1 & 1 \\ 0 & 1 \end{pmatrix}^m \begin{pmatrix} a & b \\ c & d \end{pmatrix} = \begin{pmatrix} 1 & m \\ 0 & 1 \end{pmatrix} \begin{pmatrix} a & b \\ c & d \end{pmatrix}
$$
$$
= \begin{pmatrix} a + mc & b + md \\ c & d \end{pmatrix},
$$

and $|a_1| = |a + mc| < |c| \leq |a|$, which is a reduction in the value of $|a|$.

In that way we arrive by iteration at either $c_l = 0$ or $a_l = 0$, for some l. In the first case

$$
\begin{pmatrix} 1 & b_l \\ 0 & 1 \end{pmatrix} = S(\text{product}) \begin{pmatrix} a & b \\ c & d \end{pmatrix},
$$

which yields $\left(\begin{smallmatrix} a & b \\ c & d \end{smallmatrix}\right)$ as the desired product, since $S^{-1} = S^3$. If the chain ends with $a_l = 0$, then

$$
\begin{pmatrix} 0 & -1 \\ 1 & d_l \end{pmatrix} = (\text{product}) \begin{pmatrix} a & b \\ c & d \end{pmatrix},
$$

where the expression '(product)' denotes a product of terms involving powers of T.
Since

$$
ST^{d_l} = \begin{pmatrix} 0 & -1 \\ 1 & 0 \end{pmatrix} \begin{pmatrix} 1 & d_l \\ 0 & 1 \end{pmatrix} = \begin{pmatrix} 0 & -1 \\ 1 & d_l \end{pmatrix},
$$

we have, again, expressed $\left(\begin{smallmatrix} a & b \\ c & d \end{smallmatrix}\right)$ as a product of powers of S and T.

Exercises 6.3

6.3.1 Show that the centre of $SL(2, \mathbb{Z})$ consists of the matrices $\pm \left(\begin{smallmatrix} 1 & 0 \\ 0 & 1 \end{smallmatrix}\right)$.

6.3.2 Denote by D the subset of the upper half-plane, H, consisting of the points $z \in H$ such that $|z| \geq 1$ and $-1/2 \leq \operatorname{Re} z \leq 1/2$. Sketch the region D

as a subset of H and then find and sketch the related subsets, determined by their action on D, of the set of elements of $SL_2(\mathbb{Z})/\{\pm 1\}$ defined by:

$$\{1, T, TS, ST^{-1}S, ST^{-1}, S, ST, STS, T^{-1}S, T^{-1}\}.$$

(See, for example, Copson, 1935, Chapter XV and Serre, 1970, Chapter VII, or Jones and Singerman, 1987, Chapter 5, or McKean and Moll, 1997, Chapter 4, for further background.)

6.3.3 The following table (see MacKean & Moll, 1997, p.162, but note that the numbers e_2, e_3 are interchanged in our notation) describes the action of the modular group on $\lambda(\tau) = k^2(\tau)$ in terms of the elements of the cosets of the subgroup of level 2. The permutation (ijk) is that defined by $\left(\begin{smallmatrix} 1 & 2 & 3 \\ i & j & k \end{smallmatrix}\right)$.

Coset of Γ_2 in Γ_1	Permutation of e_1, e_2, e_3	Action on $k^2(\tau)$
$\begin{pmatrix} 1 & 0 \\ 0 & 1 \end{pmatrix}$	(123)	k^2
$\begin{pmatrix} 0 & -1 \\ 1 & 0 \end{pmatrix}$	(213)	$1 - k^2$
$\begin{pmatrix} 1 & 0 \\ -1 & 1 \end{pmatrix}$	(321)	$\dfrac{1}{k^2}$
$\begin{pmatrix} 1 & -1 \\ 0 & 1 \end{pmatrix}$	(132)	$\dfrac{k^2}{k^2 - 1}$
$\begin{pmatrix} 0 & 1 \\ -1 & 1 \end{pmatrix}$	(312)	$\dfrac{1}{1 - k^2}$
$\begin{pmatrix} 1 & -1 \\ 1 & 0 \end{pmatrix}$	(231)	$1 - \dfrac{1}{k^2}$

6.4 The transformation problem

The zeros of $\theta_1(z|\tau)$ form a module

$$\Lambda = \{m\pi + n\pi\tau : m, n \in \mathbb{Z}\}.$$

Under the unimodular transformation

$$\begin{pmatrix} \omega_2 \\ \omega_1 \end{pmatrix} \mapsto \begin{pmatrix} a & b \\ c & d \end{pmatrix}\begin{pmatrix} \omega_2 \\ \omega_1 \end{pmatrix},$$

where $\omega_1 = \pi$, $\omega_2 = \pi\tau$, Λ is invariant; that is

$$\Lambda = \{m(c\tau + d)\pi + n(a\tau + b)\pi : m, n \in \mathbb{Z}\}.$$

The module

$$\Lambda' = \left\{ m\pi + n\pi \frac{a\tau + b}{c\tau + d} : m, n \in \mathbb{Z} \right\}$$

arises from Λ on multiplying by $(c\tau + d)^{-1}$. Hence

$$\theta_1 \left(\frac{z}{c\tau + d} \middle| \frac{a\tau + b}{c\tau + d} \right)$$

has the same zeros as $\theta_1(z|\tau)$. The purpose of the theory of (linear) transformations of θ-functions is to establish relations between those two functions.

In principle, we need first to establish what happens under S and T. Under $T, \tau \mapsto \tau + 1$ and we consider $G(z) = \theta_1(z|\tau + 1)/\theta_1(z|\tau)$, which is an even entire function without zeros. Now

$$G(z + \pi) = \theta_1(z + \pi|\tau + 1)/\theta_1(z + \pi|\tau) = G(z),$$

by Exercise 4.4.4. Similarly,

$$G(z + \pi\tau) = \frac{\theta_1(z + \pi\tau|\tau + 1)}{\theta_1(z + \pi\tau|\tau)} = \frac{\theta_1(z + \pi(\tau + 1)|\tau + 1)}{\theta_1(z + \pi\tau|\tau)}$$

$$= -\frac{e^{-\pi i(\tau+1)}\theta_1(z|\tau + 1)}{e^{-\pi i\tau}\theta_1(z|\tau)} = \frac{\theta_1(z|\tau + 1)}{\theta_1(z|\tau)} = G(z).$$

It follows that G is doubly periodic and entire, whence $\theta_1(z|\tau + 1) = C\theta_1(z|\tau)$, for some constant, C. Since setting $z = 0$ would lead to 0 on both sides, we divide by z and take the limit as $z \to 0$, in other words we first differentiate with respect to z and then set $z = 0$. We find that

$$\theta_1'(0|\tau + 1) = C\theta_1'(0|\tau).$$

But then, by Exercise 4.4.2 and Corollary 4.1 of Theorem 4.5, we have

$$\theta_1'(0|\tau) = 2e^{\pi i\tau/4} \prod_{n=1}^{\infty} (1 - e^{2\pi i\tau n})^3$$

and so we find that $C = e^{\pi i/4}$. It follows that

$$\theta_1(z|\tau + 1) = e^{\pi i/4}\theta_1(z|\tau). \tag{6.6}$$

The case for the transformation $S, \tau \mapsto -i/\tau$, was worked out in Section 4.7, namely

$$\theta_1(\tau^{-1}z|-\tau^{-1}) = -i(-i\tau)^{1/2} \exp(-z^2/\pi i\tau)\theta_1(z|\tau). \tag{6.7}$$

6.5   Transformation of $\theta_1(z|\tau)$, continued

We now consider the effect on $\theta_1(z|\tau)$ of the general unimodular transformation

$$\tau' = \frac{a\tau + b}{c\tau + d} \quad (ad - bc = 1). \tag{6.8}$$

We may suppose that $c > 0$, since when $c = 0$, $a = d = 1$ and (6.6) yield

$$\theta_1(z|\tau + b) = e^{\pi i b/4}\theta_1(z|\tau). \tag{6.9}$$

We shall also need, for k and l integers,

$$\theta_1(z + k\pi + l\pi\tau|\tau) = (-1)^{k+l}e^{-2ilz}e^{-il^2\pi\tau}\theta_1(z|\tau), \tag{6.10}$$

which follows from Exercise 4.4.4, by iteration.

We now observe that (6.8) implies

$$(c\tau + d)(c\tau' - a) = -1, \quad (c\tau + d)^{-1} = -c\tau' + a,$$
$$\tau(c\tau + d)^{-1} = d\tau' - b, \tag{6.11}$$

(see Exercise 6.5.1).

Proceeding as before, we note that

$$F(z) = \frac{\theta_1\left(\dfrac{z}{c\tau + d}\,\middle|\,\tau'\right)}{\theta_1(z|\tau)}$$

is an *even* entire function without zeros. Using (6.10) and (6.11) we obtain

$$F(z + \pi) = \frac{\theta_1\left(\dfrac{z}{c\tau + d} + \dfrac{\pi}{c\tau + d}\,\middle|\,\tau'\right)}{\theta_1(z + \pi|\tau)} = \frac{\theta_1\left(\dfrac{z}{c\tau + d} + (a - c\tau')\pi\,\middle|\,\tau'\right)}{\theta_1(z + \pi|\tau)}$$

$$= (-1)^{a+1-c}\exp(-ic^2\pi\tau')\exp\left(\frac{2icz}{c\tau + d}\right)\frac{\theta_1\left(\dfrac{z}{c\tau + d}\,\middle|\,\tau'\right)}{\theta_1(z|\tau)}.$$

Since a and c cannot both be even, $a + 1 - c - ac = (1 + a)(1 - c)$ is even and we can write

$$F(z + \pi) = \exp(-\pi ic(c\tau' - a))\exp\left(\frac{2icz}{c\tau + d}\right)F(z)$$

$$= \exp\left(\frac{\pi ic}{c\tau + d}\right)\exp\left(\frac{2icz}{c\tau + d}\right)F(z). \tag{6.12}$$

Similarly, we find that

$$F(z + \pi\tau) = \frac{\theta_1\left(\dfrac{z}{c\tau+d} + \dfrac{\pi\tau}{c\tau+d}\middle|\tau'\right)}{\theta_1(z+\pi\tau|\tau)}$$

$$= \frac{\theta_1\left(\dfrac{z}{c\tau+d} - b\pi + d\pi\tau'\middle|\tau'\right)}{\theta_1(z+\pi\tau|\tau)}$$

$$= (-1)^{-b+d+1}\exp(-\pi id^2\tau')\exp\left(\frac{-2idz}{c\tau+d}\right)e^{\pi i\tau}e^{2iz}F(z).$$

Since $1 - b + d - bd = (1-b)(1+d)$ is even (because b and d cannot both be even), we have

$$(-1)^{-b+d+1} = (-1)^{bd} = e^{\pi ibd},$$

whence

$$F(z+\pi\tau) = \exp(-\pi i(d(d\tau' - b) - \tau))\exp$$
$$\left(-\left(\frac{2iz}{c\tau+d}[d-(c\tau+d)]\right)\right)F(z).$$

Now, by (6.8) and (6.11) (see Exercise 6.5.2), $(d\tau' - b)(c\tau + d) = \tau$, whence $d(d\tau' - b) - \tau = -c\tau(d\tau' - b)$. Hence, finally,

$$F(z+\pi\tau) = \exp(\pi ic\tau(d\tau' - b))\exp\left(\frac{2ic\tau z}{c\tau+d}\right)F(z). \qquad (6.13)$$

Now recall the device of Section 4.7: set $f(z) = e^{az^2}$ and use again the fundamental equation

$$f(z+s) = e^{as(2z+s)}f(z). \qquad (6.14)$$

When $s = \pi$, (6.14) resembles (6.12) if $a = ic\pi^{-1}(c\tau + d)^{-1}$, and when $s = \pi\tau$ (6.14) resembles (6.13), if, as before, $a = ic\pi^{-1}(c\tau + d)^{-1}$. Hence, if we set

$$G(z) = \exp(-ic\pi^{-1}(c\tau + d)^{-1}z^2)F(z),$$

we find that $G(z + \pi) = G(z) = G(z + \pi\tau)$. It follows that the doubly periodic entire function $G(z)$ must be a constant and we conclude that

$$\theta_1\left(\frac{z}{c\tau+d}\middle|\tau'\right) = K\exp\left(\frac{icz^2}{\pi(c\tau+d)}\right)\theta_1(z|\tau), \qquad (6.15)$$

for some constant K (independent of z).

Exercises 6.5

6.5.1 (See (6.11)). Let

$$\tau' = \frac{a\tau + b}{c\tau + d} \cdot (ad - bc = 1).$$

Prove that

$$(c\tau + d)(c\tau' - a) = -1,$$
$$(c\tau + d)^{-1} = (-c\tau' + a),$$
$$\tau(c\tau + d)^{-1} = d\tau' - b.$$

6.5.2 (See the proof of (6.13).) Use the previous exercise to prove that

$$(d\tau' - b)(c\tau + d) = \tau$$

and deduce that

$$d(d\tau' - b) - \tau = -c\tau(d\tau' - b).$$

6.6 Dependence on τ

The constant K is independent of z but depends both on τ and the matrix $A = \left(\begin{smallmatrix} a & b \\ c & d \end{smallmatrix}\right)$ that induces the transformation (6.8). We now determine the dependence on τ.

Rewrite (6.15) as

$$\theta_1(z'|\tau') = K \exp(\pi^{-1}i\rho z^2)\theta_1(z|\tau), \qquad (6.16)$$

where

$$z' = \frac{z}{c\tau + d}, \qquad \rho = \frac{c}{c\tau + d}, \qquad \tau' = \frac{a\tau + b}{c\tau + d}.$$

In Exercise 4.4.6 we saw that $y = \theta_j(z|\tau)$ $(j = 1, 2, 3, 4)$ satisfies the heat equation

$$\left(4\pi^{-1}i\frac{\partial}{\partial \tau} - \frac{\partial^2}{\partial z^2}\right) y = 0.$$

Lemma 6.1 *With the foregoing notation,*

$$4\pi^{-1}i\frac{\partial}{\partial \tau'} - \frac{\partial^2}{\partial z'^2} = (c\tau + d)^2 \left(4\pi^{-1}i\frac{\partial}{\partial \tau} - \frac{\partial^2}{\partial z^2}\right)$$

$$+ 4\pi^{-1}icz(c\tau + d)\frac{\partial}{\partial z}. \qquad (6.17)$$

Proof We have

$$\frac{d\tau}{d\tau'} = \frac{1}{\frac{d\tau'}{d\tau}} = (c\tau + d)^2$$

and

$$\frac{dz}{d\tau'} = cz'\frac{d\tau}{d\tau'} = cz'(c\tau + d)^2,$$

whence

$$\frac{\partial}{\partial\tau'} = \frac{\partial\tau}{\partial\tau'}\frac{\partial}{\partial\tau} + \frac{\partial z}{\partial\tau'}\frac{\partial}{\partial z}$$

$$= (c\tau + d)^2\frac{\partial}{\partial\tau} + cz'(c\tau + d)^2\frac{\partial}{\partial z}$$

$$= (c\tau + d)^2\frac{\partial}{\partial\tau} + cz(c\tau + d)\frac{\partial}{\partial z}.$$

Moreover,

$$\frac{\partial}{\partial z'} = \frac{\partial z}{\partial z'}\frac{\partial}{\partial z} + \frac{\partial\tau}{\partial z'}\frac{\partial}{\partial\tau} = (c\tau + d)\frac{\partial}{\partial z};$$

the lemma then follows.

Now set $\psi = \exp(\pi^{-1}\mathrm{i}\rho z^2)$ to write (6.16) as

$$\theta_1(z'|\tau') = K\psi\theta_1(z|\tau)$$

and apply the operator of Lemma 6.1 to both sides. Then

$$0 = 4\pi^{-1}\mathrm{i}\frac{\partial K}{\partial\tau}\psi\theta_1 + 4\pi^{-1}\mathrm{i}K\frac{\partial\psi}{\partial\tau}\theta_1 + 4\pi^{-1}\mathrm{i}K\psi\frac{\partial\theta_1}{\partial\tau}$$

$$- K\frac{\partial^2\psi}{\partial z^2}\theta_1 - 2K\frac{\partial\psi}{\partial z}\frac{\partial\theta_1}{\partial z} - K\psi\frac{\partial^2\theta_1}{\partial z^2}$$

$$+ 4\pi^{-1}\mathrm{i}\rho z\left(K\frac{\partial\psi}{\partial z}\theta_1 + K\psi\frac{\partial\theta_1}{\partial z}\right).$$

Since $4\pi^{-1}\mathrm{i}\frac{\partial\theta_1}{\partial\tau} - \frac{\partial^2\theta_1}{\partial z^2} = 0$ and $\frac{\partial\psi}{\partial z} = (2\pi^{-1}\mathrm{i}\rho z)\psi$, we can divide by θ_1, when $0 < |z| \ll 1$, to obtain

$$4\pi^{-1}\mathrm{i}\frac{\partial K}{\partial\tau}\psi + 4\pi^{-1}\mathrm{i}K\frac{\partial\psi}{\partial\tau} - K\frac{\partial^2\psi}{\partial z^2} + 4\pi^{-1}\mathrm{i}z\rho K\frac{\partial\psi}{\partial z} = 0.$$

Let $z \to 0$ and observe that

$$\psi\bigg|_{z=0} = 1, \quad \frac{\partial\psi}{\partial z}\bigg|_{z=0} = 0, \quad \frac{\partial\psi}{\partial\tau}\bigg|_{z=0} = 0, \quad \frac{\partial^2\psi}{\partial z^2}\bigg|_{z=0} = 2\pi^{-1}\mathrm{i}\rho.$$

Whence

$$4\pi^{-1}i\frac{\partial K}{\partial \tau} - 2\pi^{-1}i\rho K = 0,$$

or

$$\frac{\partial K}{\partial \tau} - \frac{1}{2}\rho K = 0.$$

Now insert the value of ρ to obtain

(a) $\quad \dfrac{\partial K}{\partial \tau} = \dfrac{1}{2}\dfrac{c}{c\tau + d}K.$

To solve that equation, set $K = K_0(c\tau + d)^{1/2}$.
Then

$$\frac{\partial K}{\partial \tau} = \frac{1}{2}K_0 c(c\tau + d)^{-1/2} + \frac{\partial K_0}{\partial \tau}(c\tau + d)^{1/2}$$

and $\frac{1}{2}c(c\tau + d)^{-1}K = \frac{1}{2}K_0 c(c\tau + d)^{-1/2}$. It follows that $\frac{\partial K_0}{\partial \tau} = 0$, that is K_0 is independent of τ. Hence, finally,

$$\theta_1\left(\frac{z}{c\tau + d}\bigg|\tau'\right) = K_0(c\tau + d)^{1/2}\frac{icz^2}{\exp(\pi(c\tau + d))}\theta_1(z|\tau). \tag{6.18}$$

The number K_0 depends on the matrix A; we shall see in the next section that $K_0^8 = 1$.

6.7 The Dedekind η-function

If we divide both sides of (6.18) by z and let $z \to 0$, we obtain

$$\theta_1'(0|\tau') = K_0(c\tau + d)^{3/2}\theta_1'(0|\tau). \tag{6.19}$$

There is an ambiguity implicit in the power $3/2$ in (6.19), but we absorb that into K_0 and resolve the ambiguity in due course. For definiteness, we write every complex number z in the form $z = |z|e^{i\theta}$, where $-\pi \le \theta < \pi$, and for arbitrary real s we set $z^s = |z|^s e^{is\theta}$. The drawback is, of course, that $(z_1 z_2)^s \ne z_1^s z_2^s$ in general, for $s \notin \mathbb{Z}$.

Now recall from the discussion preceding (6.5) that

$$\theta_1'(0, q) = 2q^{1/4}\prod_{n=1}^{\infty}(1 - q^{2n})^3.$$

We introduce the Dedekind η-function as follows.

Definition 6.3 *The Dedekind η-function is defined by*

$$\eta(\tau) = e^{\pi i \tau/12} \prod_{n=1}^{\infty} (1 - e^{2\pi i n \tau}) = q^{1/12} \prod_{n=1}^{\infty} (1 - q^{2n}),$$

where $q = e^{\pi i \tau}$ and $\mathrm{Im}(\tau) > 0$.

It follows that

$$\eta^3(\tau) = \frac{1}{2}\theta_1'(0, \tau)$$

and so (6.19) becomes

$$\eta^3(\tau') = K_0(c\tau + d)^{3/2}\eta^3(\tau). \tag{6.20}$$

Now take cube roots and introduce the following notation. Let

$$A = \begin{pmatrix} a_{11} & a_{12} \\ a_{21} & a_{22} \end{pmatrix} \in SL(2, \mathbb{Z}),$$

$$A \circ \tau = \frac{a_{11}\tau + a_{12}}{a_{21}\tau + a_{22}},$$

to obtain

$$\eta(A \circ \tau) = \chi(A)(a_{21}\tau + a_{22})^{1/2}\eta(\tau), \tag{6.21}$$

where the square root is determined by the convention stated above and the ambiguity in the cube root is absorbed into the multiplier $\chi(A)$.

Theorem 6.7 *The multiplier $\chi(A)$ in (6.21) satisfies*

$$\chi(AB) = \pm\chi(A)\chi(B).$$

Proof Let $B = \begin{pmatrix} b_{11} & b_{12} \\ b_{21} & b_{22} \end{pmatrix}$. Then

$$C = AB = \begin{pmatrix} c_{11} & c_{12} \\ c_{21} & c_{22} \end{pmatrix} = \begin{pmatrix} a_{11}b_{11} + a_{12}b_{21} & a_{11}b_{12} + a_{12}b_{22} \\ a_{21}b_{11} + a_{22}b_{21} & a_{21}b_{12} + a_{22}b_{22} \end{pmatrix}.$$

Thus

$$\eta(AB \circ \tau) = \chi(AB)((a_{21}b_{11} + a_{22}b_{21})\tau + a_{21}b_{12} + a_{22}b_{22})^{1/2}\eta(\tau)$$

and also

$$\eta(AB \circ \tau) = \eta(A(B \circ \tau)) = \chi(A)(a_{21}B \circ \tau + a_{22})^{1/2}\eta(B \circ \tau)$$

$$= \chi(A)\chi(B)(a_{21}B \circ \tau + a_{22})^{1/2}(b_{21}\tau + b_{22})^{1/2}\eta(\tau).$$

Now divide by $\eta(\tau)$ and then square to obtain

$$\chi(AB)^2[(a_{21}b_{11}+a_{22}b_{21})\tau+(a_{21}b_{12}+a_{22}b_{22})]$$
$$=\chi(A)^2\chi(B)^2(a_{21}B\circ\tau+a_{22})(b_{21}\tau+b_{22})$$
$$=\chi(A)^2\chi(B)^2[a_{21}(b_{11}\tau+b_{12})(b_{21}\tau+b_{22})^{-1}+a_{22}](b_{21}\tau+b_{22})$$
$$=\chi(A)^2\chi(B)^2[(a_{21}b_{11}+a_{22}b_{21})\tau+a_{21}b_{12}+a_{22}b_{22}].$$

Hence $\chi(AB)^2=\chi(A)^2\chi(B)^2$ or $\chi(AB)=\pm\chi(A)\chi(B)$.

Corollary 6.1 *In the notation of the theorem,*

$$\chi(A)^{24}=1$$

for every $A\in SL(2,\mathbb{Z})$.

Proof Since S,T, as defined above in Section 6.2, generate $SL(2,\mathbb{Z})$ and the 24th roots of unity form a group, it suffices to show that $\chi(S)$ and $\chi(T)$ are 24th roots of unity.

Now the definition of $\eta(\tau)$ implies immediately that $\eta(\tau+1)=\exp(\pi i/12)\eta(\tau)$, whence $\chi(S)=\exp(\pi i/12)$. Similarly, (4.52) (or (6.7)) implies

$$\theta_1'(0|-\tau^{-1})=(-i\tau)^{3/2}\theta_1'(0|\tau),$$

whence taking a cube root yields

$$\eta(-\tau^{-1})=\omega(-i\tau)^{1/2}\eta(\tau),$$

where ω denotes some cube root of unity. To determine ω, set $\tau=i$, whence $\omega=1$. Hence, finally, $\eta(-\tau^{-1})=(-i\tau)^{1/2}\eta(\tau)$. Since now $c\tau+d=\tau$, we have $\chi(S)=(-i)^{1/2}$, again a 24th root of unity.

Corollary 6.2 *Let $K_0=K_0(A)$ be defined by (6.18), for every $A\in SL(2,\mathbb{Z})$. Then $K_0^8=1$.*

Proof It follows immediately from (6.20) and (6.21) that $K_0=K_0(A)=\pm\chi(A)^3$. Now apply Corollary 6.1 to obtain the result.

In the following exercises it should be noted that the functions $f(\tau)$, $f_1(\tau)$, $f_2(\tau)$ are not to be confused with the function $f_4(z|\tau)$ of Chapter 6 and the functions $f_\alpha(z|\tau)$, $\alpha=2,3,4$, of Chapters 5 and 7. The former are used in the solution of the quintic given in Chapter 10 (see (10.35)), the latter are important in the theory of modular forms and functions. Clearly the definitions (7.53) and those in 6.7.2 afford a connection. Again, the functions f, defined in terms of $S\tau$ and $T\tau$ in section 6.2, should not be confused with the function f. It is important to note the distinction, since we shall be concerned with Möbius transformations of f, f_1, f_2 and η.

Exercises 6.7

6.7.1 Verify, under the convention adopted in the foregoing, that $|\arg(-i\tau)| < \pi/2$, that we have

$$(-i\tau)^{1/2} = (-i)^{1/2}\tau^{1/2}.$$

6.7.2 The functions $f(\tau)$, $f_1(\tau)$ and $f_2(\tau)$ are defined by:

$$f(\tau) = q^{-1/24} \prod_{n=1}^{\infty} (1 + q^{2n-1}),$$

$$f_1(\tau) = q^{-1/24} \prod_{n=1}^{\infty} (1 - q^{2n-1}),$$

$$f_2(\tau) = \sqrt{2} q^{1/12} \prod_{n=1}^{\infty} (1 + q^{2n}).$$

Verify that

$$\theta_1' = \theta_1'(0) = 2\eta^3(\tau),$$
$$\theta_3 = \theta_3(0) = \eta(\tau) f^2(\tau),$$
$$\theta_4 = \theta_4(0) = \eta(\tau) f_1^2(\tau),$$
$$\theta_2 = \theta_2(0) = \eta(\tau) f_2^2(\tau).$$

6.7.3 Use the connection with theta functions to prove that

$$f^8 = f_1^8 + f_2^8$$

and that

$$f f_1 f_2 = \sqrt{2}.$$

6.7.4 Show that:

$$f(\tau) = \frac{e^{-\pi i/24} \eta\left(\dfrac{\tau+1}{2}\right)}{\eta(\tau)},$$

$$f_1(\tau) = \frac{\eta\left(\dfrac{\tau}{2}\right)}{\eta(\tau)},$$

$$f_2(\tau) = \sqrt{2}\frac{\eta(2\tau)}{\eta(\tau)}.$$

6.7.5 Prove that:

$$f(\tau+1) = e^{-\pi i/24} f_1(\tau),$$
$$f_1(\tau+1) = e^{-\pi i/24} f(\tau),$$
$$f_2(\tau+1) = e^{-\pi i/12} f_2(\tau).$$

6.7.6 Prove that:

$$f(\tau+2) = e^{-\pi i/12} f(\tau),$$
$$f(\tau+48) = f(\tau),$$
$$f_1\left(-\frac{1}{\tau}\right) = f_2(\tau),$$
$$f_2\left(-\frac{1}{\tau}\right) = f_1(\tau).$$

6.7.7 Use the fact that (see 6.7.3)

$$f(\tau) f_1(\tau) f_2(\tau) = \sqrt{2},$$

and the last two results in 6.7.6 to prove that

$$f\left(-\frac{1}{\tau}\right) = f(\tau).$$

6.7.8 Use the results in 6.7.4 and

$$f_1(2\tau) = e^{-\pi i/24} f(2\tau - 1), \quad f_2(\tau) = f_1\left(-\frac{1}{\tau}\right) = e^{\pi i/24} f\left(1 - \frac{1}{\tau}\right)$$

to prove that

$$f(\tau) f\left(\frac{\tau-1}{\tau+1}\right) = \sqrt{2}.$$

(For further details and background relating to those questions, see, for example, Prasolov and Solovyev, 1997, Section 7.6.)

7

The Weierstrass elliptic functions

The treatment offered here owes much to lectures by A. M. Macbeath.

7.1 Construction of the Weierstrass functions; the Weierstrass sigma and zeta functions; the Weierstrass $\wp$-function

At the end of Chapter 3 we made brief reference to the possibility of constructing an elliptic function having periods (ω_1, ω_2) and a double pole at the origin, with residue 0. We assumed its existence and anticipated its development,

$$\wp(z) = z^{-2} + \sum_{\omega \neq 0} \{(z - \omega)^{-2} - \omega^{-2}\}, \tag{7.1}$$

where ω runs through the points $\omega = n_1\omega_1 + n_2\omega_2$ of the period lattice, Λ, other than the point $\omega = 0$.

Then in Section 5.5 of Chapter 5 we showed how to construct such a function in the form

$$\wp(z; \pi, \pi\tau) = f_\alpha^2(z|\tau) + C, \tag{7.2}$$

where $\tau = iK'/K$ is purely imaginary, and where $f_\alpha(z, \tau)$ denotes a quotient of theta functions and may be written as a quotient of Jacobi elliptic functions.

In this chapter we shall give an alternative construction of the Weierstrass functions, in which the infinite products defining the theta functions are replaced by the sigma functions, which also may be defined in terms of infinite products.

It will be helpful, in trying to understand our definition of the Weierstrass functions, to begin by seeing how a similar method may be applied to the trigonometric functions (see the Appendix, where those functions are defined in terms of infinite products).

156

We shall use the method to obtain the partial fraction expansion

$$\frac{\pi^2}{\sin^2 \pi z} = \sum_{n=-\infty}^{\infty} \frac{1}{(z-n)^2},$$ (7.3)

for $z \in \mathbb{C}$.

To that end, consider the function:

$$G(z) = \frac{\pi^2}{\sin^2 \pi z} - \sum_{n=-\infty}^{\infty} \frac{1}{(z-n)^2}.$$

Clearly, $G(z+1) = G(z)$ and $G(z)$ is periodic with period 1. Moreover,

$$\lim_{z \to 0} G(z) = \lim_{z \to 0} \left(\frac{\pi^2}{\sin^2 \pi z} - \frac{1}{z^2} \right) - \lim_{z \to 0} \sum_{n \neq 0} \frac{1}{(z-n)^2} = -\sum_{n \neq 0} \frac{1}{n^2}.$$ (7.4)

By periodicity we see that $G(z)$ has a finite limit at the integers.

It follows that the function $G(z)$ may be extended to a function defined for all $z \in \mathbb{C}$ and that function is continuous everywhere. We now prove that $G(z)$ must be identically zero.

First observe that for $z = x + iy$, we have

$$|\sin \pi z|^2 = \cosh^2 y - \cos^2 x$$

and hence $\pi^2/\sin^2 \pi z$ tends uniformly to 0 as $|y| \to \infty$. It follows that the function $G(z)$ is bounded in the period strip $0 \le x \le 1$, say

$$|G(z)| \le M, \quad 0 \le \operatorname{Re} z \le 1.$$

But then $|G(z)|$ is bounded, by periodicity, for all $z \in \mathbb{C}$.

Now it is easy to prove the identity

$$G\left(\frac{1}{2}z\right) + G\left(\frac{1}{2}(z+1)\right) = 4G(z)$$

and so $|4G(z)| \le 2M$. But then $4M \le 2M$ and so $M = 0$. It follows that $G(z)$ is identically zero and the partial fraction expansion follows from (7.3).

We try to use similar ideas to deal with doubly periodic functions.

Let ω_1, ω_2 be complex numbers such that ω_2/ω_1 is not real. As in earlier chapters, we consider the lattice (that is the additive subgroup of $\mathbb{C}$) generated by ω_1 and ω_2:

$$\Lambda = \{\omega \in \mathbb{C} : \omega = n_1 \omega_1 + n_2 \omega_2, n_1, n_2 \in \mathbb{Z}\}.$$ (7.5)

We shall be concerned with functions $f(\omega)$ of ω and we shall be concerned with absolutely convergent sums,

$$\sum_{\omega \neq 0} f(\omega),$$

the sum being over all points $\omega \in \Lambda$ other than $\omega = 0$. Since the sums are absolutely convergent, they are independent of any ordering of the terms.

Let Λ_N denote the set of lattice points $n_1\omega_1 + n_2\omega_2$ for which $\max(|n_1|, |n_2|) = N$. There are $4N$ such points, all of which lie on a parallelogram, centre 0, with sides $\pm N\omega_1, \pm N\omega_2$; and if d and d' denote the greatest and least distances, respectively, from the origin to the parallelogram, then $nd' \leq |\omega| \leq nd$, for $\omega \in \Lambda_n$. So if

$$\phi_n(f) = \sum_{\omega \in \Lambda_n} |f(\omega)|$$

then the series

$$\sum_{\omega \in \Lambda, \omega \neq 0} f(\omega)$$

is absolutely convergent if and only if $\sum_{n=1}^{\infty} \phi_n(f)$ is convergent.

We shall apply that to the function $f(\omega) = \omega^{-k}, k > 0$, and note that

$$4(nd')^{-k} \geq \phi_n(f) \geq 4n(nd)^{-k}.$$

So the series $\sum_{n=1}^{\infty} \phi_n(f)$ converges if and only if the series $\sum_{n=1}^{\infty} n^{1-k}$ converges; that is, if $k > 2$. In particular, it converges if $k = 3$.

7.1.1 The Weierstrass sigma functions

Definition 7.1 *Let*

$$g(\omega, z) = \left(1 - \frac{z}{\omega}\right) \exp\left(\frac{z}{\omega} + \frac{1}{2}\left(\frac{z}{\omega}\right)^2\right). \tag{7.6}$$

Then

$$\sigma(z; \omega_1, \omega_2) = \sigma(z, \Lambda) = z \prod_{\omega \neq 0} g(\omega, z). \tag{7.7}$$

We have to prove that the definition makes sense (that is that the infinite product in (7.7) converges) and to that end we prove:

Proposition 7.1 *Let $g(\omega, z)$ be defined by (7.6), then the infinite product (7.7) converges absolutely and uniformly on every bounded subset of the z-plane. The function $\sigma(z, \Lambda)$ is an entire function with simple zeros at the points of Λ.*

Proof Given z, the set of $\omega \in \Lambda$ for which $|\omega| < 2|z|$ is finite. If ω is not in that set, we have

$$|\log g(\omega, z)| = \left| \frac{1}{3}\left(\frac{z}{\omega}\right)^3 + \frac{1}{4}\left(\frac{z}{\omega}\right)^4 + \cdots \right|$$

$$\leq \frac{1}{3}\left|\frac{z}{\omega}\right|^3 \left(1 + \frac{1}{2} + \frac{1}{4} + \cdots \right) < \left|\frac{z}{\omega}\right|^3.$$

So the infinite product converges absolutely and uniformly on bounded subsets of the z-plane. Accordingly, it defines an entire function and it is clear that its zeros are at the points of Λ and are simple.

7.1.2 The Weierstrass zeta functions

These functions, which are not to be confused with the Riemann zeta function, afford an important intermediate step between the infinite product representation of the sigma function and the partial fraction decomposition (resembling (7.3)) of the elliptic functions.

We recall that (see the Appendix for similar examples) if the product

$$\prod_{n=1}^{\infty} f_n(z)$$

converges uniformly to $F(z)$, then

$$\sum_n \log f_n(z) = \log F(z) + 2k\pi\mathrm{i},$$

and on differentiating term-by-term (permissible since we are dealing with uniformly convergent sequences) we obtain

$$\sum_n \frac{f_n'(z)}{f_n(z)} = \frac{F'(z)}{F(z)}, \tag{7.8}$$

where the dash $'$ denotes differentiation with respect to z. (Note that the integer k is a constant, since it must be a continuous function of z and so a constant.)

We apply (7.8) to the definition of $\sigma(z, \Lambda)$ to obtain:

Definition 7.2 *Define:*

$$\zeta(z, \Lambda) = \frac{\sigma'(z, \Lambda)}{\sigma(z, \Lambda)} = \frac{1}{z} + \sum_{\omega \neq 0}\left(\frac{1}{z - \omega} + \frac{1}{\omega} + \frac{z}{\omega^2}\right). \tag{7.9}$$

Proposition 7.2 *The function $\zeta(z, \Lambda)$ is meromorphic, with poles of order 1 and residue 1 at the points of Λ.*

Proof The fact that $\zeta(z, \Lambda)$ is differentiable except for the zeros of $\sigma(z, \Lambda)$ follows from the remarks preceding our definition. Clearly, the poles of $\zeta(z, \Lambda)$ are given by the zeros of $\sigma(z, \Lambda)$, which are at the points of Λ.

Note that the functions $\sigma(z, \Lambda)$, $\zeta(z, \Lambda)$ have the following homogeneity properties with respect to z and Λ. We have:

$$\sigma(\lambda z, \lambda \Lambda) = \lambda \sigma(z, \Lambda), \quad \sigma(-z, \Lambda) = -\sigma(z, \Lambda) \tag{7.10}$$

and

$$\zeta(\lambda z, \lambda \Lambda) = \lambda^{-1} \zeta(z, \Lambda), \quad \zeta(-z, \Lambda) = -\zeta(z, \Lambda). \tag{7.11}$$

The Weierstrass function $\wp(z, \Lambda)$ may now be defined as in:

Definition 7.3 *Let $\zeta(z, \Lambda)$ be defined as in Definition 7.2. Then*

$$\wp(z, \Lambda) = -\zeta'(z, \Lambda) = \frac{1}{z^2} + \sum_{\omega \neq 0} \left(\frac{1}{(z - \omega)^2} - \frac{1}{\omega^2} \right). \tag{7.12}$$

The form of $\wp(z, \Lambda)$ in (7.12) follows immediately from (7.9) by term-by-term differentiation, which is permissible by uniform convergence for $z \notin \Lambda$. The properties of $\wp(z, \Lambda)$ and related ideas will occupy our attention for the remainder of this chapter; we conclude this introductory section with a preliminary look at its basic properties.

We begin by observing that $\wp(z, \Lambda)$ satisfies the homogeneity conditions

$$\wp(\lambda z, \lambda \Lambda) = \lambda^{-2} \wp(z, \Lambda), \quad \wp(-z, \Lambda) = \wp(z, \Lambda); \tag{7.13}$$

the latter says that $\wp(z, \Lambda)$ is an even function of z.

On differentiating (7.12) with respect to z, we obtain

$$\wp'(z, \Lambda) = -\frac{2}{z^3} - \sum_{\omega \neq 0} \frac{2}{(z - \omega)^3} = -2 \sum_{\omega \in \Lambda} \frac{1}{(z - \omega)^3}, \tag{7.14}$$

and, from (7.14), we obtain

$$\wp'(\lambda z, \lambda \Lambda) = \lambda^{-3} \wp'(z, \Lambda), \quad \wp'(-z, \Lambda) = -\wp'(z, \Lambda); \tag{7.15}$$

so that $\wp'$ is an odd function of z.

In (7.14), we observe that as ω runs through the points of Λ, so also does $\omega + \omega_1$ for a fixed $\omega_1 \in \Lambda$ and therefore

$$\wp'(z + \omega_1, \Lambda) = \wp'(z, \Lambda); \tag{7.16}$$

that is, $\wp'$ is periodic, with respect to Λ, in z.

It follows that

$$\frac{\mathrm{d}}{\mathrm{d}z}\{\wp(z + \omega_1, \Lambda) - \wp(z, \Lambda)\} = 0$$

so $\wp(z + \omega_1, \Lambda) - \wp(z, \Lambda)$ must be a constant. To find the value of the constant, we take $z = -\omega_1/2$. Then, since $\wp$ is an even function, we obtain

$$\wp(z + \omega_1) - \wp(z) = \wp\left(\frac{1}{2}\omega_1\right) - \wp\left(-\frac{1}{2}\omega_1\right) = 0;$$

so $\wp(z, \Lambda)$ is also periodic with respect to Λ.

We have proved:

Proposition 7.3 *The function $\wp(z) = \wp(z, \Lambda)$ defined in (7.12) and its derivative $\wp'(z, \Lambda)$ have the following properties.*

(i) *Both $\wp(z)$ and $\wp'(z)$ are periodic with period lattice Λ.*

(ii) *The functions $\wp(z)$ and $\wp'(z)$ satisfy the homogenity conditions (7.13) and (7.15).*

(iii) *Both functions are meromorphic, with poles at the points of Λ.*

It is essential to emphasize that the functions σ, ζ, $\wp$ and $\wp'$ are defined with respect to a lattice Λ, a subgroup of the additive group of $\mathbb{C}$ defined by a pair of periods (ω_1, ω_2) such that $\mathrm{Im}(\omega_2/\omega_1) \neq 0$. Where it is clear that we are working with a particular lattice, Λ, we shall write $\wp(z)$ in place of $\wp(z, \Lambda)$, and so on.

The functions $\zeta(z, \Lambda)$ and $\sigma(z, \Lambda)$ also have properties resembling periodicity, 'pseudo-periodicity' with respect to Λ; we conclude this section by elaborating those.

We begin with the relation $\wp(z + \omega) = \wp(z)$, $\omega \in \Lambda$, and integrate it (recall the definition (7.12)) to obtain

$$\zeta(z + \omega) = \zeta(z) + \eta(\omega, \Lambda), \tag{7.17}$$

where η depends on ω but not on z. Let $\omega' \in \Lambda$ and consider

$$\zeta(z + \omega + \omega') = \zeta(z + \omega) + \eta(\omega', \Lambda)$$
$$= \zeta(z) + \eta(\omega, \Lambda) + \eta(\omega', \Lambda),$$

using (7.17). Again, using (7.17),

$$\zeta(z + \omega + \omega') = \zeta(z) + \eta(\omega + \omega', \Lambda),$$

and on comparing those two expressions we see that $\eta(\omega, \Lambda)$ is linear in the periods:

$$\eta(\omega + \omega', \Lambda) = \eta(\omega, \Lambda) + \eta(\omega', \Lambda). \tag{7.18}$$

Let $\Lambda = \{n_1\omega_1 + n_2\omega_2, n_1, n_2 \in Z\}$ and write $\eta(\omega_1) = \eta(\omega_1, \Lambda) = \eta_1$, $\eta(\omega_2) = \eta(\omega_2, \Lambda) = \eta_2$. Then, by repeated application of (7.18), we have

$$\eta(\omega) = \eta(n_1\omega_1 + n_2\omega_2) = n_1\eta_1 + n_2\eta_2. \qquad (7.19)$$

Denote[1] by ω_3 the period such that $\omega_1 + \omega_2 + \omega_3 = 0$ and write $\eta_3 = \eta(\omega_3) = \eta(\omega_3, \Lambda)$. Then, by (7.19),

$$\eta_1 + \eta_2 + \eta_3 = 0. \qquad (7.20)$$

Suppose that the point $\omega/2$ is *not* a period. Since $\zeta(z, \Lambda)$ is an odd function, we have

$$\eta(\omega, \Lambda) = \zeta\left(\frac{1}{2}\omega\right) - \zeta\left(-\frac{1}{2}\omega\right) = 2\zeta\left(\frac{1}{2}\omega\right).$$

We turn to the pseudo-periodicity of the σ-function. We shall prove that

$$\sigma(z + \omega) = \varepsilon(\omega)\left(\exp\eta\left(z + \frac{1}{2}\omega\right)\right)\sigma(z), \qquad (7.21)$$

where

$$\varepsilon(\omega) = -1, \quad \text{if } \omega \notin \frac{1}{2}\Lambda; \quad \varepsilon(\omega) = +1, \quad \text{if } \omega \in \frac{1}{2}\Lambda. \qquad (7.22)$$

Suppose that $z \notin \Lambda$ (both sides of (7.21) are zero if $z \in \Lambda$) and also that $\omega/2 \notin \Lambda$. Then integrate the relation (7.17) to obtain

$$\int_{z_0}^{z} \zeta(z + \omega)dz = \int_{z_0}^{z} \zeta(z)dz + (z - z_0)\eta(\omega). \qquad (7.23)$$

Now

$$\zeta(z) = \frac{\sigma'(z)}{\sigma(z)}$$

and $\sigma(z) \neq 0$, since $z \notin \Lambda$. So

$$\text{Log}\,\sigma(z + \omega) = \text{Log}\,\sigma(z) + z\,\eta(\omega, \Lambda) + C, \qquad (7.24)$$

where the constant C depends on the choice of z_0 in (7.23) and Log denotes the principal value. It follows that

$$\sigma(z + \omega) = K\sigma(z)\exp(z\eta(\omega, \Lambda)).$$

---

[1] See footnote 1 in Chapter 6; we have changed the notation to that referred to there. The reader should be familiar with both.

To find K, put $z = -\omega/2$ and recall that $\sigma(\omega/2) \neq 0$, since $\omega/2 \notin \Lambda$. Whence

$$\sigma\left(\frac{1}{2}\omega\right) = K\sigma\left(-\frac{1}{2}\omega\right) \exp\left(-\frac{1}{2}\omega\eta(\omega, \Lambda)\right),$$

from which we obtain

$$\sigma(z + \omega) = -\sigma(z)\left(\exp\left(z + \frac{1}{2}\omega\right)\eta(\omega, \Lambda)\right),$$

which is (7.21), with $\varepsilon(\omega) = -1$.

If $\omega/2 \in \Lambda$, then we may write $\omega = 2\omega'$, $\omega' \in \Lambda$. If now $\omega'/2 \notin \Lambda$, we iterate the result in (7.21) to obtain the result in (7.21) with $\varepsilon(\omega) = +1$, and so on.

We summarize the results proved for the pseudo-periodicity of $\zeta(z, \Lambda)$ and $\sigma(z, \Lambda)$ in:

Proposition 7.4 *With the foregoing notation:*
(a) *the function* $\zeta(z, \Lambda)$ *satisfies*

$$\zeta(z + \omega, \Lambda) = \zeta(z) + \eta(\omega, \Lambda), \quad \omega \in \Lambda,$$

where, if $\eta_1 = \eta(\omega_1)$, $\eta_2 = \eta(\omega_2)$,

$$\eta(n_1\omega_1 + n_2\omega_2) = n_1\eta_1 + n_2\eta_2, \quad n_1, n_2, \in \mathbb{Z};$$

(b) *we have, on writing*

$$\eta = \eta(\omega),$$

$$\sigma(z + \omega) = \varepsilon(\omega)\left(\exp\eta\left(z + \frac{1}{2}\omega\right)\right) \cdot \sigma(z),$$

where $\varepsilon(\omega)$ *is given by (7.22).*

Exercises 7.1

7.1.1 Let

$$G(z) = \frac{\pi^2}{\sin^2 \pi z} - \sum_{n=\alpha}^{\infty} \frac{1}{(z - n)^2},$$

(see (7.8)). Prove the result used in the text:

$$G\left(\frac{1}{2}z\right) + G\left(\frac{1}{2}(z + 1)\right) = 4G(z).$$

7.1.2 (i) Prove that in the 'harmonic case' where the lattice Λ is made up of squares and $i\Lambda = \Lambda$, then

$$\sigma(iz, \Lambda) = i\sigma(z, \Lambda).$$

(ii) The 'equianharmonic case' is that in which the lattice Λ is made up of equilateral triangles. Prove that, in that case, $\rho\Lambda = \Lambda$, where ρ is a complex cube root of unity, $\rho^3 = 1$, and then

$$\sigma(\rho z, \Lambda) = \rho\sigma(z, \Lambda).$$

7.1.3 Complete the proof of Proposition 7.4 by carrying out the details of the iteration referred to for $\omega = 2\omega'$, $\omega' \in \Lambda$.

7.1.4 Let $F(z) = \sum_{m,n\in\mathbb{Z}} (z - (m + ni))^{-3}$. Prove that $F(z)$ is a doubly periodic, meromorphic function (that is, an elliptic function) with period lattice given by $(1, i)$. Show that $F(z)$ has three zeros in the fundamental period parallelogram (that is, the square with vertices at $0, 1, i + 1, i$) and that it is an odd function. Deduce that the zeros are located at $z = 1/2, z = i/2$ and $z = (1 + i)/2$.

7.1.5 Is the function $\exp(F(z))$, where $F(z)$ is defined in 7.1.4, an elliptic function?

7.1.6 Let ω_3 be defined by $\omega_1 + \omega_2 + \omega_3 = 0$ (see footnote 1 above) and for $r = 1, 2, 3$ define

$$\sigma_r(z) = \frac{\exp\left(-\frac{1}{2}\eta_r z\right)\sigma\left(z + \frac{1}{2}\omega_r\right)}{\sigma\left(\frac{1}{2}\omega_r\right)}.$$

Prove that $\sigma_r(z)$ is an entire function with simple zeros at the points $\omega_r/2 + \Lambda$. Show further that the functions σ_r possess the pseudo-periodicity properties

$$\sigma_r(z + \omega_r) = -\exp\left(\eta_r\left(z + \frac{1}{2}\omega_r\right)\right)\sigma_r(z)$$

and, for $r \neq s$,

$$\sigma_r(z + \omega_s) = \exp\left(\eta_s\left(z + \frac{1}{2}\omega_s\right)\right)\sigma_r(z).$$

(As we shall see later, the functions σ_r are related to the theta functions and the Jacobi elliptic functions and to the preliminary definition of $\wp(z)$ given in Chapter 5, cf. (7.2). The reader should be aware that some authors use $(2\omega_1, 2\omega_2)$ as a basis for Λ and so there are corresponding changes to the definitions used here. See, for example, the books by Copson (1935) and Whittaker & Watson (1927).

7.2 The Laurent expansions: the differential equation satisfied by $\wp(z)$

We have seen, in Propositions 7.2 and 7.3, that the functions $\zeta(z, \Lambda)$ and $\wp(z, \Lambda)$ are meromorphic, with poles at the points of the period lattice, or module, Λ. We turn to the problem of finding the Laurent expansions of those functions at the poles.

Theorem 7.1 *Let* $\zeta(z) = \zeta(z, \Lambda)$ *and* $\wp(z) = \wp(z, \Lambda)$ *be defined as in Definitions 7.2 and 7.3. Then, in the neighbourhood of* $z = 0$, *we have the Laurent expansions:*

(a) $\zeta(z, \Lambda) = 1/z - S_4 z^3 - S_6 z^5 - \cdots$;
(b) $\wp(z, \Lambda) = 1/z^2 + 3S_4 z^2 + 5S_6 z^4 + \cdots$;
where:

(c) $S_m = \displaystyle\sum_{\omega \in \Lambda, \omega \neq 0} \frac{1}{\omega^m}$, *m even*,

$S_m = 0$, *m odd.*

Proof Begin by recalling that the series for S_m converges if $m \geq 3$, which is the case here.

Now

$$\frac{1}{z - \omega} = -\frac{1}{\omega} \cdot \frac{1}{1 - \dfrac{z}{\omega}}$$

$$= -\sum_{n=0}^{\infty} \frac{z^n}{\omega^{n+1}}.$$

Substitute in the expansion for $\zeta(z)$ given in (7.12) to obtain

$$\zeta(z) = \frac{1}{z} - \sum_{n=2}^{\infty} z^n \sum_{\omega \in \Lambda, \omega \neq 0} \frac{1}{\omega^{n+1}}$$

$$= \frac{1}{z} - S_4 z^3 - S_6 z^5 - S_8 z^7 \cdots,$$

the interchange of the order of the two summations being justified by absolute convergence. In the sum

$$\sum_{\omega \in \Lambda, \omega \neq 0} \frac{1}{\omega^{n+1}},$$

the terms in ω and $-\omega$ cancel if n is even, that is $n+1$ odd; so leaving the coefficients S_{2k}. That proves (a).

To prove (b), we observe that the Laurent series may be differentiated term-by-term and so $\wp(z) = -\zeta'(z)$ has Laurent expansion

$$\wp(z) = -\zeta'(z) = \frac{1}{z^2} + 3S_4 z^2 + 5S_6 z^4 + \cdots,$$

as stated in (b).

We can use Theorem 7.1 to prove that $\wp(z)$ satisfies a differential equation, whose form enables us to use the points $(x, y) = (\wp(z), \wp'(z))$ to parameterize the 'elliptic curve' (so called because of that parameterization)

$$y^2 = 4x^3 - ax - b, \tag{7.25}$$

which has profound number-theoretic applications (see Cassels, 1991) and which plays a crucial part in the proof of Fermat's Last Theorem (see the lecture by Stevens in Cornell *et al.*, 1997). The result which follows was obtained by a different method in Chapter 6, Section 6.2.

Theorem 7.2 *The function $\wp(z, \Lambda)$ satisfies the differential equation*

$$\wp'(z)^2 = 4\wp(z)^3 - g_2\wp(z) - g_3, \tag{7.26}$$

where

$$g_2 = g_2(\Lambda) = 60S_4, \quad g_3 = g_3(\Lambda) = 140 S_6.$$

Proof By Theorem 7.1

$$\wp(z) = \frac{1}{z^2} + 3S_4 z^2 + 5S_6 z^4 + \cdots$$

and we may differentiate the Laurent expansion term-by-term to obtain

$$\wp'(z) = -\frac{2}{z^3} + 6S_4 z + 20S_6 z^3 + \cdots,$$

where the coefficients S_m, m even, are given by (c) of Theorem 7.1.

Now

$$4\wp(z)^3 = \frac{4}{z^6} + \frac{36S_4}{z^2} + 60S_6 + \cdots,$$

and

$$\wp'(z)^2 = \frac{4}{z^6} - \frac{24S_4}{z^2} - 80S_6 + \cdots.$$

It follows that

$$4\wp(z)^3 - \wp'(z)^2 - 60S_4\wp(z) - 140S_6 = \phi(z), \tag{7.27}$$

where $\phi(z)$ is a power series beginning with z^2 and so is holomorphic at $z = 0$ and has a zero there. Now the left-hand side of (7.27) is a polynomial in $\wp(z)$ and $\wp'(z)$ and is doubly periodic, with period lattice Λ, since that is true of $\wp(z)$ and $\wp'(z)$.

We have seen that $\wp(z)$ and $\wp'(z)$ have no singularity except at the points of Λ (and in particular at $z = 0$).

It follows that $\phi(0) = 0$ and that $\phi(z)$ is holomorphic. That is, $\phi(z)$ is an elliptic function with period lattice Λ having no singularities. Now we saw in Theorem 3.4 of Chapter 3 that an entire elliptic function is a constant (Liouville's Theorem); so $\phi(z)$ must be a constant and, by taking $z = 0$, we see that the value of the constant is 0.

Our theorem follows.

Corollary 7.1 *In the notation of Theorem 7.2, if $\wp'(z) \neq 0$, then*

$$\wp''(z) = 6\wp(z)^2 - \frac{1}{2}g_2.$$

Proof Differentiate (7.26) to obtain

$$2\wp'(z)\wp''(z) = (12\wp(z)^2 - g_2)\wp'(z),$$

whence the result, on dividing by $\wp'(z)$.

In Corollary 7.1, we supposed that $\wp'(z) \neq 0$. The zeros of $\wp'(z)$ play an important part in what follows and it will be convenient to turn to them now.

Theorem 7.3 *Let ω_1, ω_2 be a pair of primitive periods (that is a $\mathbb{Z}$-basis for Λ) and suppose that*

$$\omega_1 + \omega_2 + \omega_3 = 0.$$

Then

$$\wp'(z)^2 = 4\left(\wp(z) - \wp\left(\frac{1}{2}\omega_1\right)\right)\left(\wp(z) - \wp\left(\frac{1}{2}\omega_2\right)\right)\left(\wp(z) - \wp\left(\frac{1}{2}\omega_3\right)\right).$$

The numbers $e_k = \wp(\omega_k/2)$, $k = 1, 2, 3$, are distinct and

$$\wp'(z)^2 = 4(\wp(z) - e_1)(\wp(z) - e_2)(\wp(z) - e_3).$$

The zeros of $\wp'(z)$ are the points $\omega_k/2$, mod Λ.

Proof We have seen that $\wp'(z)$ has a single triple pole, at $z = 0$, in the period parallellogram and so, by Theorem 3.6, takes every value three times; in particular it has three zeros, mod Λ.

By periodicity, $\wp'(z + \omega_k) = \wp'(z)$, $k = 1, 2, 3$, and so, if we take $z = -\omega_k/2$, we obtain $\wp'(\omega_k/2) = \wp'(-\omega_k/2) = -\wp'(\omega_k/2) = 0$, since $\wp'$ is an odd function of z.

It follows that there are three zeros, mod Λ, namely $\omega_1/2$, $\omega_2/2$ and $\omega_3/2$. Write $\wp(\omega_k/2) = e_k$, $k = 1, 2, 3$. It remains to show that the e_k are distinct.

To that end, write $f_k(z) = \wp(z) - e_k$. Then $f_k(z)$ has a double pole at $z = 0$ and so (by Theorem 3.5) there are two zeros in the period parallelogram. But $\omega_k/2$ is a zero of multiplicity 2, since $f_k'(z) = \wp'(z)$ vanishes there (see above). So $\omega_k/2$ is the only zero of f_k, mod Λ. Hence e_1, e_2, e_3 are distinct, because $e_1 = e_2$ implies $\wp(z) - e_1$ has a zero at $z = \omega_2/2$, a point not congruent to $\omega/2$, (mod Λ). The contradiction establishes the result and our theorem is proved.

Corollary 7.2 *Let e_1, e_2, e_3 be the numbers $\wp(\omega_1/2)$, $\wp(\omega_2/2)$ and $\wp(\omega_3/2)$ defined in the theorem. Then*

$$e_1 + e_2 + e_3 = 0.$$

Proof In the relation $(\wp'(z))^2 = 4(\wp(z))^3 - g_2\wp(z) - g_3$, put $z = \omega_k/2$, $k = 1, 2, 3$. We obtain $0 = 4e_k^3 - g_2e_k - g_3$. So the polynomial $4x^3 - g_2x - g_3$ has the three distinct zeros e_1, e_2 and e_3, and therefore

$$4x^3 - g_2x - g_3 = 4(x - e_1)(x - e_2)(x - e_3)$$

and

$$e_1 + e_2 + e_3 = 0.$$

Exercises 7.2

7.2.1 Express $\wp'''(z)$ in terms of $\wp(z)$ and $\wp'(z)$.

7.2.2 Prove that

$$\wp''(z) = 6(\wp(z))^2 - \frac{1}{2}g_2$$

by repeating the argument used in the proof of Theorem 7.2.

7.2.3 Write the Laurent expansion of $\wp(z)$ in the form

$$\wp(z) = \frac{1}{z^2} + c_2z^2 + c_3z^4 + \cdots + c_nz^{2n-2} + \cdots$$

(so that the c_n are related to the S_{2m} of Theorem 7.2). By substituting the expansion for $\wp(z)$ in the result of 7.2.2, show that

$$\{(2n - 2)(2n - 3) - 12\}c_n = 6 \sum_{\substack{r+s=n \\ r \geq 2, s \geq 2}} c_r c_s \quad (n = 4, 5, 6, \ldots)$$

and deduce that

$$c_4 = \frac{1}{3}c_2^2, \quad c_5 = \frac{3}{11}c_2 c_3, \quad c_6 = \frac{1}{13}(2c_2 c_4 + c_3^2) = \frac{1}{13}\left(\frac{2}{3}c_2^3 + c_3^2\right).$$

Hence prove that the Laurent expansion coefficients c_n of $\wp(z)$ are rational integral functions of the invariants g_2 and g_3 with positive rational coefficients. (Note that $g_2 = 20c_2$, $g_3 = 28c_3$.)

7.2.4 Prove that $\zeta(z - a) - \zeta(z - b)$ (where $\zeta(z) = \zeta(z, \Lambda)$ is the Weierstrass zeta function) is an elliptic function of order 2 with simple poles at $z = a$ and $z = b$.

7.2.5 Prove that, if $c + d = a + b$, then

$$\frac{\sigma(z - c)\sigma(z - d)}{\sigma(z - a)\sigma(z - b)}$$

is an elliptic function of order 2 with simple poles at $z = a$ and $z = b$. (Both this exercise and the result in 7.2.4 will be generalized in what follows.)

7.2.6 Starting from the result (see Chapter 3) that if $f(z)$ and $g(z)$ are elliptic functions having the same poles and the same principal parts with respect to a lattice Λ, then $f(z) - g(z)$ must be a constant, prove that

$$2\zeta(2u) - 4\zeta(u) = \frac{\wp''(u)}{\wp'(u)}.$$

(Hint: choose $f(z)$ and $g(z)$ appropriately and then, by giving u a particular value, find the constant.)

7.2.7 Prove that

$$\wp(z) = -\frac{d^2}{dz^2}\{\log \sigma(z)\} = \frac{\sigma'^2(z) - \sigma(z)\sigma''(z)}{\sigma^2(z)}.$$

7.2.8 Prove that

$$e_1 e_2 + e_2 e_3 + e_1 e_3 = -\frac{1}{2}g_2$$

$$e_1 e_2 e_3 = \frac{1}{4}g_3.$$

7.2.9 By using Liouville's Theorem, prove that if $u + v + w = 0$, then

$$\{\zeta(u) + \zeta(v) + \zeta(w)\}^2 + \zeta'(u) + \zeta'(v) + \zeta'(w) = 0.$$

(Compare with 7.2.6.)

7.3 Modular forms and functions

We return to the functions g_2 and g_3 defined in Theorem 7.2 in terms of the sums S_4 and S_6, where

$$g_2 = 60S_4, \quad S_4(\Lambda) = \sum_{\omega \in \Lambda, \omega \neq 0} \frac{1}{\omega^4}, \quad g_3 = 140S_6, \quad S_6(\Lambda) = \sum_{\omega \in \Lambda, \omega \neq 0} \frac{1}{\omega^6}.$$

$$(7.28)$$

It will be convenient in what follows to denote a sum taken over all lattice points ω, other than $\omega = 0$, by $\sum'$.

Evidently g_2 and g_3 depend on the period lattice Λ and so may be regarded as functions of the two complex numbers, ω_1 and ω_2, that form a basis for Λ. They are analytic for every pair of generators ω_1, ω_2, such that[2] $\operatorname{Im}(\omega_2/\omega_1) > 0$, or we may regard them as functions of the lattice Λ itself, since the sums (7.28) used to define g_2 and g_3 are sums over all the lattice points and do not depend on the particular pair of generators chosen.

It follows from the definitions that, for $\lambda \neq 0$,

$$g_2(\lambda \Lambda) = \lambda^{-4} g_2(\Lambda), \quad g_3(\lambda \Lambda) = \lambda^{-6} g_3(\Lambda), \tag{7.29}$$

when $\lambda \Lambda$ denotes the set of points $\lambda \omega, \omega \in \Lambda$.

A holomorphic (analytic) function of a lattice that possesses a homogeneity property like that in (7.29) is said to be a *modular form* and the powers of λ occurring are called the *weight*[3] (or the dimension) of the form; thus the forms g_2 and g_3 in (7.29) are of dimensions $+4$ and $+6$, respectively.

A particularly important function is the *discriminant*, Δ, of the polynomial $4x^3 - g_2 x - g_3$, whose roots are the numbers e_1, e_2 and e_3 defined in Theorem 7.3. The discriminant is defined by

$$\Delta = 16(e_1 - e_2)^2(e_2 - e_3)^2(e_1 - e_3)^2. \tag{7.30}$$

Note that $\Delta \neq 0$, since the numbers e_1, e_2, e_3 are always distinct (by Theorem 7.3).

If we write

$$f(x) = 4x^3 - g_2 x - g_3 = 4(x - e_1)(x - e_2)(x - e_3),$$

---

[2] Some authors interchange ω_1 and ω_2; so that $\operatorname{Im}(\omega_1/\omega_2) > 0$; there are then some changes in the following formulae, to which we shall draw attention.

[3] Some authors write 'of weight (or dimension), -4 or -6', others 'of weight $+2$ or $+3$'. We follow the usage of McKean and Moll (1997) and Serre (1970) (for example). See also the preceding footnote (2).

then

$$f'(e_1) = 4(e_1 - e_2)(e_1 - e_3),$$

with similar expressions for $f'(e_2)$ and $f'(e_3)$. It follows that

$$\Delta = -\frac{1}{4} f'(e_1) f'(e_2) f'(e_3)$$

and so, since $f'(x) = 12x^2 - g_2$,

$$\Delta = -\frac{1}{4} (12e_1^2 - g_2)(12e_2^2 - g_2)(12e_3^2 - g_2) = g_2^3 - 27g_3^2, \qquad (7.31)$$

on using the elementary symmetric functions of e_1, e_2, e_3:

$$e_1 + e_2 + e_3 = 0, \quad e_1e_2 + e_2e_3 + e_3e_1 = -\frac{g_2}{4}, \quad e_1e_2e_3 = \frac{g_3}{4}.$$

The discriminant is accordingly a modular form of weight 12, since

$$\Delta(\lambda\Lambda) = \lambda^{-12}\Delta(\Lambda), \qquad (7.32)$$

on using (7.28) and (7.31).

We now introduce a function, $J(\tau)$, of $\tau = \omega_2/\omega_1$, of such importance that it, or rather the function $j(\tau) = 1728J(\tau)$, is called *the* modular function. The function J is actually a function of the lattice, Λ, defined in terms of generators (ω_1, ω_2), with $\operatorname{Im} \tau = \operatorname{Im}(\omega_2/\omega_1) > 0$, by

$$J(\Lambda) = J(\omega_1, \omega_2) = \frac{g_2^3}{\Delta}. \qquad (7.33)$$

Since each of g_2^3 and Δ has weight 12, it follows that $J(\Lambda)$ is homogeneous of weight 0. Thus

$$J(\omega_1, \omega_2) = J(\omega_1, \tau\omega_1) = J(1, \tau),$$

and we see that J depends only on τ and so may be denoted without ambiguity by $J(\tau)$. The function $J(\tau)$ is an analytic (holomorphic) function of τ for all values of $\tau = \omega_2/\omega_1$ that are not real.

Since J depends only on the lattice Λ, we may replace the ordered pair of generating periods (ω_1, ω_2) by any other similarly ordered pair (ω_1', ω_2'): thus

$$\omega_1' = a\omega_1 + b\omega_2,$$
$$\omega_2' = c\omega_1 + d\omega_2, \qquad (7.34)$$

where a, b, c, d are integers such that $ad - bc = +1$. (That is, the matrix

$$\begin{pmatrix} a & b \\ c & d \end{pmatrix} \in SL_2(Z),$$

a subgroup of the special linear group $SL_2(\mathbb{R})$. The condition that the determinant should be $+1$ is required in order to ensure that $\text{Im}(\omega_2'/\omega_1') > 1$.)

If we write

$$\tau = \frac{\omega_2}{\omega_1}, \quad \tau' = \frac{\omega_2'}{\omega_1'} = \frac{c\omega_1 + d\omega_2}{a\omega_1 + b\omega_2} = \frac{c + d\tau}{a + b\tau},$$

then[4]

$$J\left(\frac{c + d\tau}{a + b\tau}\right) = J(\tau). \tag{7.35}$$

In the notation employed in (7.34) and (7.35), the condition for a function f to be a *modular form* of weight $2n$ is that it should be holomorphic for all τ with $\text{Im}\,\tau > 0$, including the points at ∞, and

$$f(\tau) = (a + b\tau)^{-2n} f\left(\frac{c + d\tau}{a + b\tau}\right). \tag{7.36}$$

A *modular function* is a function defined on the upper half-plane H, $\text{Im}\,\tau > 0$, that is meromorphic on H and at ∞ and satisfies (7.36).[5]

We illustrate those definitions by considering the example of $g_2(\tau)$. We recall that

$$g_2(\omega_1, \omega_2) = 60 \sideset{}{'}\sum \frac{1}{(m\omega_1 + n\omega_2)^4}$$

$$= 60\omega_1^{-4} \sideset{}{'}\sum_{m_1 n} \frac{1}{(m + n\tau)^4},$$

where $\tau = \omega_2/\omega_1$. Now suppose

$$\tau' = \frac{c + d\tau}{a + b\tau}, \quad ab - cd = 1.$$

Then

$$g_2(\tau') = g_2\left(\frac{c + d\tau}{a + b\tau}\right) = \sideset{}{'}\sum_{m', n'} \frac{1}{\left(m' + n'\dfrac{c + d\tau}{a + b\tau}\right)^4}$$

---

[4] Some authors write $J(\frac{a+b\tau}{c+d\tau}) = J(\tau)$ in accordance with their usage $\tau = \omega_1/\omega_2$. The reader should be familiar with both formulations and their consequences.

[5] For further details and the applications of modular forms to the theory of numbers, see the books by Hardy and Wright (1979), McKean and Moll (1997) Serre, (1970) and some of the work in Chapter 10. In the notation of footnote (4), the condition reads
$$f(z) = (cz + d)^{-2n} f\left(\frac{az + b}{cz + d}\right), z = \omega_1/\omega_2$$

$$= (a + b\tau)^4 \sum_{m',n'}' \frac{1}{(m'(a + b\tau) + n'(c + d\tau))^4}$$

$$= (a + b\tau)^4 \sum_{m',n'}' \frac{1}{((m'a + n'c) + (m'b + n'd)\tau)^4}$$

$$= (a + b\tau)^4 \sum_{m,n}' \frac{1}{(m + n\tau)^4},$$

where

$$m = m'a + n'c, \quad n = m'b + n'd.$$

We see that

$$g_2\left(\frac{c + d\tau}{a + b\tau}\right) = (a + b\tau)^4 g_2(\tau) \tag{7.37}$$

(which agrees with (7.36)) and, from the infinite series that defines it, $g_2(\tau)$ is holomorphic on $\operatorname{Im}\tau > 0$ and at ∞. So $g_2(\tau)$ is a modular form of dimension 4.

The functions $\Delta(\tau)$ and $j(\tau)$ have great significance and remarkable arithmetical properties and we shall refer to them in Chapter 10. We conclude this section by obtaining series expansions for those two functions in $q = e^{\pi i \tau}$, $\operatorname{Im}\tau > 0$, and also product expansions and expressions in terms of theta functions.

For the present (and for notational reasons that will emerge), we begin by replacing τ by z ($\operatorname{Im} z > 0$) and write the definition of the functions $g_2(z)$, $g_3(z)$ in the forms

$$g_2(z) = 60\, G_2(z) = 60\, S_4(z), \quad g_3(z) = 140\, G_3(z) = 140\, S_6(z),$$

where $S_l(z)$ is defined by the *Eisenstein series*

$$S_l(z) = \sum_{\omega \in \Lambda, \omega \neq 0} \frac{1}{\omega^l}, \quad l \text{ even},$$

and on choosing our lattice Λ to have basis $(1, z)$, $z = \frac{\omega_2}{\omega_1}$ (so that $\omega_1(1, z) = (\omega_1, \omega_2)$), and writing $l = 2k$, we have

$$G_k(z) = \sum_{(m,n) \neq (0,0)} \frac{1}{(m + nz)^{2k}}. \tag{7.38}$$

We are looking for an expansion of $G_k(z)$ in terms of $q = e^{2\pi i z}$ (later we replace $e^{2\pi i z}$ by $e^{\pi i \tau}$ and then q will be replaced by[6] q^2.

---

[6] Some authors (for example Serre (1970) and Prasolov & Solovyev (1997)) use $q = e^{2\pi i z}$ where later we use $q = e^{\pi i \tau}$; so that their 'q' is replaced by q^2.

Proposition 7.5 *With the foregoing notation,*

$$G_k(z) = 2\zeta(2k) + \frac{2(2\pi i)^{2k}}{(2k-1)!} \sum_{n=1}^{\infty} \sigma_{2k-1}(n) q^n, \tag{7.39}$$

where $q = e^{2\pi i z}$ and ζ denotes the Riemann zeta function (see Appendix) and where $\sigma_{2k-1}(n)$ denotes the sum of the powers d^{2k-1} of the positive divisions of n, thus $\sigma_{2k-1}(n) = \sum_{d|n} d^{2k-1}$.

Proof We begin with the definition (see (7.38))

$$G_k(z) = \sum_{(m,n)\neq(0,0)} \frac{1}{(m+nz)^{2k}},$$

and then we separate out those terms for which $n = 0$ to obtain

$$G_k(z) = 2 \sum_{m=1}^{\infty} \frac{1}{m^{2k}} + 2 \sum_{n=1}^{\infty} \sum_{m\in\mathbb{Z}} \frac{1}{(m+nz)^{2k}}$$

$$= 2\zeta(2k) + 2\frac{(-2\pi i)^{2k}}{(2k-1)!} \sum_{n=1}^{\infty} \sum_{a=1}^{\infty} a^{2k-1} q^{an}$$

(on replacing z by nz in the result proved in the Appendix, A.5.4). It follows that

$$G_k(z) = 2\zeta(2k) + \frac{2(2\pi i)^{2k}}{(2k-1)!} \sum_{n=1}^{\infty} \sigma_{2k-1}(n) q^n,$$

as was to be proved.

Now use Proposition 7.5 to write $G_k(z)$ in the form

$$G_k(z) = 2\zeta(2k) E_k(z), \tag{7.40}$$

where (the notation $E_k(z)$ is suggested by the term 'Eisenstein series')

$$E_k(z) = 1 + \gamma_k \sum_{n=1}^{\infty} \sigma_{2k-1}(n) q^n, \tag{7.41}$$

with

$$\gamma_k = (-1)^k \frac{4k}{B_k}, \tag{7.42}$$

where $B_k (k = 1, 2, 3, \ldots)$ denotes the Bernoulli numbers, defined by

$$\zeta(2k) = \frac{2^{2k-1}}{(2k)!} B_k \pi^{2k}$$

(see Appendix, Section A.6, Equation (A.12)). In particular $B_1 = 1/6$, $B_2 = 1/30$, $B_3 = 1/42$, $B_4 = 1/30$.

It follows that[7]

$$E_2(z) = 1 + 240 \sum_{n=1}^{\infty} \sigma_3(n)q^n, \quad q = e^{2\pi i z}, \tag{7.43}$$

$$E_3(z) = 1 - 504 \sum_{n=1}^{\infty} \sigma_5(n)q^n, \quad q = e^{2\pi i z}, \tag{7.44}$$

whence

$$
\begin{aligned}
g_2(z) &= (2\pi)^4 \frac{1}{2^2 \cdot 3} \cdot E_2(z) \\
&= \frac{4}{3}\pi^4 \left(1 + 240 \sum_{n=1}^{\infty} \sigma_3(n)q^n \right),
\end{aligned}
\tag{7.45}
$$

and

$$
\begin{aligned}
g_3(z) &= (2\pi)^6 \cdot \frac{1}{2^3 \cdot 3^3} \cdot E_3(z) \\
&= \frac{2^3}{3^3}\pi^6 \left(1 - 504 \sum_{n=1}^{\infty} \sigma_5(n)q^n \right), \quad q = e^{2\pi i z}.
\end{aligned}
\tag{7.46}
$$

Proposition 7.6 *Let $\Delta(z)$ denote the discriminant $\Delta(z) = g_2(z)^3 - 27g_3{}^2(z)$ (see (7.30) and (7.31)), then $\Delta(z)$ has the series expansion in $q = e^{2\pi i z}$ given by:*

$$\Delta(z) = (2\pi)^{12}\{q - 24q^2 + 252q^3 - 1472q^4 + \cdots\}. \tag{7.47}$$

Proof We have, using (7.45) and (7.46),

$$
\begin{aligned}
\Delta(z) &= g_2(z)^3 - 27g_3(z)^2 \\
&= (2\pi)^{12} \cdot 2^{-6} \cdot 3^{-3}(E_2(z)^3 - E_3(z)^2) \\
&= (2\pi)^{12} 2^{-6} 3^{-3} \left\{ \left(1 + 240 \sum_{n=1}^{\infty} \sigma_3(n)q^n \right)^3 - \left(1 - 504 \sum_{n=1}^{\infty} \sigma_5(n)q^n \right)^2 \right\} \\
&= (2\pi)^{12} 2^{-6} \cdot 3^{-3} \left\{ 3 \cdot 240 \sum_{n=1}^{\infty} \sigma_3(n)q^n + \cdots + 2 \cdot 504 \sum_{n=1}^{\infty} \sigma_5(n)q^n + \cdots \right\} \\
&= (2\pi)^{12}\{q - 24q^2 + 252q^3 - 1472q^4 + \cdots\}.
\end{aligned}
$$

---

[7] There are remarkable relations between the $E_k(z)$, for example $E_2 E_3 = E_5$, which imply no less remarkable identities involving the divisor functions $\sigma_k(n)$; the proofs require results concerning the vector space of modular forms of weight $2k$ – see Serre (1970) for details.

(The coefficients of the terms involving $q^2, q^3, \cdots$ are not obvious, but are readily obtained from writing out the expansions of the powers of the series in the penultimate line.)

Corollary 7.3 *Let $q = e^{\pi i \tau}$, Im, $\tau > 0$. Then the expansion of $\Delta(\tau)$ as a series in powers of q is*

$$\Delta(\tau) = q^2 \{1 - 24q^2 + 252q^4 - 1472q^6 + \cdots\}. \tag{7.48}$$

Proof The proof is obvious; replacing $e^{2\pi i z}$ by $e^{\pi i \tau}$ (where the underlying lattice consists of the points $m + nz = m + n\tau$) effectively replaces q by q^2.

Theorem 7.4 *The modular function $j(z)$ has a series expansion in powers of $q = e^{2\pi i z}$ given by*

$$j(z) = \frac{1}{q} + 744 + \sum_{n=1}^{\infty} c_n q^n, \tag{7.49}$$

where Im $z > 0$ and the $c(n)$ are integers.

Proof We have (see (7.33)),

$$j(z) = \frac{1728 g_2(z)^3}{\Delta(z)} = \frac{1728 g_2(z)^3}{(2\pi)^{12} q (1 - 24q + 252q^2 - \cdots)}$$

$$= \frac{1728 \cdot (2\pi)^{12} (12)^{-3} \left(1 + 240 \sum_{n=1}^{\infty} \sigma_3(n) q^n\right)^3}{(2\pi)^{12} q (1 - 24q + 252q^2 - \cdots)}$$

$$= \frac{\left(1 + 240 \sum_{n=1}^{\infty} \sigma_3(n) q^n\right)^3}{q (1 - 24q + 252q^2 - \cdots)}$$

$$= \frac{1}{q} (1 + 720q + \cdots)(1 + 24q + \cdots)$$

$$= \frac{1}{q} \left(1 + 744q + \sum_{n=1}^{\infty} c(n) q^{n+1}\right)$$

$$= q^{-1} + 744 + \sum_{n=1}^{\infty} c(n) q^n.$$

(It is clear, from the binomial expansions, that the $c(n)$ are integers, though their values are less readily accessible. For example, $c(1) = 2^3 \cdot 3^3 \cdot 1823 = 196884$, $c(2) = 2^{11} \cdot 5.2099 = 21493760$. The $c(n)$ have beautiful and surprising, indeed astonishing, divisibility properties, first proved by A. O. L. Atkin and J. N. O'Brien; see Serre (1970), p.147, for references.)

Corollary 7.4 *The function $j(\tau)$ has the series expansion in $q = e^{\pi i \tau}$, $\operatorname{Im} \tau >$ 0, given by*

$$j(\tau) = q^{-2} + 744 + \sum_{n=1}^{\infty} c(n) q^{2n},$$

where the $c(n)$ are integers.

Proof As with Corollary 7.3, the proof consists in replacing $q = e^{2\pi i z}$ by $q = e^{\pi i \tau}$ and so q by q^2 (and of course using τ in place of z in the underlying lattice).

We conclude this section by obtaining an expression for $\Delta(\tau)$ as an infinite product involving the $q = e^{\pi i \tau}$. This remarkable formula brings together the work of this section and much of our earlier work on theta functions and the Weierstrass functions.

Theorem 7.5 (Jacobi) *Let $q = e^{\pi i \tau}$, $\operatorname{Im} \tau > 0$, then*

$$\Delta(\tau) = (2\pi)^{12} q^2 \prod_{n=1}^{\infty} (1 - q^{2n})^{24}. \tag{7.50}$$

Proof We begin by observing that the infinite product term is G^{24}, where $G = \prod_{n=1}^{\infty} (1 - q^{2n})$ is the constant occurring in the expression for the theta functions as infinite products, and indeed

$$2^8 q^2 \prod_{n=1}^{\infty} (1 - q^{2n})^{24} \tag{7.51}$$

is the eighth power of

$$\theta_1' = 2q^{1/4} \prod_{n=1}^{\infty} (1 - q^{2n})^3. \tag{7.52}$$

Now $\theta_1' = \theta_2 \theta_3 \theta_4$, which suggests that we should try to express the discriminant, Δ, in terms of the theta functions $\theta_2, \theta_3, \theta_4$. (For the foregoing, see Chapter 4, Sections 4.4–4.6, Chapter 5, Section 5.5, and Chapter 6, Section 6.2.)

Now recall the definition of the *root functions*, in Section 5.5, namely

$$f_\alpha(z|\tau) = \frac{\theta_1'}{\theta_\alpha} \cdot \frac{\theta_\alpha(z|\tau)}{\theta_1(z|\tau)}, \quad \alpha = 2, 3, 4. \tag{7.53}$$

and (whence the name)

$$(\wp(z) - e_j)^{1/2} = f_{j+1}(z|\tau), \quad j = 1, 2, 3. \tag{7.54}$$

We recall from Chapter 6, Section 6.2, Equation (6.4) that

$$\wp(z; \omega_1, \omega_2) = A\theta_4^2\left(\left.\frac{\pi z}{\omega_1}\right|\tau\right) + B, \quad \tau = \frac{\omega_2}{\omega_1} \tag{7.55}$$

and (Theorem 6.2)

$$\lambda(\tau) = \frac{\theta_2^4(0|\tau)}{\theta_3^4(0|\tau)} = \frac{e_3 - e_2}{e_1 - e_2},$$

where (see the remarks following Theorem 6.3)

$$A = \frac{\pi^2}{\omega_1^2}\theta_1^2\theta_3^2, \tag{7.56}$$

$$B = -\frac{\pi^2}{\omega_1^2} \cdot \frac{1}{3} \cdot (\theta_3^4 + \theta_2^4). \tag{7.57}$$

In an appeal to that last, we shall also require the result (4.26) from Chapter 4 that

$$\theta_3^4 = \theta_2^4 + \theta_4^4. \tag{7.58}$$

Now we recall from (7.30) that the discriminant, Δ, is defined by

$$\Delta = 16(e_1 - e_2)^2(e_2 - e_3)^2(e_1 - e_3)^2,$$

where the numbers e_1, e_2, e_3 are distinct and are the values of $\wp$ at the half-periods $\omega_1/2, \omega_2/2, \omega_3/2$. It follows from (7.55) that (taking $\alpha = 1$ in the formulae for e_1, e_2, e_3)

$$e_1 = \frac{\pi^2}{\omega_1^2}\left(\theta_3^4 - \frac{1}{3}(\theta_3^4 + \theta_2^4)\right),$$

$$e_2 = \frac{\pi^2}{\omega_1^2}\left(-\frac{1}{3}(\theta_3^4 + \theta_2^4)\right),$$

$$e_3 = \frac{\pi^2}{\omega_1^2}\left(\theta_2^4 - \frac{1}{3}(\theta_3^4 + \theta_2^4)\right).$$

So

$$e_1 - e_2 = \frac{\pi^2}{\omega_1^2}\theta_3^4,$$

$$e_2 - e_3 = -\frac{\pi^2}{\omega_1^2}\theta_2^4,$$

$$e_1 - e_3 = \frac{\pi^2}{\omega_1^2} (\theta_3^4 - \theta_2^4) = \frac{\pi^2}{\omega_1^2} \theta_4^4, \tag{7.59}$$

on using (7.58).

On substituting from (7.59) in the definition of the discriminant, we obtain

$$
\begin{aligned}
\Delta &= 16(e_1 - e_2)^2 (e_2 - e_3)^2 (e_1 - e_3)^2 \\
&= 16 \frac{\pi^{12}}{\omega_1^{12}} \theta_2^8 \theta_3^8 \theta_4^8 \\
&= 16 \left(\frac{\pi}{\omega_1} \right)^{12} (\theta_1')^8 \\
&= 16 \left(\frac{\pi}{\omega_1} \right)^{12} 2^8 q^2 \prod_{n=1}^{\infty} (1 - q^{2n})^{24}.
\end{aligned}
$$

For the lattice with basis $(1, \tau)$, we have $\omega_1 = 1$ and so, finally, we obtain the desired result.

We conclude with yet another remarkable application of Theorem 7.5 and Proposition 7.6. From (7.48) and (7.50) we obtain an identity of the form

$$(2\pi)^{-12} \Delta(z) = \sum_{n=1}^{\infty} \tau(n) q^n = q \prod_{n=1}^{\infty} (1 - q^n)^{24}, \quad q = e^{2\pi i z}$$

(the τ in $\tau(n)$ has nothing to do with the τ in $q = e^{\pi i \tau}$!), where the coefficients $\tau(n)$ may be obtained from the coefficients in (7.48). The function $n \mapsto \tau(n)$ that gives the coefficients $\tau(n)$ is called the *Ramanujan function*, whose properties are outlined in Serre (1970), pp.156ff, together with open questions concerning the values of $\tau(n)$, and in particular the value of $\tau(p)$, for prime numbers p, for which Ramanujan conjectured the upper bound $|\tau(p)| < 2p^{11/2}$.

Exercises 7.3

7.3.1 (i) In the harmonic case, the lattice generated by $(1, i)$ is the same as the lattice generated by $(-i, 1) = -i(1, i)$. Deduce that $g_3(1, i) = g_3(-i, 1) = (-i)^6 g_3(1, i)$ and hence $g_3(1, i) = 0$. Show that $\Delta(1, i) = (g_2(1, i))^3$ and $J(i) = 1$.

(ii) In the equianharmonic case, let ρ denote an imaginary cube root of 1. Show that the lattice generated by $(1, \rho)$ is the same as the lattice generated by $(\rho, \rho^2) = \rho(1, \rho)$ and deduce that

$$g_2(1, \rho) = g_2(\rho, \rho^2) = \rho^{-4} g_2(1, \rho),$$

whence

$$g_2(1, \rho) = 0 \quad \text{and} \quad J(\rho) = 0.$$

7.3.2 Complete the details of the proof of (7.31).

7.3.3 Given that

$$\wp\left(z + \frac{1}{2}\omega_1\right) = e_1 + \frac{(e_1 - e_2)(e_1 - e_3)}{\wp(z) - e_1},$$

with similar formulae for the other half-periods (see Exercise 7.4.2), prove that the discriminant Δ, defined in (7.31), is

$$\wp'(z)\wp'\left(z + \frac{1}{2}\omega_1\right)\wp'\left(z + \frac{1}{2}\omega_2\right)\wp'\left(z + \frac{1}{2}\omega_3\right).$$

(Hint: differentiate the three expressions of the type $\wp\left(z + \frac{1}{2}\omega_1\right)$ and then use

$$\wp'(z)^2 = 4\prod_{i=1}^{3}(\wp(z) - e_i).)$$

7.3.4 Prove that $g_3(\tau)$ is a modular form of weight 6 by imitating the argument used to obtain (7.37).

7.3.5 Prove that if f_1 and f_2 are modular forms of weight $2n_1$, $2n_2$, respectively, then $f_1 f_2$ is a modular form of weight $2n_1 + 2n_2$.

7.3.6 Prove that if $g = \begin{pmatrix} a & b \\ c & d \end{pmatrix}$ and $z \in \mathbb{C}$, then if $g\tau = \frac{c+d\tau}{a+b\tau}$, $\operatorname{Im}(g\tau) = \frac{\operatorname{Im}\tau}{|a+b\tau|^2}$.

7.3.7 Prove that

$$e_1 = \frac{1}{3}\pi^2\left(\theta_3^4 + \theta_4^4\right),$$

$$e_2 = \frac{1}{3}\pi^2\left(\theta_2^4 - \theta_4^4\right),$$

$$e_3 = \frac{1}{3}\pi^2\left(\theta_2^4 + \theta_3^4\right).$$

7.3.8 Prove that

$$g_2 = 2\left(e_1^2 + e_2^2 + e_3^2\right)$$

$$= \frac{2}{3}\pi^4\left(\theta_2^8 + \theta_3^8 + \theta_4^8\right).$$

7.3.9 Prove that

$$j(\tau) = \frac{g_2^3}{\Delta} = \frac{1}{54} \frac{\left(\theta_2^8 + \theta_3^8 + \theta_4^8\right)^3}{\left(\theta_2^8 \theta_3^8 \theta_4^8\right)}.$$

7.3.10 Prove that

$$k^2 = \frac{\theta_2^4}{\theta_3^4} = 16q \prod_{n=1}^{\infty} \left(\frac{1 + q^{2n}}{1 + q^{2n-1}}\right)^8, \quad q = e^{\pi i \tau},$$

and

$$k'^2 = \frac{\theta_4^4}{\theta_3^4} = \prod_{n=1}^{\infty} \left(\frac{1 - q^{2n-1}}{1 + q^{2n-1}}\right)^8, \quad q = e^{\pi i \tau}.$$

7.3.11 Show that

$$\frac{k'}{(k/4\sqrt{q})^{1/2}} = \prod_{n=1}^{\infty} (1 - q^{2n-1})^4 (1 + q^n)^{-2}$$

$$= \prod_{n=1}^{\infty} (1 - q^{2n})^8, \quad q = e^{\pi i \tau}.$$

7.3.12 Given that $E_2^2 = E_4$, deduce that

$$\sigma_7(n) = \sigma_3(n) + 120 \sum_{m=1}^{n-1} \sigma_3(m)\sigma_3(n - m);$$

and, given that $E_2 E_3 = E_5$, show that

$$11\,\sigma_9(n) = 21\sigma_5(n) - 10\,\sigma_3(n) + 5040 \sum_{m=1}^{n-1} \sigma_3(n)\sigma_5(n - m).$$

(See Serre 1970, and McKean & Moll 1997, for further background details).

7.3.13 Compute the Ramanujan numbers

$$\tau(1) = 1, \tau(2) = -24, \tau(3) = 252, \tau(4) = -1472, \tau(5) = 4830.$$

7.3.14 Given that $B_6 = \dfrac{691}{2730}$, show that

$$E_6(z) = 1 + \frac{65520}{691} \sum_{n=1}^{\infty} \sigma_{11}(n)q^n, q = e^{2\pi i z}.$$

(For an application of that result to congruence properties of the Ramanujan function $\tau(n)$, see McKean & Moll, 1997, Chapter 4, pp. 176–177, where references to other results are given; see also Serre, 1970 and the references given there.)

7.4 Construction of elliptic functions with given zeros and poles; the addition theorem for $\wp(z)$

7.4.1 Construction of elliptic functions with given zeros and poles

In this section we consider the following problem: suppose we are given a lattice Λ, with basis $(\omega_1, \omega_2,)$ and two sets of points in the complex plane $a_1, \ldots, a_n$ and $b_1, \ldots, b_n$ where those points are counted according to multiplicities – that is we allow repetitions in which $a_1 = a_2 = a_3 = \cdots = a_k$, which then counts k times, and so on. We know from Theorem 3.6 that if the $a_1, \ldots, a_n$ denote the zeros and $b_1, \ldots, b_n$ the poles of an elliptic function, then the number of points (counted according to multiplicities) in each set must be the same (that is, n) and also that

$$a_1 + \cdots + a_n \equiv b_1 + \cdots + b_n (\mathrm{mod}\ \Lambda). \tag{7.60}$$

By replacing one of the points a_k or b_k by one which differs from it by a period we may, and in what follows we shall, suppose that

$$a_1 + \cdots + a_n = b_1 + \cdots + b_n. \tag{7.61}$$

We now ask the question: given two sets of points $a_1, \ldots, a_n$ and $b_1, \ldots, b_n$ satisfying (7.60), is it possible to construct an elliptic function having those points as zeros and poles, respectively, and is such a function essentially unique? We shall prove:

Theorem 7.1 *Let $a_1, \ldots, a_n$ and $b_1, \ldots, b_n$ be points of the complex plane satisfying (7.61). Then the function*

$$f(z) = \frac{\sigma(z - a_1)\sigma(z - a_2)\cdots\sigma(z - a_n)}{\sigma(z - b_1)\sigma(z - b_2)\cdots\sigma(z - b_n)}, \tag{7.62}$$

where σ denotes the Weierstrass function (Definition 7.1), is an elliptic function with zeros at the points a_k and poles at the points b_k. The function $f(z)$ is unique in the sense that any other elliptic function with the same property must be a constant multiple of f.

Proof Consider the function

$$f(z) = \frac{\sigma(z - a_1)\cdots\sigma(z - a_n)}{\sigma(z - b_1)\cdots\sigma(z - b_n)}. \tag{7.62}$$

Since both the numerator and denominator have zeros at the points a_k, b_k, respectively, it follows that $f(z)$ has zeros and poles of the right multiplicities at the given points. Moreover, the transformation law for the sigma function

$$\sigma(z + \omega_1) = -\exp\left(\eta_1\left(z + \frac{1}{2}\omega_1\right)\right)\sigma(z)$$

(see (7.21)), gives

$$f(z + \omega_1) = f(z) \exp\left(\eta_1 \sum_{k=1}^{n} \left(z + \frac{1}{2}\omega_1 - a_k \right) - \eta_1 \sum_{k=1}^{n} \left(z + \frac{1}{2}\omega_1 - b_k \right) \right)$$
$$= f(z),$$

by the relation (7.39). Similarly, $f(z + \omega_2) = f(z)$ and so f has both ω_1 and ω_2 as periods and is an elliptic function.

If $g(z)$ is any function with zeros and poles at the same points of the same multiplicities, then $f(z)/g(z)$ would be an elliptic function without zeros or poles and therefore must be a constant, by Liouville's Theorem.

An important example Consider the function $\wp(z) - \wp(v)$. That function has a double pole at $z = 0$ and zeros at $z = \pm v$. Since the number of zeros of an elliptic function is equal to the number of poles, it follows that those are simple zeros and the only ones. So we may apply the theorem with $a_1 = v, a_2 = -v, b_1 = b_2 = 0$ and $z = u$ to show that

$$\wp(u) - \wp(v) = C(v) \frac{\sigma(u - v)\sigma(u + v)}{(\sigma(u))^2},$$

where $C(v)$ is a constant. To evaluate $C(v)$, we appeal to the Laurent expansion of $\wp(u)$ and the infinite product for $\sigma(z)$ to show that

$$1 = \lim_{u \to 0}[u^2(\wp(u) - \wp(v))] = C(v)\sigma(v)\sigma(-v) \lim_{u \to 0}\left\{ \frac{u}{\sigma(u)} \right\}^2 = -C(v)(\sigma(v))^2.$$

Whence $C(v) = -1/(\sigma(v))^2$ and we conclude that

$$\wp(u) - \wp(v) = -\frac{\sigma(u - v)\sigma(u + v)}{\sigma^2(u)\sigma^2(v)}. \tag{7.63}$$

7.4.2 The addition theorem for $\wp(z)$, first version

We saw in Theorem 2.1 that the Jacobian elliptic functions *sn*, *cn* and *dn* possess addition theorems similar to those for sine and cosine (to which they reduce when $k = 0$) and we shall now obtain the addition theorem for the Weierstrass function $\wp(z)$, by using (7.63).

In (7.63), take logarithmic derivatives with respect to u to obtain

$$\frac{\wp'(u)}{\wp(u) - \wp(v)} = \zeta(u - v) + \zeta(u + v) - 2\zeta(u). \tag{7.64}$$

Now interchange u and v and repeat the argument using differentiation with respect to v to obtain

$$\frac{\wp'(v)}{\wp(v) - \wp(u)} = \zeta(v - u) + \zeta(u + v) - 2\zeta(v). \tag{7.65}$$

On adding (7.42) and (7.43), and recalling that $\zeta(z)$ is an odd function, we have

$$\frac{\wp'(u) - \wp'(v)}{\wp(u) - \wp(v)} = 2\zeta(u + v) - 2\zeta(u) - 2\zeta(v). \tag{7.66}$$

On differentiating with respect to u, we find

$$\frac{1}{2}\frac{\partial}{\partial u}\left(\frac{\wp'(u) - \wp'(v)}{\wp(u) - \wp(v)}\right) = -\wp(u + v) + \wp(u)$$

or

$$\wp(u + v) = \wp(u) - \frac{1}{2}\frac{\partial}{\partial u}\left\{\frac{\wp'(u) - \wp'(v)}{\wp(u) - \wp(v)}\right\}, \tag{7.67}$$

which is one form of the addition theorem for the $\wp$-function.

Another form (the one usually referred to as the addition theorem) is provided by:

Theorem 7.2 *The Weierstrass function $\wp(z) = \wp(z, \Lambda)$ satisfies the addition theorem:*

$$\wp(u + v) = \frac{1}{4}\left\{\frac{\wp'(u) - \wp'(v)}{\wp(u) - \wp(v)}\right\}^2 - \wp(u) - \wp(v). \tag{7.68}$$

Proof Consider the function

$$\frac{1}{4}\left(\frac{\wp'(u) - \wp'(v)}{\wp(u) - \wp(v)}\right)^2,$$

regarded as a function of the complex variable u. Its singularities occur at the point $u = 0$ (where $\wp(u)$ and $\wp'(u)$ have poles) and at the points $u = \pm v$ (where the denominator vanishes).

At $u = 0$ we have the Laurent expansion (in terms of u)

$$\frac{\left(\dfrac{1}{u^3} + \dfrac{1}{2}\wp'(v) + \cdots\right)^2}{\left(\dfrac{1}{u^2} - \wp(v) + \cdots\right)^2} = \frac{1}{u^2}\left(1 + \frac{1}{2}\wp'(v)u^3 + \cdots\right)^2$$

$$(1 - \wp(v)u^2 + \cdots) = \frac{1}{u^2} + 2\wp(v) + \cdots.$$

At $u = v$ the function is analytic (holomorphic), certainly if v is not a half-period, since the zero in the denominator is cancelled by the one in the numerator. (The case when v is a half-period may be dealt with similarly.) At $u = -v$, there is a double pole with residue 0 (since that is the only pole). Since $\wp(u)$ is even and $\wp'(u)$ is odd, we have

$$\lim_{u \to -v} (u + v)^2 \cdot \frac{1}{4} \left\{ \frac{\wp'(u) - \wp'(v)}{\wp(u) - \wp(v)} \right\}^2$$

$$= \lim_{u \to -v} \left(\frac{u + v}{\wp(u) - \wp(-v)} \right)^2 \cdot \frac{1}{4} (\wp'(-v) - \wp'(v))^2$$

$$= \left(\frac{1}{\wp'(-v)} \right)^2 \cdot \frac{1}{4} (-2\wp'(-v))^2 = 1.$$

It follows that the function

$$\frac{1}{4} \left(\frac{\wp'(u) - \wp'(v)}{\wp(u) - \wp(v)} \right)^2 - \wp(u) - \wp(u + v)$$

has no poles and is accordingly a constant. To find the value of the constant, we consider the limit as $u \to 0$ and we find that the value is $-\wp(v)$. We have obtained (7.44) and the theorem is proved.

Corollary 7.1

$$\wp(2u) = \frac{1}{4} \left\{ \frac{\wp''(u)}{\wp'(u)} \right\}^2 - 2\wp(u).$$

Exercises 7.4

7.4.1 Prove that

$$\frac{\sigma(z - c)\sigma(z - d)}{\sigma(z - a)\sigma(z - b)}, c + d = a + b,$$

is an elliptic function of order 2, with simple poles at $z = a$ and $z = b$.

7.4.2 Let $\wp(\omega_k/2) = e_k$. Prove that

$$\wp\left(z + \frac{1}{2}\omega_1\right) = e_1 + \frac{(e_1 - e_2)(e_1 - e_3)}{\wp(z) - e_1}.$$

7.4.3 Prove Corollary 7.5.

7.4.4 Show that:

$$\wp\left(\frac{1}{4}\omega_1\right) = e_1 \pm \{(e_1 - e_2)(e_1 - e_3)\}^{1/2},$$

and, using the previous question with ω_2 in place of ω_1, show that

$$\wp\left(\frac{1}{4}\omega_1 + \frac{1}{2}\omega_2\right) = e_1 \mp \{(e_1 - e_2)(e_1 - e_3)\}^{1/2}$$

7.4.5 Prove that

$$\wp(u + v) - \wp(u - v) = -\frac{\wp'(u)\wp'(v)}{(\wp(u) - \wp(v))^2}.$$

7.4.6 Show that

$$\wp(z) - \wp(a) = -\frac{\sigma(z - a)\sigma(z + a)}{\sigma(z)^2\sigma(a)^2}$$

and deduce that

$$\wp'(z) = 6\frac{\sigma(z + a)\sigma(z - a)\sigma(z + b)\sigma(z - b)}{\sigma(z)^4\sigma(a)^2\sigma(b)^2},$$

where a and b are constants satisfying

$$\wp(a)^2 = \frac{g_2}{12}, \qquad \wp(b) = -\wp(a).$$

7.4.7 Prove the addition theorem for $\wp(z, \Lambda)$ by investigating the poles of

$$\wp(u + a) + \wp(u), \qquad a \notin \Lambda.$$

(Hint: construct another elliptic function with the same poles and principal parts; the difference must be a constant; and then find the constant by choosing u suitably.)

7.4.8 Given that the sum of the zeros of a non-constant elliptic function $f(z)$ is congruent, mod Λ, to the sum of the poles, counted according to multiplicities, consider the zeros of the functions of the form

$\lambda\wp(z)^2 + \mu\wp'(z) + \nu\wp(z) + \rho$ and deduce that if $u+v+w+x \equiv 0 (\text{mod } \Lambda)$,

then

$$\begin{vmatrix} \wp'(u) & \wp(u)^2 & \wp(u) & 1 \\ \wp'(v) & \wp(v)^2 & \wp(v) & 1 \\ \wp'(w) & \wp(w)^2 & \wp(w) & 1 \\ \wp'(x) & \wp(x)^2 & \wp(x) & 1 \end{vmatrix} = 0.$$

7.4.9 Show that

$$\frac{\wp'(z) - \wp'(\alpha)}{\wp(z) - \wp(\alpha)} = 2\zeta(z + \alpha) - 2\zeta(z) - 2\zeta(\alpha).$$

Deduce that, if $u + v + w = 0$, then

$$(\zeta(u) + \zeta(v) + \zeta(w))^2 + \zeta'(u) + \zeta'(v) + \zeta'(w) = 0.$$

7.4.10 Using the fact that every elliptic function may be expressed as a quotient of σ-functions, prove that

$$\begin{vmatrix} 1 & \wp(u) & \wp'(u) \\ 1 & \wp(v) & \wp'(v) \\ 1 & \wp(w) & \wp'(w) \end{vmatrix} = 2\sigma(u+v+w) \cdot \frac{\sigma(u-v)\sigma(v-w)\sigma(w-u)}{(\sigma(u)\sigma(v)\sigma(w))^3}.$$

Deduce the addition theorem for $\wp(z)$.
Deduce also that

$$12\wp(u)(\wp'(u))^2 - (\wp''(u))^2 = 4\frac{\sigma(3u)}{(\sigma(u))^9}.$$

7.4.11 Prove that

$$\wp'(z) = 2\frac{\sigma\left(z - \frac{1}{2}\omega_1\right)\sigma\left(z - \frac{1}{2}\omega_2\right)\sigma\left(z - \frac{1}{2}\omega_3\right)}{\sigma\left(\frac{1}{2}\omega_1\right)\sigma\left(\frac{1}{2}\omega_2\right)\sigma\left(\frac{1}{2}\omega_3\right)\sigma^3(z)}.$$

(Hint: $\wp'(z)$ has a triple pole at $z = 0$ and simple zeros at $z = \omega_k/2$, $k = 1, 2, 3$.)

7.4.12 Prove that

$$\wp'(z) = -\frac{\sigma(2z)}{(\sigma(z))^4}.$$

Let

$$\phi_n(z) = \frac{\sigma(nz)}{(\sigma(z))^{n^2}}.$$

By using the result for $\wp'(z)$ and the first part of 7.4.6, or otherwise, prove that $\phi_n(z)$ is an elliptic function satisfying the recurrence relation

$$\phi_{n+1}(z) = -\frac{(\phi_n(z))^2}{\phi_{n-1}(z)}(\wp(nz) - \wp(z)).$$

7.4.13 By considering the zeros and poles of the functions on either side, prove that

$$\left(\frac{\wp\left(z + \frac{1}{2}\omega_1\right)}{\wp'(z)}\right) = -\left\{\frac{\wp\left(\frac{1}{4}\omega_1\right) - \wp\left(\frac{1}{2}\omega_1\right)}{\wp(z) - \wp\left(\frac{1}{2}\omega_1\right)}\right\}^2.$$

7.5 The field of elliptic functions

7.5.1 The field of elliptic functions

The ideas of this section are of special significance in arithmetic, algebra and analysis and in the theory of elliptic integrals (see Chapter 8), which has applications in dynamics and physics and in the theory of probability.

We consider elliptic functions having a given, fixed lattice, Λ, of periods. Let $f(z)$ be an *even elliptic function,* that is, $f(-z) = f(z)$ (recall that $\wp(z)$ is an even function). We denote by $(\alpha_1, -\alpha_1), (\alpha_2, -\alpha_2), \ldots, (\alpha_r, -\alpha_r)$ the points, other than the origin, that are zeros or poles of $f(z)$ (they occur in pairs because we are supposing that $f(-z) = f(z)$). With each pair $(\alpha_k, -\alpha_k), k = 1, 2, \ldots, r$, we associate an integer m_k, according to the following rule: if α_k is not a half-period and is a zero, then m_k is the multiplicity of the zero. If α_k is a half-period and a zero, then $2m_k$ is the multiplicity of the zero.

Similarly, if α_k is not a half-period and is a pole, then m_k is negative and $-m_k$ is the multiplicity of the pole. If α_k is a half-period and a pole, then $-2m_k$ is the multiplicity of the pole.

We have tacitly assumed that m_k is an integer and in particular that if α_k is a half-period then the multiplicity there must be even. To justify that, we argue as follows. The function

$$\varphi(z) = f\left(z + \frac{1}{2}\omega_k\right)$$

is an even function, since (using the fact that f is even and then appealing to periodicity)

$$\varphi(-z) = f\left(-z + \frac{1}{2}\omega_k\right) = f\left(z - \frac{1}{2}\omega_k\right) = f\left(z + \frac{1}{2}\omega_k\right) = \varphi(z).$$

So $\varphi(z)$ must have a zero or pole of even multiplicity at $z = 0$ and accordingly $f(z)$ must have even multiplicity at $\omega_k/2$.

Our object now is to prove that every elliptic function with period lattice Λ may be obtained from the elliptic functions $\wp(z)$ and $\wp'(z)$ in the sense that every such function is a rational function (that is a quotient of two polynomials) of $\wp$ and $\wp'$ with coefficients in $\mathbb{C}$. More precisely, we prove:

Theorem 7.8 *Let Λ be a given period lattice and let $\wp(z) = \wp(z, \Lambda), \wp'(z) = \wp'(z, \Lambda)$ be the Weierstrass functions for Λ.*
Then:

(a) *every even elliptic function with period lattice Λ is a rational function of $\wp(z)$;*

(b) *every elliptic function with period lattice* Λ *is a rational function of* $\wp(z)$ *and* $\wp'(z)$.

Proof Let $f(z)$ be an even elliptic function with respect to Λ. Consider the function

$$F(z) = \prod_{k=1}^{n} (\wp(z) - \wp(\alpha_k))^{m_k}, \tag{7.69}$$

where α_k and m_k are the numbers defined above, with respect to the even elliptic function $f(z)$. The function $F(z)$ has exactly the same zeros and poles as $f(z)$, with the same orders, apart from possible poles or zeros at the origin (which was excluded from our earlier definition). Since the number of poles is equal to the number of zeros for both of the elliptic functions $F(z)$ and $f(z)$, we must have agreement at the origin also. It follows that $f(z) = kF(z)$.

We conclude that every even elliptic function for Λ is a rational function of $\wp(z)$.

To prove (b), we observe that $\wp'(z)$ is an odd function and so, if $g(z) = g(z, \Lambda)$ is an odd elliptic function the quotient $g(z)/\wp'(z)$ is an even function and therefore, by (a), a rational function of $\wp(z)$.

Now let $h(z) = h(z, \Lambda)$ be any elliptic function with period lattice Λ. Then

$$h(z) = \frac{1}{2}\{h(z) + h(-z)\} + \frac{1}{2}\{h(z) - h(-z)\}$$
$$= f(z) + g(z),$$

where $f(z)$ is even and $g(z)$ is odd. So, by the foregoing argument,

$$h(z) = P(\wp(z)) + \wp'(z)Q(\wp(z)),$$

where P and Q are rational functions.

We conclude that the field of elliptic functions with period lattice Λ is obtained from $\mathbb{C}$ by adjoining the transcendental elements $\wp(z)$ and $\wp'(z)$; so that we obtain the field

$$\mathbb{C}(\wp(z), \wp'(z))$$

of rational functions of $\wp(z)$ and $\wp'(z)$. (In arithmetical applications one is usually interested in sub-fields, such as the field $\mathbb{Q}(\wp, \wp')$; see Cassels, 1991.)

7.5.2 Elliptic functions with given points for poles

We have seen how to construct elliptic functions with given zeros and poles, using the σ-functions. Suppose now that nothing is known about the position of the zeros of an elliptic function $f(z)$, with period lattice Λ, but that it is known that $f(z)$ has poles at the points $u_1, u_2, \ldots, u_r$ and that the principal part at u_k is

$$\sum_{r=1}^{n_k} \frac{a_{r,k}}{(z - u_k)^r};$$ (7.70)

so that $f(z)$ has a pole of order n_k at $z = u_k$, and the residue there is $a_{1,k}$. Since the sum of the residues at the poles must be zero, a *necessary* condition for the expressions in (7.65) to be the principal parts of an elliptic function for Λ is that

$$\sum_{k=1}^{r} a_{1,k} = 0.$$ (7.71)

We shall suppose that (7.71) is satisfied and we shall use the functions $\zeta(z)$, $\wp(z)$ and the derivatives of $\wp(z)$ to construct the required function.

The function

$$\sum_{k=1}^{r} a_{1,k}\, \zeta(z - u_k)$$ (7.72)

is an elliptic function (using the fact that $\zeta(z + \omega) = \zeta(z) + \eta(\omega)$) with the correct residue at $z = u_k$, and the function

$$\sum_{k=1}^{r} a_{2,k}\, \wp(z - u_k)$$ (7.73)

has the correct terms of the second order (using the Laurent expansion of $\wp$).

Similarly,

$$-\frac{1}{2} \sum_{k=1}^{r} a_{3,k}\, \wp'(z - u_k)$$ (7.74)

has the correct terms of the third order (using the Laurent expansion of $\wp'$) to agree with (7.70), and so on, using the higher derivatives of $\wp$.

If we add together all such expressions, we obtain a function with the same singularities as the principal parts in (7.70). The elliptic function so obtained must differ from the elliptic function $f(z)$ we seek by an elliptic function with no singularities and so must be a constant. The freedom afforded by that arbitrary constant makes up for the linear relation (7.71) between the $a_{1,k}$, and so the dimension of the vector space of elliptic functions, having poles at the points u_k and the order of whose pole at u_k does not exceed n_k, is precisely

$$n_1 + n_2 + \cdots + n_r.$$ (7.75)

The result just obtained is the genus one case of a general theorem in algebraic geometry called the 'Riemann–Roch Theorem'. We state it as:

Theorem 7.9 *The elliptic functions with period lattice* Λ, *having given principal parts and poles of orders at most* $n_1, \ldots n_r$ *at the points* $u_1, \ldots, u_r$ *respectively, form a vector space over* $\mathbb{C}$ *of dimension* $n_1 + n_2 + \cdots + n_r$.

Exercises 7.5

7.5.1 Prove that an elliptic function with only simple poles can be expressed as a constant plus a linear combination of functions of the form $\zeta(z - z_k)$.

Prove that

$$
\frac{1}{\wp'(z)} = \frac{\zeta\left(z - \frac{1}{2}\omega_1\right) + \zeta\left(\frac{1}{2}\omega_1\right)}{\wp''\left(\frac{1}{2}\omega_1\right)} + \frac{\zeta\left(z - \frac{1}{2}\omega_2\right) + \zeta\left(\frac{1}{2}\omega_2\right)}{\wp''\left(\frac{1}{2}\omega_2\right)}
$$

$$
+ \frac{\zeta\left(z - \frac{1}{2}\omega_1 - \frac{1}{2}\omega_2\right) + \zeta\left(\frac{1}{2}\omega_1 + \frac{1}{2}\omega_2\right)}{\wp''\left(\frac{1}{2}\omega_1 + \frac{1}{2}\omega_2\right)}.
$$

7.5.2 Prove that

$$
\wp(z-a)\wp(z-b) = \wp(a)\wp(b) + \wp(a-b)\{\wp(z - a) + \wp(z - b) - \wp(a) - \wp(b)\}
$$
$$
+ \wp'(a - b)\{\zeta(z - a) - \zeta(z - b) + \zeta(a) - \zeta(b)\}.
$$

7.5.3 Show that

$$
\zeta(z - a) - \zeta(z - b) - \zeta(a - b) + \zeta(2a - 2b)
$$

is an elliptic function with period lattice Λ.

Deduce that it is equal to

$$
\frac{\sigma(z - 2a + b)\sigma(z - 2b + a)}{\sigma(2b - 2a)\sigma(z - a)\sigma(z - b)},
$$

7.5.4 Show that

$$
\wp(z - a) - \wp(z + a) = \frac{\wp'(z)\wp'(a)}{(\wp(z) - \wp(a))^2}.
$$

7.5.5 Prove that, if n is an integer, then $\wp(n\,u)$ can be expressed as a rational function of $\wp(u)$.

Show, in particular, that (writing $\wp = \wp(u)$ etc.)

(i) $\wp(2u) = \dfrac{\wp^4 + \dfrac{1}{2}g_2\wp^2 + 2g_3\wp + \dfrac{1}{16}g_2^2}{4\wp^3 - g_2\wp - g_3};$

(ii) $\wp(3u) = \wp(u) + \dfrac{\wp'^2(\wp'^4 - \psi\wp'')}{\psi^2},$

where $\psi = \wp'^2(u)(\wp(u) - \wp(2u))$.

7.5.6 Use the method of Theorem 7.9 to prove that

$$\wp^2(z) = \frac{1}{6}\wp''(z) + \frac{1}{12}g_2.$$

7.6 Connection with the Jacobi functions

In Exercise 7.1.6, we introduced the functions $\sigma_r(z)$ and promised to explain how they provide a connection between the Weierstrass and Jacobi theories; we conclude this chapter by developing those ideas and, finally, by saying a little more about the modular functions and the inversion problem.

We recall from Section 7.4, (7.63) the representation

$$\wp(z) - \wp(\alpha) = -\frac{\sigma(z - \alpha)\sigma(z + \alpha)}{\sigma^2(z)\sigma^2(\alpha)}, \tag{7.76}$$

in terms of σ-functions.

We begin by using (7.76) to obtain a representation of the square root $(\wp(z) - e_r)^{1/2}$, where $e_r = \wp(\omega_r/2)$, $\omega_1 + \omega_2 + \omega_3 = 0$ and the pair (ω_1, ω_2) is a basis for the period lattice Λ. We must make clear which square root we are using and we do that by defining $(\wp(z) - e_r)^{1/2}$ to be that square root of $\wp(z) - e_r$ having a simple pole with residue 1 at $z = 0$ (recall from Theorem 7.1 that $\wp(z) = 1/z^2 + 3S_4z^2 + 5S_6z^4 + \cdots$).
Since $\wp'(z) = -2/z^3 + 6S_4z + 20S_6z^3 + \cdots$, in the neighbourhood of $z = 0$, we have

$$\wp'(z) = -2\{\wp(z) - e_1\}^{1/2}\{\wp(z) - e_2\}^{1/2}\{\wp(z) - e_3\}^{1/2} \tag{7.77}$$

and by writing $\alpha = \omega_r/2$, $r = 1, 2, 3$, in (7.76), we obtain

$$\wp(z) - e_r = -\frac{\sigma\left(z - \dfrac{1}{2}\omega_r\right)\sigma\left(z + \dfrac{1}{2}\omega_r\right)}{\sigma^2(z)\sigma^2\left(\dfrac{1}{2}\omega_r\right)}.$$

Now

$$\sigma\left(z - \frac{1}{2}\omega_r\right) = \sigma\left(z + \frac{1}{2}\omega_r - \omega_r\right),$$

and so (by Proposition 7.4)

$$\sigma\left(z - \frac{1}{2}\omega_r\right) = -e^{\eta(-\omega_r)\left(z + \frac{1}{2}\omega_r - \frac{1}{2}\omega_r\right)}\sigma\left(z + \frac{1}{2}\omega_r\right),$$

and by substituting in (7.76) we obtain

$$\wp(z) - e_r = +e^{-\eta_r z} \frac{\sigma^2\left(z + \frac{1}{2}\omega_r\right)}{\sigma^2(z)\sigma^2\left(\frac{1}{2}\omega_r\right)}.$$

It follows that

$$(\wp(z) - e_r)^{1/2} = \pm e^{-\frac{1}{2}\eta_r z} \frac{\sigma\left(z + \frac{1}{2}\omega_r\right)}{\sigma(z)\sigma\left(\frac{1}{2}\omega_r\right)}, \tag{7.78}$$

and by considering the behaviour of each side in the neighbourhood of $z = 0$ we see that the plus sign in (7.78) is to be chosen.

Hence

$$(\wp(z) - e_r)^{\frac{1}{2}} = e^{-\frac{1}{2}\eta_r z} \frac{\sigma\left(z + \frac{1}{2}\omega_r\right)}{\sigma(z)\sigma\left(\frac{1}{2}\omega_r\right)}, \tag{7.79}$$

and we recall that[8]

$$\eta_1 = \eta(\omega_1) = 2\zeta\left(\frac{1}{2}\omega_1\right), \eta_2 = \eta(\omega_2) = 2\zeta\left(\frac{1}{2}\omega_2\right), \eta_3 = \eta(\omega_3) = 2\zeta\left(\frac{1}{2}\omega_3\right).$$

It follows from (7.79) that $(\wp(z) - e_r)^{1/2}$ has simple poles at the points of Λ and simple zeros at the points of $\omega_r/2 + \Lambda$.

---

[8] The work of this section is based on the treatment of the Jacobi functions given by Copson (1935); but note that Copson uses the notation $(2\omega_1, 2\omega_2)$ for the basis of Λ and some significant changes result.

We are now in a position to introduce the functions $\sigma_r(z)$, and to begin with we define $(e_1 - e_2)^{1/2}$ to be the value of $(\wp(z) - e_2)^{1/2}$ at $z = \omega_1/2$ and

$$(e_r - e_s)^{1/2} = e^{-\eta_s \omega_r/4} \frac{\sigma\left(\dfrac{1}{2}\omega_r + \dfrac{1}{2}\omega_s\right)}{\sigma\left(\dfrac{1}{2}\omega_r\right)\sigma\left(\dfrac{1}{2}\omega_s\right)}. \tag{7.80}$$

Now the expression for $(\wp(z) - e_r)^{1/2}$ suggests the introduction of the following functions we already encountered in Exercise 7.1.6.

Definition 7.4 *We define, for $r = 1, 2, 3$,*

$$\sigma_r(z) = e^{-\frac{1}{2}\eta_r z} \frac{\sigma\left(z + \dfrac{1}{2}\omega_r\right)}{\sigma\left(\dfrac{1}{2}\omega_r\right)}.$$

In that notation we have

$$(\wp(z) - e_r)^{1/2} = \frac{\sigma_r(z)}{\sigma(z)},$$

and

$$(e_r - e_s)^{\frac{1}{2}} = \frac{\sigma_s\left(\dfrac{1}{2}\omega_r\right)}{\sigma\left(\dfrac{1}{2}\omega_r\right)}.$$

We prove now that: (a) $\sigma_r(z)$ is an even function of z; (b) $\sigma_r(z)$ is an entire (integral) function of z; and (c) $\sigma_r(z)$ has pseudo-periodicity properties, resembling those of $\sigma(z)$. Our main results are contained in the following three propositions.

Proposition 7.7 *The function $\sigma_r(z)$ is an even function of z.*

Proof We have, using Definition (7.4) and the properties of $\sigma(z)$,

$$\sigma_r(-z) = e^{\eta_r z/2} \frac{\sigma\left(-z + \dfrac{1}{2}\omega_r\right)}{\sigma\left(\dfrac{1}{2}\omega_r\right)}$$

$$= e^{\eta_r z/2} \frac{\left(-\sigma\left(z - \dfrac{1}{2}\omega_r\right)\right)}{\sigma\left(\dfrac{1}{2}\omega_r\right)}$$

$$= -e^{\eta_r z/2} \frac{\sigma\left(z + \frac{1}{2}\omega_r - \omega_r\right)}{\sigma\left(\frac{1}{2}\omega_r\right)}$$

$$= -e^{\eta_r/2}\left(-e^{\eta(-\omega_r)(z+\frac{1}{2}\omega_r-\frac{1}{2}\omega_r)}\right) \cdot \frac{\sigma\left(z + \frac{1}{2}\omega_r\right)}{\sigma\left(\frac{1}{2}\omega_r\right)}$$

$$= e^{-\eta_r z/2} \frac{\sigma\left(z + \frac{1}{2}\omega_r\right)}{\sigma\left(\frac{1}{2}\omega_r\right)};$$

$$= \sigma_r(z),$$

so $\sigma_r(z)$ is an even function of z.

Proposition 7.8 *The function $\sigma_r(z)$ is an entire (integral) function, with simple zeros at the points $\omega_r/2 + \Lambda, r = 1, 2, 3$.*

Proof By definition,

$$\sigma_r(z) = e^{-\frac{1}{2}\eta_r z}\left(z + \frac{1}{2}\omega_r\right) \prod_{\omega \in \Lambda}' \left(1 - \frac{z + \frac{1}{2}\omega_r}{\omega}\right) \cdot e^{\frac{z}{\omega}+\frac{1}{2}\left(\frac{z}{\omega}\right)^2} \bigg/ \sigma\left(\frac{1}{2}\omega_r\right),$$

where the $'$ denotes that the infinite product is over all $\omega \in \Lambda$, other than $\omega = 0$. It follows that the function is an entire function and has zeros at the points $z \equiv \omega_r/2 \,(\text{mod } \Lambda)$.

Proposition 7.9 *The functions $\sigma_r(z)$ have the pseudo-periodicity properties*

(i) $\sigma_r(z + \omega_r) = -e^{\eta_r\left(z + \frac{1}{2}\omega_r\right)}\sigma_r(z)$;
    and if $r \neq s$,
(ii) $\sigma_r(z + \omega_s) = +e^{\eta_s\left(z + \frac{1}{2}\omega_s\right)}\sigma_r(z)$.

Proof (i) By definition,

$$\sigma_r(z + \omega_r) = e^{-\eta_r(z+\omega_r)/2} \frac{\sigma\left(z + \omega_r + \frac{1}{2}\omega_r\right)}{\sigma\left(\frac{1}{2}\omega_r\right)}$$

$$= e^{-\eta_r(z+\omega_r)/2}\left\{-e^{\eta_r(z+\frac{1}{2}\omega_r+\frac{1}{2}\omega_r)}\frac{\sigma\left(z+\frac{1}{2}\omega_r\right)}{\sigma\left(\frac{1}{2}\omega_r\right)}\right\}$$

$$= -e^{\eta_r(z+\omega_r)/2}\frac{\sigma\left(z+\frac{1}{2}\omega_r\right)}{\sigma\left(\frac{1}{2}\omega_r\right)}$$

$$= -e^{\eta_r(z+\omega_r)/2}e^{\eta_r z/2}\sigma_r(z)$$

$$= -e^{\eta_r(z+\frac{1}{2}\omega_r)}\sigma_r(z).$$

(ii) Again, if $r \neq s$,

$$\sigma_r(z+\omega_s) = e^{-\eta_r(z+\omega_s)/2}\frac{\sigma\left(z+\omega_s+\frac{1}{2}\omega_r\right)}{\sigma\left(\frac{1}{2}\omega_r\right)}$$

$$= e^{-\eta_r(z+\omega_s)/2}\left\{-e^{\eta_s(z+\frac{1}{2}\omega_r+\frac{1}{2}\omega_s)}\frac{\sigma\left(z+\frac{1}{2}\omega_r\right)}{\sigma\left(\frac{1}{2}\omega_r\right)}\right\}$$

$$= -e^{-\eta_r z/2}e^{\eta_s z}e^{(\eta_s\omega_r-\eta_r\omega_s)/2}e^{\eta_s\omega_s/2}\frac{\sigma\left(z+\frac{1}{2}\omega_r\right)}{\sigma\left(\frac{1}{2}\omega_r\right)}$$

$$= +e^{-\eta_s z}e^{-\eta_r z/2}e^{\eta_s\omega_s/2}\frac{\sigma\left(z+\frac{1}{2}\omega_r\right)}{\sigma\left(\frac{1}{2}\tilde{\omega}_r\right)}$$

$$= e^{\eta_s z}e^{-\eta_r z/2}e^{\eta_s\omega_s/2}\frac{\sigma\left(z+\frac{1}{2}\omega_r\right)}{\sigma\left(\frac{1}{2}\omega_r\right)}$$

$$= e^{\eta_s z}e^{-\eta_r z/2}e^{\eta_s\omega_s/2}\sigma_r(z)e^{\eta_r z/2}$$

$$= e^{\eta_s(z+\frac{1}{2}\omega_s)}\sigma_r(z).$$

On the basis of those three propositions, we can now establish the connection between our approach to the Weierstrass $\wp$-function developed in this Chapter and the Jacobi functions and theta functions; so reversing the order of things in our earlier chapters.

We begin by defining three functions, $S(z)$, $C(z)$ and $D(z)$, which will, eventually, yield the Jacobi functions $sn(u)$, $cn(u)$ and $dn(u)$, as follows.

Definition 7.5 *With the foregoing notations, define*

$$S(z) = (e_1 - e_2)^{1/2} \frac{\sigma(z)}{\sigma_2(z)} = \left\{ \frac{e_1 - e_2}{\wp(z) - e_2} \right\}^{1/2},$$

$$C(z) = \frac{\sigma_1(z)}{\sigma_2(z)} = \left\{ \frac{\wp(z) - e_1}{\wp(z) - e_2} \right\}^{1/2},$$

$$D(z) = \frac{\sigma_3(z)}{\sigma_2(z)} = \left\{ \frac{\wp(z) - e_3}{\wp(z) - e_2} \right\}^{1/2},$$

where the square root is so chosen in C and D that each has the value $+1$ at $z = 0$. (It follows from the definition that $S(0) = 0$.)

Theorem 7.10 *The functions $S(z)$, $C(z)$ and $D(z)$ are meromorphic and periodic, with periods given by the following table (for example $S(z + \omega_r) = \pm S(z)$, $r = 1, 2, 3$).*

	ω_1	ω_2	ω_3
$S(z)$	-1	$+1$	-1
$C(z)$	-1	-1	$+1$
$D(z)$	$+1$	-1	-1

Corollary 7.6 *The functions $S(z)$, $C(z)$ and $D(z)$ are elliptic functions with period lattices generated by $(2\omega_1, \omega_2)$, $(2\omega_1, \omega_3)$ and $(2\omega_3, \omega_1)$, respectively.*

Proof The functions are quotients of analytic functions with poles at the points where the denominator vanishes, or where the numerator has poles, in each case. It follows that they are meromorphic.

To verify the entries in the table, we appeal to the pseudo-periodicity properties of the σ-functions; for example

$$S(z + \omega_1) = (e_1 - e_2)^{1/2} \frac{\sigma(z + \omega_1)}{\sigma_2(z + \omega_1)}$$

$$= \frac{(e_1 - e_2)^{1/2}(-1)e^{\eta_1(z + \frac{1}{2}\omega_1)}\sigma(z)}{+ e^{\eta_1(z + \frac{1}{2}\omega_1)}\sigma_2(z)}$$

$$= (e_1 - e_2)^{1/2}(-1)\frac{\sigma(z)}{\sigma_2(z)} = -S(z).$$

Corollary 7.6 follows immediately, by repeated application of the entries in the table, where necessary.

If we define

$$k = -\left\{\frac{e_3 - e_2}{e_1 - e_2}\right\}^{1/2},$$

then it follows that

$$C(z) = (1 - S^2(z))^{1/2}, \quad D(z) = (1 - k^2 S^2(z))^{1/2}$$

and

$$S'(z) = (e_1 - e_2)^{1/2} C(z) D(z)$$
$$C'(z) = -(e_1 - e_2)^{1/2} D(z) S(z)$$
$$D'(z) = -(e_1 - e_2)^{1/2} k^2 S(z) C(z).$$

The proofs are immediate consequences of the definitions and are left as exercises (see Exercise 7.6.3).

The idea now is to define the Jacobi function $sn(u, k)$ as $S((e_1 - e_2)^{-1/2}u)$, with similar definitions for $cn(u, k)$ and $dn(u, k)$. But before we do that we need to complete our verification of the usual properties of sn, cn, dn by proving:

Theorem 7.11 *With the notation of the preceding theorem, the elliptic function $S(z)$ has the following properties.*

(a) *The period lattice Λ of $S(z)$ has basis $(2\omega_1, \omega_2)$.*
(b) *$S(z)$ is an odd function of z.*
(c) *$S(z)$ has zeros at the points of Λ and is of order 2.*
(d) *The poles of $S(z)$ are at the points $\equiv \omega_2/2$ or $\equiv \omega_1 + \omega_2/2 \pmod{\Lambda}$.*
(e) *The residue of $S(z)$ at $\omega_2/2$ is $-(e_3 - e_2)^{-1/2}$; the residue at $\omega_1 + \omega_2/2$ is $+(e_3 - e_2)^{-1/2}$.*

(The reader should compare the assertions of the theorem with the results in Chapter 2.)

Proof Part (a) is Corollary 7.6
To prove (b) we observe that

$$S(-z) = (e_1 - e_2)^{1/2} \frac{\sigma(-z)}{\sigma_2(-z)} = -(e_1 - e_2)^{1/2} \frac{\sigma(z)}{\sigma_2(z)} = -S(z),$$

since $\sigma(z)$ is an odd function and $\sigma_2(z)$ is even.

(c) The function $S(z)$ has zeros at the points where $\sigma(z) = 0$, that is at the points of Λ. In the period parallelogram with basis $(2\omega_1, \omega_2)$ there are two such points, namely 0 and ω_1, and so $S(z)$ has order 2.

(d) The poles of $S(z)$ are given by the points where $\sigma_2(z) = 0$ (that is $(\wp(z) - e_2)^{1/2} = 0$) and so the poles in the fundamental parallelogram are simple poles at the points $\omega_2/2, \omega_1 + \omega_2/2$. (Note there are two poles, in accordance with the result proved in (c).)

(e) The residue of $S(z)$ at $\omega_2/2$ is

$$\lim_{z \to \frac{1}{2}\omega_2} \left\{ (e_1 - e_2)^{1/2} \sigma(z) \frac{z - \frac{1}{2}\omega_2}{\sigma_2(z)} \right\} = (e_1 - e_2)^{1/2} \sigma\left(\frac{1}{2}\omega_2\right) \cdot \frac{1}{\sigma_2'\left(\frac{1}{2}\omega_2\right)}.$$

Now

$$\sigma_2(z) = e^{-\frac{1}{2}\eta_2 z} \sigma\left(z + \frac{1}{2}\omega_2\right) \Big/ \sigma\left(\frac{1}{2}\omega_2\right)$$

and to find $\sigma_2'(\omega_2/2)$ we put $z = \omega_2/2$ in

$$\sigma_2'(z) = -\frac{\eta_2}{2} e^{-\frac{1}{2}\eta_2 z} \frac{\sigma\left(z + \frac{1}{2}\omega_2\right)}{\sigma\left(\frac{1}{2}\omega_2\right)} + e^{-\frac{1}{2}\eta_2 z} \frac{\sigma'\left(z + \frac{1}{2}\omega_2\right)}{\sigma\left(\frac{1}{2}\omega_2\right)},$$

whence

$$\sigma_2'\left(\frac{1}{2}\omega_2\right) = e^{-\frac{1}{4}\eta_2\omega_2} \frac{\sigma'(\omega_2)}{\sigma\left(\frac{1}{2}\omega_2\right)} = -\frac{e^{\frac{1}{4}\eta_2\omega_2}}{\left(\frac{1}{2}\omega_2\right)}$$

(where we have used the fact that

$$\frac{\sigma'(z)}{\sigma(z)} = \zeta(z)$$

and

$$\sigma'(z) = \frac{d}{dz}\left\{ z \prod_{\omega}' \left(1 - \frac{z}{\omega}\right) e^{\frac{z}{\omega} + \frac{1}{2}\frac{z^2}{\omega^2}} \right\}.$$

Now recall that

$$(e_1 - e_2)^{\frac{1}{2}} = e^{-\frac{1}{4}\eta_2\omega_1} \frac{\sigma\left(\frac{1}{2}\omega_1 + \frac{1}{2}\omega_2\right)}{\sigma\left(\frac{1}{2}\omega_1\right) \sigma\left(\frac{1}{2}\omega_2\right)}$$

$$= \frac{e^{-\frac{1}{4}\eta_2\omega_1} \sigma\left(-\frac{1}{2}\omega_3\right)}{\sigma\left(\frac{1}{2}\omega_1\right) \sigma\left(\frac{1}{2}\omega_2\right)}$$

$$= -\frac{e^{-\frac{1}{4}\eta_2\omega_2} \sigma\left(\frac{1}{2}\omega_3\right)}{\sigma\left(\frac{1}{2}\omega_1\right) \sigma\left(\frac{1}{2}\omega_2\right)}.$$

It follows that the residue at $z = \omega_2/2$ is

$$-\frac{e^{-\frac{1}{4}\eta_2\omega_1}\sigma\left(\frac{1}{2}\omega_3\right)}{\sigma\left(\frac{1}{2}\omega_1\right)\sigma\left(\frac{1}{2}\omega_2\right)}\sigma\left(\frac{1}{2}\omega_2\right)\cdot\frac{1}{\sigma_2'\left(\frac{1}{2}\omega_2\right)}$$

$$=\frac{e^{-\frac{1}{4}\eta_2\omega_1}\sigma\left(\frac{1}{2}\omega_3\right)}{\sigma\left(\frac{1}{2}\omega_1\right)\sigma\left(\frac{1}{2}\omega_2\right)}\sigma\left(\frac{1}{2}\omega_2\right)\frac{\sigma\left(\frac{1}{2}\omega_1\right)}{e^{\frac{1}{4}\eta_2\omega_2}}$$

$$=-e^{\frac{1}{4}\eta_2\omega_3}\frac{\sigma\left(\frac{1}{2}\omega_2\right)\sigma\left(\frac{1}{2}\omega_3\right)}{\sigma\left(\frac{1}{2}\omega_2+\frac{1}{2}\omega_3\right)}$$

(using $\omega_1 + \omega_2 + \omega_3 = 0$). Finally, we obtain the residue at $z = \omega_2/2$ to be $-(e_3 - e_2)^{-1/2}$.

Since the sum of the residues in the period parallelogram is 0, the residue at $(\omega_1 + \omega_2/2)$ is $+(e_3 - e_2)^{-1/2}$.

That completes the proof of the theorem.

The proofs of the following two theorems are similar to the proof of the theorem just proved and are left as an exercise.

Theorem 7.12

(a) *The function $C(z)$ is an even function with periods $(2\omega_2, \omega_3)$.*
(b) *The zeros of $C(z)$ are simple zeros at the points $\omega_1/2 + \Lambda$; so there are two in each period parallelogram.*
(c) *The poles of $C(z)$ are at the points congruent to $\omega_2/2$ or $3\omega_2/2$ (mod $2\omega_2, \omega_3$).*
(d) *The residue at $\omega_2/2$ is $i(e_3 - e_2)^{-1/2}$, the residue at $3\omega_2/2$ is $-i(e_3 - e_2)^{-1/2}$.*

Theorem 7.13

(a) *The function $D(z)$ is an even function with periods $(2\omega_3, \omega_1)$.*
(b) *The function has simple zeros at the points $\omega_3/2 + \Lambda$ and is of order 2.*
(c) *The function has poles at the points $\omega_2/2$ and $\omega_2/2 + \omega_3$(mod $2\omega_3, \omega_1$).*
(d) *The residue at $\omega_2/2$ is $-i(e_1 - e_2)^{-1/2}$; the residue at $\omega_2/2 + \omega_3$ is $+i(e_1 - e_2)^{-1/2}$.*

We have already defined the modulus k by

$$k = -\frac{(e_3 - e_2)^{1/2}}{(e_1 - e_2)^{1/2}},$$

and we define the complementary modulus, k', by

$$k' = \frac{(e_1 - e_3)^{1/2}}{(e_1 - e_2)^{1/2}}.$$

One readily verifies that

$$k^2 + k'^2 = \frac{e_3 - e_2}{e_1 - e_2} + \frac{e_1 - e_3}{e_1 - e_2} = 1.$$

Finally, we may now define the Jacobi functions of Chapter 2 as follows. Write

$$u = (e_1 - e_2)^{1/2} z.$$

Then

$$sn(u) = S((e_1 - e_2)^{-1/2} u),$$
$$cn(u) = C((e_1 - e_2)^{-1/2} u),$$
$$dn(u) = D((e_1 - e_2)^{-1/2} u). \tag{7.81}$$

The properties of the Jacobi functions proved in Chapter 2 may now be derived (see Exercises 7.6 for an outline and Chapter 14 of Copson (1935) for a more detailed account).

Exercises 7.6

7.6.1 Prove that $\sigma_r(z)$ is an even function of z and that its Taylor expansion is

$$\sigma_r(z) = 1 - \frac{1}{2} e_r z^2 + \frac{1}{48} (g_2 - 6e_r^2) z^4 + \cdots.$$

7.6.2 Show that:

(i) $\sigma_r^2(z) - \sigma_s^2(z) = (e_s - e_r)\sigma^2(z)$;

(ii) $(e_2 - e_3)\sigma_1^2(z) + (e_3 - e_1)\sigma_2^2(z) + (e_1 - e_2)\sigma_3^2(z) = 0$.

7.6.3 Prove that

$$C(z) = (1 - S^2(z))^{1/2}, \quad D(z) = (1 - k^2 S^2(z))^{1/2},$$

where $k = -\left\{\dfrac{e_3 - e_2}{e_1 - e_2}\right\}^{1/2}$. Show further that

$$S'(z) = (e_1 - e_2)^{1/2} C(z) D(z),$$
$$C'(z) = -(e_1 - e_2)^{1/2} D(z) S(z),$$
$$D'(z) = -(e_1 - e_2)^{1/2} k^2 S(z) C(z).$$

7.6.4 Complete the proofs of Theorem 7.10. and 7.11. by imitating the proof of Theorem 7.9.

7.6.5 Using the definitions in (7.81) and the results in 7.6.3, prove that, if $sn\,(u) = sn(u, k)$ etc., then

$$\frac{d}{du} sn\,(u) = cn\,(u)\,dn\,(u), \qquad \frac{d}{du} cn\,(u) = -dn\,(u)\,sn\,(u),$$
$$\frac{d}{du} dn\,(u) = -k^2 sn\,(u)\,cn\,(u).$$

7.6.6 The numbers

$$K = \frac{1}{2}\omega_1(e_1 - e_2)^{1/2}, \quad iK' = \frac{1}{2}\omega_2(e_1 - e_2)^{1/2}$$

are called the *quarter-periods*. Show that when u is increased by a *half-period*, $2K, 2iK'$ or $2K + 2iK'$, then the functions $sn\,(u)$, $cn\,(u)$, $dn\,(u)$, are reproduced according to the scheme shown below, with multipliers ± 1.

	$2K$	$2iK'$	$2K + 2iK'$
$sn\,(u)$	-1	$+1$	-1
$cn\,(u)$	-1	-1	$+1$
$dn\,(u)$	$+1$	-1	-1

(Thus the entry in the first row of the first column reads: $sn(u + 2K) = -sn\,(u)$, etc.)

7.6.7 Verify that

$$sn\left(Kz \middle/ \frac{1}{2}\omega_1\right) = \frac{K}{\omega_1/2} \frac{\sigma(z)}{\sigma_2(z)},$$
$$cn\left(Kz \middle/ \frac{1}{2}\omega_1\right) = \frac{\sigma_1(z)}{\sigma_2(z)},$$
$$dn\left(Kz \middle/ \frac{1}{2}\omega_1\right) = \frac{\sigma_3(z)}{\sigma_2(z)}.$$

7.6.8 Show that

$$sn\ (0) = 0, \qquad sn\ (K) = 1, \qquad sn(K + \mathrm{i}K') = 1/k,$$
$$cn\ (0) = 1, \qquad cn\ (K) = 0, \qquad cn(K + \mathrm{i}K') \quad = -\mathrm{i}k'/k,$$
$$dn\ (0) = 1, \qquad dn\ (K) = k', \qquad dn(K + \mathrm{i}K') = 0.$$

7.6.9 Prove that the Jacobi elliptic functions in the neighbourhood of the origin have the Taylor expansions:

$$sn\ (u) = u - \frac{1}{3!}(1 + k^2)u^3 + \frac{1}{5!}(1 + 14k^2 + k^4)u^5 - \cdots ;$$

$$cn\ (u) = 1 - \frac{1}{2!}u^2 + \frac{1}{4!}(1 + 4k^2)u^4 - \cdots ;$$

$$dn\ (u) = 1 - \frac{1}{2!}k^2 u^2 + \frac{1}{4!}(4k^2 + k^4)u^4 - \cdots .$$

7.6.10 Prove that

$$\begin{vmatrix} \sigma_1(2u) & \sigma_1(2v) & \sigma_1(2w) \\ \sigma_2(2u) & \sigma_2(2v) & \sigma_2(2w) \\ \sigma_3(2u) & \sigma_3(2v) & \sigma_3(2w) \end{vmatrix} = 4\mathrm{i}e^{\frac{1}{4}(\eta_1\omega_1 + \eta_2\omega_2 + \eta_3\omega_3)}$$

$$\times \frac{\sigma(u - v)\sigma(u + v)\sigma(v + w)\sigma(v - w)\sigma(w + u)\sigma(w - u)}{\sigma^2\left(\frac{1}{2}\omega_1\right)\sigma^2\left(\frac{1}{2}\omega_2\right)\sigma^2\left(\frac{1}{2}\omega_3\right)}.$$

7.7 The modular functions $j(\tau)$, $\lambda(\tau)$ and the inversion problem

We saw in Section 7.3 that the function

$$j(\tau) = 1728\ J(\tau) = 1728\ \frac{g_2^3}{\Delta}, \tag{7.82}$$

where Δ is given by (7.30), is a modular function (indeed *the* modular function), invariant under the action of the modular group $SL(2, \mathbb{Z})$ generated by S, T (see Chapter 6, Section 6.3).

We had already found the fundamental region for that group in Chapter 3 (see Figure 3.1) and other transformation properties associated with it in Chapter 6. In Chapter 10, on the solution of the general quintic equation in terms of elliptic functions, we shall need a fundamental property of $j(\tau)$, namely that if $c \in \mathbb{C}$, then the equation $j(\tau) = c$ has exactly one solution in the fundamental region $\mathfrak{R}$. That will be proved in Theorem 7.14.

In Chapter 5, we solved the inversion problem for the function

$$\lambda(\tau) = m = k^2 = \frac{\theta_2^4(0|\tau)}{\theta_3^4(0|\tau)} = \frac{e_3 - e_2}{e_1 - e_2}, \tag{7.83}$$

(cf. Chapter 6, Definition 6.1), which is fundamental in the study of the Jacobian elliptic functions and which, again, arises in connection with the solution of the quintic.

We shall obtain the fundamental region for λ and we shall show that

$$J(\tau) = \frac{4}{27} \frac{(1 - \lambda + \lambda^2)^3}{\lambda^2(1 - \lambda)^2}, \quad \lambda = \lambda(\tau). \tag{7.84}$$

We begin with the inversion problem for $j(\tau)$ (see Copson, 1935, Serre, 1970, Chapter VII, for a variation on the theme; see also, for example, the books, Dutta & Debnath, 1965, Prasolov & Solovyev 1997) and Whittaker & Watson, 1927).

Theorem 7.14 *The equation*

$$J(\tau) = c, \ c \in \mathbb{C},$$

has a unique solution in the fundamental region $\mathfrak{F}$.

Proof We refer to Figure 3.1 in Chapter 3.

Since $J(\tau) \to \infty$ as $\operatorname{Im} \tau \to \infty$ (cf. the expansion in (7.49)), we can choose $t_0 > 0$ such that $|J(\tau)| > |c|$, provided that $\operatorname{Im} \tau \geq t_0$. Denote by D, D', respectively, the points on the lines $\operatorname{Re}(\tau) = -1/2$, $\operatorname{Re}(\tau) = +1/2$ such that $D = -1/2 + it_0$, $D' = 1/2 + it_0$ and denote by C the contour bounded by $ABA'D'DA$ (see Figure 3.1).

Let N denote the number of zeros of $J(\tau) - c$ within C, then

$$N = \frac{1}{2\pi i} \int_C \frac{J'(\tau)}{J(\tau) - c} d\tau,$$

where the dash, $'$, denotes differentiation with respect to τ. That is,

$$N = \frac{1}{2\pi i} \int_C d\{\log(J(\tau) - c)\}. \tag{7.85}$$

We begin by supposing that $J(\tau) - c$ has no zeros on the boundary ∂C of C. Then

$$N = \frac{1}{2\pi i} \left[\int_A^B + \int_B^{A'} + \int_{A'}^{D'} + \int_{D'}^D + \int_D^A \right] d\{\log(J(\tau) - c)\}. \tag{7.86}$$

The transformation T sends $A'D'$ to AD and S sends the arc AB to the arc $A'B$ and so the integrals along those parts of the contour cancel and we are left with

$$N = \frac{1}{2\pi i} \int_{D'}^{D} d\{\log(J(\tau) - c)\}. \tag{7.87}$$

Now write $z = \exp(2\pi i \tau)$ and observe that, as τ goes from D' to D, z describes a circle of radius $e^{-2\pi t}$. If we choose t so large that the circle $|z| = e^{-2\pi t}$ contains no poles or zeros of $\log\{J(\tau) - c\}$ except the simple pole at $z = 0$ (that is, $\tau - i\infty$), we see that the integral (7.87) is 1 and so Theorem 7.14 is true in that case.

Suppose now that there are zeros of $J(\tau) - c$ on the boundary ∂C of C. Then, by appealing to the transformations T and S, we may now suppose that there are similarly placed zeros on $A'D'$ and AD and also on $A'B$ and AB. If semi-circles are drawn with those zeros as centres, then we may replace the boundary ∂C by excluding those points and replacing the portions of the original lines by arcs of circles for which the new path has no zeros on its boundary. By appealing to the previous case we obtain that $N = 1$.

Finally, suppose that there are zeros within C and on the boundary, as above, and also at the vertices A, B.

We recall from Figure 3.1 that, at the point A, $\tau = \exp(2\pi i/3) = \alpha$, say, and at B, $\tau = i$. Evidently $\alpha^3 = 1$.

Consider

$$g_2(1, \alpha) = 60 \sum_{\substack{m,n=-\infty \\ m,n\neq 0,0}}^{\infty} \frac{1}{(m + n\alpha)^4}$$

$$= 60 \sum_{\substack{m,n=-\infty \\ m,n\neq 0,0}}^{\infty} \frac{1}{\alpha\{(m - n) + m\alpha\}^4}$$

$$= \frac{1}{\alpha} g_2(1, \alpha),$$

whence $g_2(1, \alpha) = 0$ and so $J(\alpha) = 0$. So $J(\tau) - c$ has a zero only at the point A where $\tau = \alpha$; that is when $c = 0$.

At the point B, $\tau = i$, we find

$$g_3(1, i) = 140 \sum_{\substack{m,n=-\infty \\ m,n\neq 0,0}}^{\infty} \frac{1}{(m + ni)^6}$$

$$= -140 \sum_{\substack{m,n=-\infty \\ m,n\neq 0,0}}^{\infty} \frac{1}{(n - mi)^6}$$

$$= -g_3(1, i).$$

So, as before, $g_3(1, 1) = 0$ and $J(i) = 1$.

Those considerations lead us to consider the cases $J(\tau) = 1$ and $J(\tau) = 0$. The details of the concluding stages of the argument are outlined in Exercise 7.7.1 and are left (with hints) to the reader.

We can use Theorem 7.14 to prove that every function $g(\tau)$ defined on the upper half-plane and invariant under the modular group can be expressed in the form $G(j(\tau))$ for some function G.

It follows from Theorem 7.14 that $j(\tau)$ determines a one-to-one map of $\Im \cup \infty$ on to $\mathbb{C} \cup \infty$, since to any point in $\mathbb{C} \cup \infty$ there corresponds a unique point c in $\Im \cup \infty$ and then, by the theorem, there exists τ such that $j(\tau) = c$. It follows that there exists an inverse map:
$j^{-1} : \mathbb{C} \cup \infty \mapsto D \cup \infty$, and if we write $G(z) = g(j^{-1}(z))$, then $G(j(\tau)) = g(j^{-1}(j(\tau))) = g(\tau)$.

We can prove further that G is a rational function if $g(\tau)$ is a meromorphic function of $q = e^{\pi i \tau}$ in the disc $|q| \leq 1$. For if g is meromorphic, then the finite singularities must be poles. The value $j(\tau) = \infty$ corresponds to $q = 0$; hence by our hypothesis concerning g, the point ∞ cannot be an essential singularity. It follows that in $\mathbb{C} \cup \infty$ the function G has no singular points other than poles. If we remove from G the principal parts of its Laurent series expansions at the singular points, we obtain a function without singular points in $\mathbb{C} \cup \infty$, that is, a constant. Since the principal parts that we removed consist of finitely many terms, it follows that G is a rational function.

We turn now to the *elliptic modular function*, $\lambda(\tau)$, defined by (7.83).

We have already seen, in Chapter 5, Theorem 5.1, that one can find τ in the upper half-plane such that, for given m, $\lambda(\tau) = m$ and we also stated that (Theorem 5.2), any two solutions τ_1, τ_2 of $\lambda(\tau) = m$ were related by

$$\tau_2 = \frac{a\tau_1 + b}{c\tau_1 + d}, \quad a, b, c, d \in \mathbb{Z}, \tag{7.88}$$

where $ad - bc = 1$, $a, d \equiv 1 \pmod 2$, and $b, c \equiv 0 \pmod 2$. That suggests that we should begin by constructing explicitly the fundamental region of the λ-group, as defined implicitly by (7.88), and to that we now turn.

Theorem 7.15 *The fundamental region of the λ-group is defined by the set of points, τ, in the upper half-plane such that*

$$-1 \leq \operatorname{Re} \tau \leq 0, \left| \tau - \frac{1}{2} \right| \geq \frac{1}{2} \quad \text{or} \quad 0 < \operatorname{Re} \tau < 1, \left| \tau + \frac{1}{2} \right| \geq \frac{1}{2}.$$

Proof We saw in Chapter 6 that the points τ_1, τ_2 are congruent with respect to the λ-group, (7.88), if they are connected by a transformation of the form

$$\begin{pmatrix} \tau_2 \\ 1 \end{pmatrix} = \begin{pmatrix} a & b \\ c & d \end{pmatrix} \begin{pmatrix} \tau_1 \\ 1 \end{pmatrix},$$

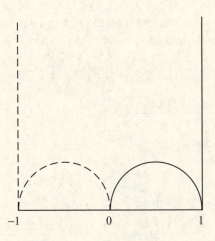

Figure 7.1 Fundamental region of the λ-group.

where $a, d \equiv 1(\mathrm{mod}\ 2)$ and $b, c \equiv 0(\mathrm{mod}\ 2)$ and $ad - bc = 1$. It is clear that every point τ' in the upper half-plane is congruent to $\tau \in H$ with respect to the λ-group, where

$$|2\tau \pm 1| > 1, \quad |\tau| \le |\tau \pm 2|. \tag{7.89}$$

Moreover, there cannot be two congruent points τ and τ' such that

$$-1 \le \mathrm{Re}\ \tau \le 0, \quad \left|\tau - \frac{1}{2}\right| \ge \frac{1}{2} \tag{7.90}$$

or

$$0 < \mathrm{Re}\ \tau < 1, \quad \left|\tau + \frac{1}{2}\right| \ge \frac{1}{2} \tag{7.91}$$

The shape of the fundamental region follows immediately

It follows from the work in Chapter 6 and from the foregoing that under the transformations of the modular group the function $\lambda(\tau)$ assumes one of the values:

$$\lambda(\tau), \quad 1 - \lambda(\tau), \quad \frac{1}{\lambda(\tau)}, \quad \frac{1}{1 - \lambda(\tau)}, \quad 1 - \frac{1}{\lambda(\tau)}, \quad \frac{\lambda(\tau)}{\lambda(\tau) - 1}. \tag{7.92}$$

We obtain the following connection between the functions $j(\tau)$ and $\lambda(\tau)$ (the result stated in the theorem is the same as (7.84) on using (7.82)).

Theorem 7.16 *The functions $j(\tau)$ and $\lambda(\tau)$ are connected by the relation*

$$j(\tau) = 64\frac{(1 - \lambda + \lambda^2)^3}{\lambda^2(1 - \lambda)^2}.$$

Proof The values of $\lambda(\tau)$ under the modular group given in (7.92) suggest that we should consider the following function of τ: where $\lambda = \lambda(\tau)$:

$$\Phi(\tau) = (\lambda+1)\left(\frac{1}{1-\lambda}+1\right)\left(\frac{\lambda-1}{\lambda}+1\right)\left(\frac{1}{\lambda}+1\right)\left(\frac{\lambda}{\lambda-1}+1\right)(1-\lambda+1)$$

$$= -\frac{(\lambda+1)^2(\lambda-2)^2(2\lambda-1)^2}{\lambda^2(1-\lambda)^2}. \tag{7.93}$$

We recall (from (7.83)) that

$$\lambda(\tau) = \frac{e_3 - e_2}{e_1 - e_2}$$

and, on substituting, we obtain

$$\Phi(\tau) = -\frac{(e_1 + e_3 - 2e_2)^2(e_2 + e_3 - 2e_1)^2(e_2 + e_1 - 2e_3)^2}{(e_1 - e_2)^2(e_2 - e_3)^2(e_1 - e_3)^2}.$$

We recall that

$$e_1 + e_2 + e_3 = 0,$$
$$e_1 e_2 e_3 = \frac{g_3}{4}$$

and

$$\Delta = 16(e_1 - e_2)^2(e_2 - e_3)^2(e_1 - e_3)^2 = g_2^3 - 27g_2^3,$$

and, after some calculations, we obtain

$$\Phi(\tau) = -\frac{3^6 g_3^2}{g_2^3 - 27g_3^2}$$

$$= 27\{1 - J(\tau)\}. \tag{7.94}$$

Whence

$$J(\tau) = 1 - \frac{\Phi(\tau)}{27}$$

$$= 1 + \frac{(\lambda+1)^2(\lambda-2)^2(2\lambda-1)^2}{27\lambda^2(1-\lambda)^2}$$

$$= \frac{4(1-\lambda+\lambda^2)^3}{27\lambda^2(1-\lambda)^2}. \tag{7.95}$$

Exercises 7.7

7.7.1 Complete the proof of Theorem 7.14 by carrying out the details of the following calculations.

Suppose that $J(\tau) = 1$ and exclude the point B from ∂C by an arc of a semi-circle, $\gamma(B)$. We take $c = 1$ and note that $J(\tau) - 1$ has a zero at B. Then

$$N = \frac{1}{2\pi i} \int_C d\{\log(J(\tau) - 1)\}$$

$$= \frac{1}{2\pi i} \left[\int_{D^1}^{D} + \int_{\gamma(B)} \right] d\{\log(J(\tau) - 1)\}.$$

Now let $\gamma'(B)$ be obtained from $\gamma(B)$ by the transformation defined by S and note that $J(-1/\tau) = J(\tau)$. Hence show that

$$\int_{\gamma(B)} d\{\log(J(\tau) - 1)\} = \frac{1}{2} \int_{\gamma(B)+\gamma'(B)} d\{\log(J(\tau) - 1)\}$$

$$= -2\pi i \nu,$$

where 2ν denotes the order of the zero at B; that is the order of the zero of $g_3(1, \tau)$.

Deduce that $N = 1 - \nu$ and conclude that $\nu = 1$, $N = 0$. Show that the function $J(\tau) - 1$ has one double zero at $\tau = i$ and that the number of zeros of $J(\tau) - 1$ within C is zero. Use similar arguments to show that $J(\tau)$ has a triple zero at the point $\tau = \exp(2\pi i/3)$ and no other zeros within C.

7.7.2 Prove the assertions in (7.89), (7.80) and (7.91).

7.7.3 Prove that, under the action of the modular group, the function $\lambda(\tau)$ assumes the values in (7.92).

7.7.4 Supply the details of the calculations leading to (7.94) and (7.95).

8

Elliptic integrals

8.1 Elliptic integrals in general

The treatment of the elliptic integrals given in this chapter (and later work on their applications, for example in Chapter 12) owes much to the books by Bowman (1961), Dutta and Debnath (1965), Lawden (1989) and Whittaker and Watson (1927). Detailed references are given as appropriate in what follows.

An integral of the type

$$\int R(x, y)\mathrm{d}x, \tag{8.1}$$

where y^2 is a cubic or quartic polynomial in x, and R denotes a rational function of x and y, is called an *elliptic integral*[1] (though, strictly speaking, it is an elliptic integral only if it cannot be integrated using elementary functions). Such integrals include the fundamental Legendre standard forms of Chapter 1, the integral

$$\int \frac{\mathrm{d}t}{(Q(t, k))^{1/2}} \tag{8.2}$$

used in Chapter 2 to define the Jacobi functions of a complex variable and the integrals:

$$\int f(u)\mathrm{d}u, \quad \int f^2(u)\,\mathrm{d}u, \tag{8.3}$$

where $f(u)$ denotes one of the twelve Jacobian elliptic functions $sn(u)$, $cn(u)$, $dn(u)$, $sc(u)$ etc. introduced in Chapter 2 (see also Chapter 5, Section 5.6).

---

[1] As we saw in Chapter 1, the name is derived from the problem of the rectification of an ellipse – see also Chapter 9 on applications in geometry.

The integrals

$$\int \frac{\mathrm{d}w}{v}, \qquad \int \frac{w\,\mathrm{d}w}{v}, \qquad \int \frac{v + v_r}{w - w_r} \frac{\mathrm{d}w}{v} \qquad (8.4)$$

(the Weierstrass normal integrals of the first, second and third kind, respectively), where $w = \wp(z)$, $v = \wp'(z)$, and v_r, w_r denote special values of those functions, are also elliptic integrals, which may be thought of as corresponding to the Legendre normal forms.

Our main aim in this chapter is to show that the general elliptic integral, (8.1), can, by means of suitable linear transformations and reduction formulae, be expressed in terms of the integrals (8.3) (or 8.4) and theta functions. In particular, we shall begin by proving that the general elliptic integral may be expressed as a finite sum of elementary integrals and the three types of integral given by Legendre's normal forms.

8.2 Reduction in terms of Legendre's three normal forms

We recall the three normal forms:

$$\int \frac{\mathrm{d}x}{\sqrt{(1 - x^2)}\sqrt{1 - k^2 x^2}}, \qquad (8.5)$$

$$\int \frac{\sqrt{(1 - k^2 x^2)}}{\sqrt{(1 - x^2)}}\mathrm{d}x, \qquad (8.6)$$

$$\int \frac{\mathrm{d}x}{(1 + nx^2)\sqrt{(1 - x^2)}\sqrt{(1 - k^2 x^2)}} \qquad (8.7)$$

of the first, second and third kinds, respectively. We shall see that if the coefficients in the polynomial y in (8.1) are real, then the reduction may be effected in such a way that the modulus k is real and $0 < k < 1$ (cf. Chapter 1).

Again we recall from Chapter 1 that if, in (8.5), (8.6) and (8.7), we put $x = \sin \phi$, we obtain:

$$F(k, \phi) = \int_o^\phi \frac{\mathrm{d}\phi}{\sqrt{(1 - k^2 \sin^2 \phi)}}, \qquad (8.8)$$

$$E(k, \phi) = \int_o^\phi \sqrt{(1 - k^2 \sin^2 \phi)}\,\mathrm{d}\phi, \qquad (8.9)$$

$$\prod(k, \varphi) = \int_o^\varphi \frac{\mathrm{d}\varphi}{(1 + n \sin^2 \varphi)\sqrt{(1 - k^2 \sin^2 \varphi)}}, \qquad (8.10)$$

respectively (cf. Chapter 1, (1.18), (1.19), (1.20) and Exercise 1.5.1).

Finally, if we write

$$sn(u) = x = \sin\varphi \qquad (8.11)$$

(see Chapter 1, Section 1.7), then we obtain:

$$F(k, \varphi) = u = sn^{-1}x, \qquad (8.12)$$

$$E(k, \varphi) = E(u) = \int_o^u dn^2(u)\,du, \qquad (8.13)$$

$$\prod(k, n, \varphi) = \int_o^u \frac{du}{1 + n\,sn^2(u)}. \qquad (8.14)$$

The definite integrals K and E defined by

$$K = F\left(k, \frac{1}{2}\pi\right) = \int_o^{\pi/2} \frac{d\varphi}{\sqrt{(1 - k^2\sin^2\varphi)}} \qquad (8.15)$$

$$= \int_o^1 \frac{dx}{\sqrt{(1 - x^2)}\sqrt{(1 - k^2x^2)}};$$

and

$$E = E\left(k, \frac{1}{2}\pi\right) = \int_o^{\pi/2} \sqrt{(1 - k^2\sin^2\phi)}\,d\varphi$$

$$= \int_o^1 \frac{\sqrt{(1 - k^2x^2)}}{\sqrt{(1 - x^2)}}\,dx \qquad (8.16)$$

$$= \int_o^K dn^2(u)\,du;$$

are called *complete* elliptic integrals, of the first and second kinds respectively.

We recall from Chapter 5, Section 5.6, that the three kinds of elliptic integral may be evaluated with the aid of theta functions.

In the later chapters and especially in Chapter 12 (on physical applications) we shall encounter significant applications of those integrals and of similar integrals involving the Weierstress functions $\wp(z)$, $\wp'(z)$. The following examples and the related exercises provide a preliminary list of basic integrals, which we shall use in the sequel and which arise in applications.

Examples 8.2

1 On putting $x = sn(u)$ we obtain

$$\int sn(u)\,du = \int \frac{x\,dx}{\sqrt{(1 - x^2)}\sqrt{(1 - k^2x^2)}}$$

$$= \int \frac{\frac{1}{2} dt}{\sqrt{(1-t)}\sqrt{1-k^2 t}} \qquad \text{(put } x^2 = t\text{)}$$
$$= k^{-1} \log\{\sqrt{(1-k^2 t)} - k\sqrt{(1-t)}\}$$
$$= k^{-1} \log\{dn\,(u) - k\,sn\,(u)\}.$$

(Although the final line involves elliptic functions, the preceding line expresses the integral in terms of elementary functions of t and so it is not, strictly speaking, an elliptic integral.)

2
$$\int sn^2 u\, du = \int \frac{1 - dn^2 u}{k^2}\, du = \frac{u - E(u)}{k^2}.$$

(Squares of the other basic elliptic functions may also be expressed in terms of $E(u)$ and it is for that reason that (8.13) is taken to be the fundamental integral of the second kind. See Chapter 5, Section 5.6 and Exercises 8.2.1 and 8.2.3, for example, below.)

3　We have

$$\frac{d}{du}\left\{\frac{sn(u)cu(u)}{dn(u)}\right\} = \frac{dn^4(u) - k'^2}{k^2 dn^2(u)}$$
$$= \frac{1}{k^2}\left(dn^2 u - \frac{k'^2}{dn^2 u}\right).$$

So, by integrating and multiplying through by k^2, we obtain

$$k'^2 \int \frac{du}{dn^2 u} = E(u) - k^2 \frac{sn\,(u)\,cn\,(u)}{dn\,(u)}.$$

4　Recall the results given in Chapter 5, Exercises 5.6; for example:

$$\int sn\,(u)\,du = \frac{1}{2}k^{-1} \log \frac{1 - k\,cd\,(u)}{1 + k\,cd\,(u)}$$

(cf. Example 1);

$$\int dn\,(u)\,du = am\,(u);$$

and

$$\int dc\,(u)\,du = \frac{1}{2}\log \frac{1 + sn\,(u)}{1 - sn\,(u)}.$$

Exercises 8.2

8.2.1 Prove, by differentiation, the equivalence of the following twelve expressions.

(Thus, for example, (b) and (l) give

$$\int dc^2u \, du = u + dn\,(u)\,sc\,(u) - \int dn^2u \, du.)$$

(a) $u - k^2 \int sn^2u \, du;$

(b) $\int dn^2u \, du;$

(c) $k'^2u + dn\,(u)\,sc\,(u) - k'^2 \int nc^2u \, du;$

(d) $dn\,(u)\,sc\,(u) - k'^2 \int sc^2u \, du;$

(e) $u + k^2 sn\,(u)\,cd\,(u) - k^2 \int cd^2u \, du;$

(f) $k'^2u - dn\,(u)\,cs\,(u) - \int ds^2u \, du;$

(g) $k'^2u + k^2 \int cn^2u \, du;$

(h) $u - dn\,(u)\,cs\,(u) - \int ns^2\,u \, du;$

(i) $k^2 sn\,(u)\,cd\,(u) + k'^2 \int nd^2u \, du;$

(j) $k'^2u + k^2 sn\,(u)\,cd\,(u) + k^2 k'^2 \int sd^2u \, du;$

(k) $-dn\,(u)\,cs\,(u) - \int cs^2u \, du;$

(l) $u + dn\,(u)\,sc\,(u) - \int dc^2u \, du.$

(See Whittaker & Watson, 1927, p. 516, Example 3.)

8.2.2 Verify the following:

(a) $\int cn\,(u)\,du = k^{-1}\sin^{-1}(k\,sn\,(u));$

(b) $\int dn\,(u)\,du = \sin^{-1}(sn\,(u))$

compare (a) and (b) with (b) and (c) of Exercise 5.6.3.

(c) $\int \frac{du}{sn\,(u)} = \log\frac{sn(u)}{dn(u)+cn(u)} = \log\frac{sn(u/2)}{cn(u/2)\,dn(u/2)};$

(d) $\int sn^{-1}x \, dx = x\,sn^{-1}x - k^{-1}\log\{\sqrt{(1-k^2x^2)} - k\sqrt{(1-x^2)}\};$

(e) $\int \frac{du}{sn^2u} = u - E(u) - \frac{cn(u)dn(u)}{sn(u)}$

(compare with (h) of question 8.2.1);

(f) $3\int \frac{du}{sn^4 u}$

$$= (2+k^2)u - 2(1+k^2)E(u)$$
$$- 2(1+k^2)\frac{cn\,(u)\,dn\,(u)}{sn\,(u)} - \frac{cn\,(u)\,dn\,(u)}{sn^3 u};$$

(g) $\int_o^K \frac{du}{1+cn\,(u)} = K - E + k';$

(h) $\int_o^K \frac{cn\,(u)}{1 - k\,sn\,(u)}\,du = \frac{1+k-k'}{kk'}.$ See Bowman (1961), p.18.

8.2.3   Compose a table of integrals of the following differentials (in Glaisher's notatation where appropriate).

(a) $sn\,(u)\,du$;   (b) $cn\,(u)\,du$;   (c) $dn\,(u)\,du$;

(d) $ns\,(u)\,du$;   (e) $nc\,(u)\,du$;   (f) $nd\,(u)\,du$;

(g) $sn^2 u\,du$;   (h) $cn^2 u\,du$;   (i) $dn^2 u\,du$;

(j) $\frac{1}{sn^2 u}du$;   (k) $\frac{1}{cn^2 u}du$;   (l) $\frac{1}{dn^2 u}du$;

(m) $sn^4 u\,du$;   (n) $cn^4 u\,du$;   (o) $dn^4 u\,du$;

(p) $cn^2 u\,dn^2 u\,du$;   (q) $sn^2 u\,cn^2 u\,du$;   (r) $sn^2 u\,dn^2 u\,du$.

See Bowman (1961), Ex. 4, p.18.

8.2.4   Let $V = V(u) = 1 + n\,sn^2 u$. Show that

$$\frac{d}{du}\left\{ \frac{sn\,(u)\,cn\,(u)\,dn\,(u)}{V^{m+1}} \right\} = \frac{A}{V^m} + \frac{B}{V^{m-1}} + \frac{C}{V^{m-2}} + \frac{D}{V^{m-3}},$$

where A, B, C and D are constants.

Deduce that, if m is a positive integer, then the integral

$$\int \frac{du}{(1 + n\,sn^2 u)^m}$$

is expressible in terms of u, $E(u)$ and $\prod(u)$, where

$$\prod(u) = \int \frac{du}{(1 + n\,sn^2 u)}.$$

See Bowman (1961), Ex. 3, p.19.

8.2.5   Show that

$$\frac{d^2}{du^2} sn^n u = n(n-1)sn^{n-2}u - n^2(1+k^2)sn^n u + n(n+1)k^2 sn^{n+2}u.$$

Obtain similar formulae in the cases when $sn^n u$ is replaced by $f^n(u)$ where $f(u)$ is one of the twelve Jacobian elliptic functions introduced above. (Compare with the reduction formulae for J_n given in Exercise 8.3.14) (See Whittaker & Watson, 1927, Example 4, p.516.)

8.2.6   This and the following exercises provide a list of integrals which may be reduced to integrals of the first kind that are used in applications and which the reader is advised to work through. (See Lawden, 1989 pp. 52ff, for details and other examples.)

Recall that

$$\int_o^x \{(1 - t^2)(1 - k^2 t^2)\}^{-1/2} dt = u = sn^{-1}(x, k), \quad 0 \leq x \leq 1,$$

is an expression of the fact that

$$\frac{dx}{du} = cn\,(u) \cdot dn\,(u) = \sqrt{(1 - x^2)(1 - k^2 x^2)}.$$

The substitution $t = \sin\theta$ gives the standard form

$$F(\phi, k) = sn^{-1}(\sin\phi, k) = \int_o^\phi (1 - k^2 \sin^2\theta)^{-1/2} d\theta.$$

where $x = \sin\phi$.

Let $u = cn^{-1} x$; so that $x = cn\,(u)$. Given that

$$\frac{dx}{du} = -sn\,(u)\,dn\,(u) = -\sqrt{(1 - x^2)(k'^2 + k^2 x^2)},$$

show that

$$cn^{-1}(x, k) = \int_x^1 \frac{dt}{\sqrt{(1 - t^2)(k'^2 + k^2 t^2)}}, \quad 0 \leq x \leq 1.$$

By means of the substitutions $t = s/b$, $k = b/\sqrt{a^2 + b^2}$, deduce that, for $0 \leq x \leq b$,

$$\int_x^b \frac{dt}{\sqrt{(a^2 + t^2)(b^2 - t^2)}} = \frac{1}{\sqrt{(a^2 + b^2)}} cn^{-1}\left[\frac{x}{b}, \frac{b}{\sqrt{a^2 + b^2}}\right].$$

Verify that $cn^{-1}(x) = sn^{-1}(1 - x^2)^{1/2}$ and hence obtain

$$\frac{1}{\sqrt{(a^2 + b^2)}} sn^{-1}\left[\frac{\sqrt{(b^2 - x^2)}}{b}, \frac{b}{\sqrt{(a^2 + b^2)}}\right] = \int_x^b \frac{dt}{\sqrt{(a^2 + t^2)(b^2 - t^2)}}.$$

The foregoing suggests that the substitution $s = \sqrt{(b^2 - t^2)}/b$ yields

$$\int_x^b \frac{dt}{\sqrt{(a^2 + t^2)(b^2 - t^2)}} = \frac{1}{\sqrt{(a^2 + b^2)}} \int_o^{\sqrt{(1 - x^2/b^2)}} \frac{ds}{\sqrt{(1 - s^2)(1 - k^2 s^2)}},$$

which is the Legendre normal form (8.5) of the integral of the first kind.

8.2.7 Obtain the following elliptic integrals in terms of the appropriate inverse functions, by using the changes of variable indicated. (See Question 8.2.6 for the cases leading to sn^{-1} and cn^{-1}. For further details see Lawden (1989) Chapter 3. We use Glaisher's notation throughout; e.g. $sd\,(u) = \dfrac{sn(u)}{dn(u)}$ etc.)

(a) $\int_x^b \{(a^2 - t^2)(b^2 - t^2)\}^{-1/2} dt = \frac{1}{a}cd^{-1}\left[\frac{x}{b}, \frac{b}{a}\right], \quad 0 \le x \le b < a.$

$\left(\text{Use } s^2 = \frac{a^2(b^2 - t^2)}{b^2(a^2 - t^2)}.\right)$

(b) $\int_0^x \{(a^2 + t^2)(b^2 - t^2)\}^{-1/2} dt$

$= \frac{1}{\sqrt{a^2 + b^2}} sd^{-1}\left[\frac{\sqrt{a^2 + b^2}x}{ab}, \frac{b}{\sqrt{a^2 + b^2}}\right], \quad 0 \le x \le b.$

$\left(\text{Use } s^2 = \frac{(a^2 + b^2)t^2}{b^2(a^2 + t^2)}.\right)$

(c) $\int_a^x \{(t^2 - a^2)(t^2 - b^2)\}^{-1/2} dt = \frac{1}{a}dc^{-1}\left(\frac{x}{a}, \frac{b}{a}\right), \quad b < a \le x.$

$\left(\text{Use } s^2 = \frac{t^2 - a^2}{t^2 - b^2}.\right)$

(d) $\int_x^\infty \{(t^2 - a^2)(t^2 - b^2)\}^{-1/2} dt = \frac{1}{a}ns^{-1}\left(\frac{x}{a}, \frac{b}{a}\right), \quad b < a \le x.$

$\left(\text{Use } s = \frac{a}{t}.\right)$

(e) $\int_b^x \{(a^2 - t^2)(t^2 - b^2)\}^{-1/2} dt = \frac{1}{a}nd^{-1}\left[\frac{x}{b}, \frac{\sqrt{a^2 - b^2}}{a}\right], \quad b \le x \le a.$

$\left(\text{Use } s^2 = \frac{a^2(t^2 - b^2)}{(a^2 - b^2)t^2}\right).$

(f) $\int_x^a \{(a^2 - t^2)(t^2 - b^2)\}^{-1/2} dt = \frac{1}{a}dn^{-1}\left[\frac{x}{a}, \frac{\sqrt{a^2 - b^2}}{a}\right], \quad b \le x \le a.$

$\left(\text{Use } s^2 = \frac{a^2 - t^2}{a^2 - b^2}.\right)$

(g) $\int_a^x \{(t^2 - a^2)(t^2 + b^2)\}^{-1/2} dt = \frac{1}{\sqrt{a^2 + b^2}} nc^{-1}\left[\frac{x}{a}, \frac{b}{\sqrt{a^2 + b^2}}\right], \quad a \le x.$

$\left(\text{Use } s^2 = 1 - \frac{a^2}{t^2}.\right)$

(h) $\int_x^\infty \{(t^2 - a^2)(t^2 + b^2)\}^{-1/2} dt$

$= \frac{1}{\sqrt{a^2 + b^2}} ds^{-1}\left[\frac{x}{\sqrt{a^2 + b^2}}, \frac{b}{\sqrt{a^2 + b^2}}\right], \quad a \le x.$

$\left(\text{Use } s^2 = \frac{a^2 + b^2}{t^2 + b^2}.\right)$

(i) $\int_0^x \{(t^2 + a^2)(t^2 + b^2)\}^{-1/2} dt$

$= \frac{1}{a}sc^{-1}\left[\frac{x}{b}, \frac{\sqrt{a^2 - b^2}}{a}\right], \quad 0 < b < a, \quad 0 \le x.$

$\left(\text{Use } s^2 = \frac{t^2}{t^2 + a^2}.\right)$

(j) $\int_x^\infty \{(t^2 + a^2)(t^2 + b^2)\}^{-1/2} dt$

$= \frac{1}{a} cs^{-1} \left[\frac{x}{a}, \frac{\sqrt{a^2 - b^2}}{a} \right] \quad 0 < b < a, \quad 0 \le x.$

$\left(\text{Use } s^2 = \frac{a^2}{t^2 + a^2}. \right)$

8.2.8   This question gives a list of integrals of the second kind reducible to integrals of the form $\int f^2(u) du$, where $f(u)$ denotes one of the elliptic functions, similar to the list in Question 8.2.7. See Chapter 5, Section 5.6, and Exercise 8.2.1 above. See Lawden (1989), pp.60–63 for further details.

As an example, consider 8.2.7(b), and note that

$$\{(a^2 + t^2)(b^2 - t^2)\}^{-1/2} = \frac{1}{\sqrt{a^2 + b^2}} \frac{d}{dt} sd^{-1} \left[\frac{\sqrt{a^2 + b^2}}{ab} t, \frac{b}{\sqrt{a^2 + b^2}} \right],$$

whence

$$\int t^2 \{(a^2 + t^2)(b^2 - t^2)\}^{-1/2} dt$$

$$= (a^2 + b^2)^{-1/2} \int t^2 \frac{d}{dt} sd^{-1} \left[\frac{\sqrt{a^2 + b^2}}{ab} t, \frac{b}{\sqrt{(a^2 + b^2)}} \right] dt$$

$$= a^2 b^2 (a^2 + b^2)^{-3/2} \int sd^2 u \, du,$$

where $t = ab(sd(u))/\sqrt{a^2 + b^2}$.

Now obtain the following integrals of the second kind, using the substitutions indicated. Suppose that $a > b$ throughout except possibly in the cases of (g) and (h). In each case the modulus, k, is given. (For further details see Lawden, 1989, Chapter 3, Section 3.4d, from where these exercises are taken.)

(a) $\int t^2 \{(a^2 - t^2)(b^2 - t^2)\}^{-1/2} dt = \frac{b^2}{a} \int sn^2 u \, du,$
    $t = b \, sn \, (u), \quad k = \frac{b}{a};$

(b) $\int t^2 \{(a^2 - t^2)(b^2 - t^2)\}^{-1/2} dt =$
    $-\frac{b^2}{a} \int cd^2 u \, du, \quad t = b \, cd \, (u), \quad k = \frac{b}{a};$

(c) $\int t^2 \{(t^2 - a^2)(t^2 - b^2)\}^{-1/2} dt = a \int dc^2 u \, du, \quad t = a \, dc \, (u),$
    $k = \frac{b}{a};$

(d) $\int t^2 \{(t^2 - a^2)(t^2 - b^2)\}^{-1/2} dt =$
    $-a \int ns^2 u \, du, \quad t = a \, ns \, (u), \quad k = \frac{b}{a};$

(e) $\int t^2 \{(a^2 - t^2)(t^2 - b^2)\}^{-1/2} dt = \frac{b^2}{a} \int nd^2 u \, du, \quad t = b \, nd \, (u),$
$k' = \frac{b}{a};$

(f) $\int t^2 \{(a^2 - t^2)(t^2 - b^2)\}^{-1/2} dt = -a \int dn^2 u \, du, \quad t = a \, dn \, (u),$
$k' = \frac{b}{a};$

(g) $\int t^2 \{(t^2 - a^2)(t^2 + b^2)\}^{-1/2} dt = \frac{a^2}{\sqrt{a^2 + b^2}} \int nc^2 u \, du, \quad t =$
$a \, nc \, (u),$
$k = \frac{b}{\sqrt{a^2 + b^2}};$

(h) $\int t^2 \{(t^2 - a^2)(t^2 + b^2)\}^{-1/2} dt = -\sqrt{a^2 + b^2} \int ds^2 u \, du,$
$t = \sqrt{a^2 + b^2} \, ds \, (u), \quad k = \frac{b}{\sqrt{a^2 + b^2}};$

(i) $\int t^2 \{(t^2 + a^2)(t^2 + b^2)\}^{-1/2} dt = \frac{b^2}{a} \int sc^2 u \, du, \quad t = b \, sc \, (u),$
$k' = \frac{b}{a};$

(j) $\int t^2 \{(t^2 + a^2)(t^2 + b^2)\}^{-1/2} dt = -a \int cs^2 u \, du, \quad t = a \, cs \, (u),$
$k' = \frac{b}{a}.$

8.3 Reduction to the standard form

We have seen in Section 8.2 how integrals of the first and second kinds may be reduced to particular integrals of the forms (8.5), (8.6) and (8.7), and (8.8), (8.9) and (8.10). There remains the problem of reducing the general elliptic integral (8.1) to one of those standard forms. We follow the discussion given in Whittaker & Watson (1927), pp.512–523, and see also Bowman (1961), Chapters 2 and 9 and Lawden (1989), Chapter 3.

Since $R(x, y)$ is a rational function of x and y we may write

$$R(x, y) = P(x, y)/Q(x, y), \tag{8.17}$$

where P and Q denote polynomials in two variables and so, by obvious manipulations, the reasons for which will immediately emerge,

$$R(x, y) = \frac{y}{y} \cdot \frac{P(x, y)}{Q(x, y)} \cdot \frac{Q(x, -y)}{Q(x, -y)}.$$

Now $Q(x, y) \cdot Q(x, -y)$ is a rational function of x and y^2 and so it is a rational function of x. So we may multiply out the numerator, replacing y^2, wherever it occurs, by the polynomial y^2 in x, to obtain a polynomial in x and y, which is linear in y. So, finally,

$$R(x, y) = \{R_1(x) + y R_2(x)\}/y, \tag{8.18}$$

where $R_1(x)$ and $R_2(x)$ are rational functions of x.

The integral in (8.1) accordingly reduces to an integral involving a rational function of x only, which can be evaluated using elementary functions, and an integral of the form

$$\int R_1(x) y^{-1} \mathrm{d}x.$$

We shall suppose that y is a quartic in x and that the coefficients in y^2 are real[2]. Then the roots of the equation $y^2 = 0$ as a quartic in x will be: (i) all real; (ii) two real and a pair of conjugate complex numbers; (iii) two pairs of conjugate complex numbers. It follows that the quartic y^2 can be expressed as product of two quadratic factors with real coefficients:

$$y^2 = X_1 X_2 = (a_1 x^2 + b_1 x + c_1)(a_2 x^2 + b_2 x + c_2), \tag{8.19}$$

where $a_1, b_1, c_1, a_2, b_2, c_2$ are all real (see footnote 2). We try to express the right hand side of (8.19) in the form

$$\{A_1(x - \alpha)^2 + B_1(x - \beta)^2\}\{A_2(x - \alpha)^2 + B_2(x - \beta)^2\} \tag{8.20}$$

with a view to obtaining something resembling the Legendre forms (8.5), (8.6) and (8.7) by means of a substitution of the form

$$t = (x - \alpha)/(x - \beta); \tag{8.21}$$

so that

$$\frac{\mathrm{d}x}{y} = \pm \frac{(\alpha - \beta)^{-1} \mathrm{d}t}{\{(A_1 t^2 + B_1)(A_2 t^2 + B_2)\}^{1/2}}. \tag{8.22}$$

To that end, consider the polynomial $X_1 - \lambda X_2$. We look to find values of λ that make a perfect square in x, from which we shall derive (8.20). Now $X_1 - \lambda X_2$ is a perfect square if

$$(a_1 - \lambda a_2)(c_1 - \lambda c_2) - (b_1 - \lambda b_2)^2 = 0. \tag{8.23}$$

Let the roots of that quadratic (in λ) be λ_1 and λ_2; so there exist α and β such that

$$X_1 - \lambda_1 X_2 = (a_1 - \lambda_1 a_2)(x - \alpha)^2, \quad X_1 - \lambda_2 X_2 = (a_1 - \lambda_2 a_2)(x - \beta)^2,$$

so obtaining (8.20).

---

[2] The case in which y^2 is a cubic in x can be included in this discussion by taking one root to be infinite, that is by regarding a cubic as a quartic in which the coefficient of x^4 is 0. The assumption that the coefficients in y^2 are real is not used in an essential way until the paragraph immediately following (8.23), below. The changes in what follows up to that point if the coefficients are not real will be obvious.

The discussion thus far has not used the fact that the quartic has real coefficients in an essential way. If it has real coefficients, then λ_1 and λ_2 are real and distinct, since

$$(a_1 - \lambda a_2)(c_1 - \lambda c_2) - (b_1 - \lambda b_2)^2 \qquad (8.24)$$

is positive when $\lambda = 0$ and negative when $\lambda = a_1/a_2$ (unless $a_1 b_2 = a_2 b_1$, in which case $X_1 = a_1(x - \alpha)^2 + B_1$, $X_2 = a_2(x - \alpha)^2 + B_2$).

If X_1 and X_2 have real factors $\left(x - \xi_1^{(1)}\right)\left(x - \xi_2^{(1)}\right)$ and $\left(x - \xi_1^{(2)}\right)\left(x - \xi_2^{(2)}\right)$, say, then λ_1 and λ_2 are real if

$$\left(\xi_1^{(1)} - \xi_1^{(2)}\right)\left(\xi_2^{(1)} - \xi_1^{(2)}\right)\left(\xi_1^{(1)} - \xi_2^{(2)}\right)\left(\xi_2^{(1)} - \xi_2^{(2)}\right) > 0,$$

which holds if the zeros of X_1 and those of X_2 do not interlace, which can always be arranged.

Now let α, β be defined as in (8.24) and (as already forecast in (8.21)) write:

$$t = (x - \alpha)/(x - \beta), \qquad (8.21')$$

$$\frac{dx}{y} = \pm \frac{(\alpha - \beta)^{-1}}{\{(A_1 t^2 + B_1)(A_2 t^2 + B_2)\}^{1/2}} dt. \qquad (8.22')$$

From (8.18) that substitution will yield

$$R_1(x) = \pm(\alpha - \beta) R_3(t),$$

where $R_3(t)$ is a rational function of t, and so

$$\int \frac{R_1(x)}{y} dx = \int \frac{R_3(t) dt}{\{(A_1 t^2 + B_1)(A_2 t^2 + B_2)\}^{1/2}}.$$

By considering odd and even functions of t, we have

$$R_3(t) + R_3(-t) = 2R_4(t^2),$$

$$R_3(t) - R_3(-t) = 2t\, R_5(t^2),$$

say, where R_4 and R_5 are rational functions of t^2.
Whence

$$R_3(t) = R_4(t^2) + t\, R_5(t^2).$$

Now the integral

$$\int \{(A_1 t^2 + B_1)(A_2 t^2 + B_2)\}^{-1/2} t\, R_5(t^2) dt$$

can be evaluated by writing $u = t^2$; so, on putting $R_4(t^2)$ into partial fractions, the problem of evaluating $\int R(x, y)\mathrm{d}x$ is reduced to the evaluation of the integrals

$$I_{2m} = \int t^{2m}\{(A_1 t^2 + B_1)(A_2 t^2 + B_2)\}^{-1/2}\mathrm{d}t, \qquad (8.25)$$

where m is an integer; and

$$J_m = \int (1 + Nt^2)^{-m}\{(A_1 t^2 + B_1)(A_2 t^2 + B_2)\}^{-1/2}\mathrm{d}t, \qquad (8.26)$$

where m is a positive integer and $N \neq 0$. We conclude our discussion by showing that the evaluation of I_{2m} reduces to elliptic integrals of the first and second kinds, whilst that of J_m involves elliptic integrals of the third kind.

Write

$$C = C(t) = \{(A_1 t^2 + B_1)(A_2 t^2 + B_2)\}^{1/2}.$$

Then

$$\frac{\mathrm{d}}{\mathrm{d}t}\{t^{2m-1}C\} = (2m - 1)t^{2m-2}(A_1 t^2 + B_1)(A_2 t^2 + B_2)C^{-1}$$

$$+ t^{2m}\{2A_1 A_2 t^2 + A_1 B_2 + A_2 B_1\}C^{-1}$$

and so, on integration (that is, using integration by parts) we obtain a reduction formula connecting I_{2m+2}, I_{2m} and I_{2m-2} and so, by repeated application, we can express I_{2m} in terms of I_0 and I_2. The integral I_0 is an elliptic integral of the first kind. (For example, if A_1, B_1, A_2, B_2 are all positive and $A_2 B_1 > A_1 B_2$, put

$$A_1^{1/2}t = B_1^{1/2}cs(u, k), \quad k'^2 = (A_1 B_2)/(A_2 B_1), \qquad (8.27)$$

and similarly for the other possibilities; (see Exercises 8.3 below and Exercises 8.2.)

So it remains to evaluate

$$I_2 = \int t^2\{(A_1 t^2 + B_1)(A_2 t^2 + B_2)\}^{-1/2}\mathrm{d}t. \qquad (8.28)$$

We use the substitution (8.27), and similar ones depending on the signs of A_1, B_1, A_2, B_2, to express I_2 in terms of integrals of the squares of the Jacobian elliptic functions (which can be expressed in terms of $E(u)$, as in Chapter 5 and Exercises 8.2. The details are given in Exercises 8.2, above, which, as already observed, will also serve as a reference for subsequent applications.

Finally, we have to consider J_m, the integral of the third kind.

Again, we use variations of the substitution (8.27) to reduce the integral to the form:

$$\int \frac{\alpha + \beta \, sn^2 u}{1 + v \, sn^2 u} du = \alpha u + (\beta - \alpha v) \int \frac{sn^2 u}{1 + v \, sn^2 u} du, \tag{8.29}$$

where α, β and v are constants. If $v = 0, -1, \infty$ or $-k^2$ the integral can be expressed in terms of integrals of the first or second kinds. For other values of v, determine a by $v = -k^2 sn^2 a$ and then the fundamental integral of the third kind is

$$\prod (u, a) = \int_0^u \frac{k^2 sn\,(a)\,cn\,(a)\,dn\,(a)\,sn^2 u}{1 - k^2 sn^2 a \; sn^2 u} du. \tag{8.30}$$

The integral $\Pi(u, a)$ is expressible in terms of the Jacobi theta functions and zeta functions as follows (see Chapter 5).

The integrand may be written

$$\frac{1}{2} k^2 sn\,(u)\,sn\,(a)\{sn(u + a) + sn(u - a)\} \tag{8.31}$$

(on using the addition theorem

$$sn(u + v) + sn(u - v) = \frac{2sn\,(u)\,cn\,(v)\,dn\,(v)}{1 - k^2 sn^2 u \; sn^2 v}),$$

and in its turn (8.31) may be written

$$\frac{1}{2}\{Z(u - a) - Z(u + a) + 2Z(a)\},$$

by the addition theorem

$$Z(u) + Z(v) - Z(u + v) = k^2 sn\,(u)\,sn\,(v)\,sn(u + v)$$

for the Jacobi zeta function. Since $Z(u) = \Theta'(u)/\Theta(u)$, we obtain

$$\prod (u, a) = \frac{1}{2} \log \frac{\Theta(u - a)}{\Theta(u + a)} + u Z(a), \tag{8.32}$$

a result which shows that $\Pi(u, a)$ is many-valued with logarithmic singularities at the zeros of $\Theta(u \pm a)$.

In practical calculations, a more useful form is

$$\Lambda(u, \alpha, k) = \int_0^u \frac{du}{1 - \alpha^2 sn^2 u} \tag{8.33}$$

(see Lawden, 1989, p.69 and Exercises 8.3).

8.3.1 Another way of obtaining the reduction of the general integral

The approach given here is based on that of Dutta and Debnath (1965), Chapter 9. The method used above relies on properties of the polynomial y^2 and in particular included the case of a cubic polynomial as a special case of a quartic, in which one root is infinite (see footnote 2). The modification of that approach adopted here deals with the quartic and cubic cases separately and has some affinity with the methods used for the integrals in Exercises 8.2.

We return to (8.18) and recall that it suffices to consider integrals of the form

$$\int R_1(x) y^{-1} dx, \tag{8.34}$$

where $R_1(x)$ denotes a rational function of x. We consider the cases of a quartic and of a cubic polynomial separately.

Case 1: y is a quartic polynomial

Suppose that

$$y^2 = (x - \alpha)(x - \beta)(x - \gamma)(x - \delta), \tag{8.35}$$

where α, β, γ, δ are distinct.

Write

$$z^2 = \frac{(\beta - \delta)(x - \alpha)}{(\alpha - \delta)(x - \beta)} \tag{8.36}$$

and then

$$x = \frac{\alpha + \beta}{2} + \frac{\beta - \alpha}{2} \left\{ 1 + \frac{2(\beta - \delta)}{(\alpha - \delta)z^2 - (\beta - \delta)} \right\},$$

and

$$2z\,dz = \frac{(\beta - \delta)(\alpha - \beta)}{\alpha - \delta} \frac{dx}{(x - \beta)^2}.$$

We find that

$$x - \alpha = \frac{(\beta - \alpha)(\alpha - \delta)z^2}{(\alpha - \delta)z^2 - (\beta - \delta)}$$

and by obtaining similar expressions for $x - \beta$, $x - \gamma$ and $x - \delta$ and multiplying out, one finds that

$$y^2 = \frac{(\beta - \alpha)^2(\beta - \delta)^2(\alpha - \delta)^2(\alpha - \gamma)(\beta - \delta)z^2(1 - z^2)(1 - k^2 z^2)}{((\alpha - \delta)z^2 - (\beta - \delta))^4}, \tag{8.37}$$

where

$$k^2 = \frac{(\alpha - \delta)(\beta - \gamma)}{(\beta - \delta)(\alpha - \gamma)}. \tag{8.38}$$

Finally,

$$\sqrt{(\alpha - \gamma)(\beta - \delta)} \cdot \frac{dx}{y} = \frac{2dz}{\sqrt{(1 - z^2)(1 - k^2 z^2)}}. \tag{8.39}$$

We shall return to (8.34), but for the present we turn to the case when y^2 is a cubic.

Case 2: y is a cubic polynomial

Suppose that

$$y^2 = (x - \alpha)(x - \beta)(x - \gamma), \tag{8.40}$$

where α, β, γ are distinct.

Let

$$z^2 = \frac{x - \alpha}{x - \beta} = 1 - \frac{\alpha - \beta}{x - \beta}. \tag{8.41}$$

As before, we find that

$$(x - \alpha) = \frac{(\alpha - \beta)z^2}{(1 - z^2)},$$

and there are similar expressions for $(x - \beta)$, $(x - \gamma)$. On multiplying out, as before, we obtain

$$y^2 = \frac{(\alpha - \beta)z^2(\alpha - \beta)(\alpha - \gamma)(1 - k^2 z^2)}{(1 - z^2)^3}, \tag{8.42}$$

where

$$k^2 = \frac{\beta - \gamma}{\alpha - \gamma}. \tag{8.43}$$

Again,

$$2zdz = \frac{(\alpha - \beta)}{(x - \beta)^2}dx,$$

and so

$$\sqrt{\alpha - \gamma}\frac{dx}{y} = \frac{2dz}{\sqrt{(1 - z^2)(1 - k^2 z^2)}}. \tag{8.44}$$

So in both Case 1 and Case 2 we see that

$$\int R_1(x)\frac{dx}{y} = \int \frac{Q(z^2)dz}{\sqrt{(1-z^2)(1-k^2z^2)}}, \qquad (8.45)$$

where $Q(z^2)$ is a rational function of z^2 and k^2 is a constant, not equal to 0 or 1.

The function $Q(z^2)$ may be expressed in partial fractions in the form

$$Q(z^2) = \sum_r a_r z^{2r} + \sum_{p,n} \frac{b_{p,n}}{(1+a_p z^2)^n},$$

and so

$$\int \frac{Q(z^2)dz}{\sqrt{(1-z^2)(1-k^2z^2)}} = \sum_r a_r \int \frac{z^{2r} dz}{\sqrt{(1-z^2)(1-k^2z^2)}}$$

$$+ \sum_{p,n} b_{p,n} \int \frac{dz}{(1+a_p z^2)^n \sqrt{(1-z^2)(1-k^2z^2)}}. \qquad (8.46)$$

As before, those integrals may be reduced to sums of integrals of the forms (8.5), (8.6) and (8.7).

Exercises 8.3

(Questions 8.3.1–8.3.6 are based on Bowman (1961), Chapter IX; questions 8.3.7, 8.3.8 and 8.3.9 on Dutta & Debnath (1965), Chapter 9; and questions 8.3.10–8.3.12 on Whittaker & Watson (1927), Chapter XXII. The remaining exercises are based on Whittaker & Watson (1927) and references are given in each question. Further examples may be found in those books.)

8.3.1 Show that, if

$$w = \int_a^t \frac{dt}{\{(t-a)(b-t)(c-t)(d-t)\}^{1/2}},$$

where a, b, c and d are real, then

$$w = \frac{2}{\{(c-a)(d-b)\}^{1/2}} sn^{-1} \left\{ \frac{(d-b)(t-a)}{(b-a)(d-t)} \right\}^{1/2},$$

where the modulus, k, of the Jacobi function is given by

$$k^2 = \frac{(b-a)(d-c)}{(d-b)(c-a)}.$$

8.3.2 Given that

$$u = \int_1^x \frac{dx}{(5x^2 - 4x - 1)^{1/2}(12x^2 - 4x - 1)^{1/2}},$$

show that

$$x = \frac{dn\,(4u)}{3cn\,(4u) - 2dn\,(4u)},$$

where the modulus, k, of the Jacobi function is $k = 3/4$.
(Hint: put $x = 1/y$.)

8.3.3 Let $T = 1 + 2t^2 \cos 2\alpha + t^4$. Show that

$$\int_0^x \frac{dt}{\sqrt{T}} = \int_{1/x}^\infty \frac{dt}{\sqrt{T}} = \frac{1}{2} sn^{-1} \frac{2x}{1+x^2},$$

where the modulus is $k = \sin \alpha$.
(Hint: write $t' = 1/t$ to show that the integrals are equal and then put $t = \tan \theta$ and then $y = \sin 2\theta$ in the first integral.)

8.3.4 Given that $0 < x < a$, show that

$$\int_x^a \frac{dt}{(a^4 - t^4)^{1/2}} = \int_a^{a^2/x} \frac{dt}{(t^4 - a^4)^{1/2}} = \frac{1}{a\sqrt{2}} cn^{-1} \left(\frac{x}{a}, \frac{1}{\sqrt{2}} \right).$$

8.3.5 Show that

$$\int_1^x \frac{dt}{((t-1)(t^2+1))^{1/2}} = 2^{-1/4} sn^{-1} \left(\frac{2^{5/4}(x-1)^{1/2}}{x - 1 + \sqrt{2}}, \sin \left(\frac{\pi}{8} \right) \right).$$

8.3.6 By putting $x^2 = y$, $y = 1/t$, show that

$$\int (a + bx^2 + cx^4 + dx^6)^{-1/2} dx = -\frac{1}{2} \int (at^3 + bt^2 + ct + d)^{-1/2} dt$$

and then evaluate the integral (cf. Section 8.4).

8.3.7 By using the substitution $z = a \operatorname{cosec} \phi$, prove that, if $a < z < \infty$, then

$$\int_z^\infty \frac{dz}{\{(z^2 - a^2)(z^2 - b^2)\}^{1/2}} = \frac{1}{a} sn^{-1} \left(\frac{a}{z}, \frac{b}{a} \right).$$

8.3.8 Show that, if $0 < z < b < a$, then

$$\int_z^b \frac{dz}{\{(a^2 + z^2)(b^2 - z^2)\}^{1/2}} = \frac{1}{(a^2 + b^2)^{1/2}} cn^{-1} \left(\frac{z}{b}, \frac{b}{(a^2 + b^2)^{1/2}} \right),$$

$$\int_b^z \frac{dz}{\{(a^2 + z^2)(z^2 - b^2)\}^{1/2}} = \frac{1}{(a^2 + b^2)^{1/2}} cn^{-1} \left(\frac{b}{z}, \frac{a}{(a^2 + b^2)^{1/2}} \right).$$

8.3.9 Show that, if $0 < b < z < a$, then

$$\int_z^a \frac{dz}{\{(a^2 - z^2)(z^2 - b^2)\}^{1/2}} = \frac{1}{a} dn^{-1}\left(\frac{z}{a}, \frac{(a^2 - b^2)^{1/2}}{a}\right).$$

8.3.10 Express

$$\int_0^2 \{(2x - x^2)(4x^2 + 9)\}^{-1/2} dx$$

in terms of a complete elliptic integral of the first kind with a real modulus.

8.3.11 Given that

$$u = \int_x^\infty \{(t + 1)(t^2 + t + 1)\}^{-1/2} dt,$$

express x in terms of Jacobian elliptic functions with a real modulus.

8.3.12 Given that

$$u = \int_0^x (1 + t^2 - 2t^4)^{-1/2} dt,$$

express x in terms of u by means of either Jacobian or Weierstrassian elliptic functions. (See Section 8.4.)

8.3.13 Show that

$$\int \{(x^2 - a)(x^2 - b)\}^{-1/4} dx = -\frac{1}{2}\log\frac{\sigma(z - z_0)}{\sigma(z + z_0)} + \frac{i}{2}\log\frac{\sigma(z - iz_0)}{\sigma(z + iz_0)},$$

where σ denotes the Weierstrass σ-function and where, in the notation of the Weierstrass function, $\wp$,

$$x^2 = a + \frac{1}{6}\frac{1}{\wp^2(z) - \wp^2(z_0)}, \quad g_2 = \frac{2b}{3a(a - b)}, \quad g_3 = 0, \quad \wp^2(z_0) = \frac{1}{6(a - b)}.$$

(See Section 8.4.). See Whittaker & Watson (1927), Chapter XX, p.461.

8.3.14 (See Whittaker & Watson, 1927, p. 515).

By differentiating expressions of the form

$$t^{2m-1}\{(A_1t^2 + B_1)(A_2t^2 + B_2)\}^{\frac{1}{2}},$$
$$t(1 + Nt^2)^{1-m}\{(A_1t^2 + B_1)(A_2t^2 + B_2)\}^{\frac{1}{2}},$$

obtain reduction formulae for integrals of the form

$$\int t^{2m}\{(A_1t^2 + B_1)(A_2t^2 + B_2)\}^{-\frac{1}{2}} dt$$

and

$$\int (1 + Nt^2)^{-m}\{(A_1t^2 + B_1)(A_2t^2 + B_2)\}^{-\frac{1}{2}}dt,$$

where m is an integer in the former case and a positive integer in the latter, $N \neq 0$. Use those reduction formulae to obtain the three canonical forms

(i) $\int\{(A_1t^2 + B_1)(A_2t^2 + B_2)\}^{-\frac{1}{2}}dt,$

(ii) $\int t^2\{(A_1t^2 + B_1)(A_2t^2 + B_2)\}^{-\frac{1}{2}}dt,$

(iii) $\int(1 + Nt^2)^{-1}\{(A_1t^2 + B_1)(A_2t^2 + B_2)\}^{-\frac{1}{2}}dt$

8.3.15 What is the connection between the formulae in 8.3.14 and the formulae obtained for $f^n(u)$ in Exercise 8.2.5? See Whittaker & Watson (1927), p. 516.

8.4 Reduction to the Weierstrass normal forms

In this section we shall see how the general elliptic integral (8.18) may be reduced to the Weierstrass normal forms, (8.4). For further details, see Dutta & Debnath (1965), Chapter 9.

As before, we begin with the general elliptic integral

$$\int R(x, y)dx$$

where y^2 is a cubic or quartic polynomial in x and $R(x, y)$ is a rational function of x and y.

We saw in Chapter 7, Theorem 7.8, that y and x can be expressed as rational functions of $\wp(z)$ and $\wp'(z)$ and so $R(x, y)$ is an elliptic function of z. Suppose that

$$R(x, y) = f(z).$$

Then

$$\int R(x, y)dx = \int f(z)dx = \int g(z)dz,$$

where $g(z)$ denotes an elliptic function of z.

Now we appeal to the ideas involved in Theorem 7.9 to express $g(z)$ in terms of the Weierstrass functions $\zeta(z)$ and $\wp(z)$ and obtain

$$g(z) = C + \sum_r \{a_{r,1}\zeta(z - \beta_r) + a_{r,2}\wp(z - \beta_r) + \cdots + a_{r,k-1}\wp^{(k-2)}(z - \beta_r)\}$$

$$(8.47)$$

for suitable constants $C, a_{r,n}$ and β_r.

It follows that

$$\int g(z)dz = D + Cz + \sum_r \left\{ \begin{matrix} a_{r,1} \log \sigma(z - \beta_r) + a_{r,2}\, \zeta(z - \beta_r) \\ + a_{r,3}\wp(z - \beta_r) + \cdots + a_{r,k-1}\, \wp^{(k-3)}(z - \beta_r) \end{matrix} \right\}. \tag{8.48}$$

Recall that

$$\zeta(z - \beta_r) - \zeta(z) + \zeta(\beta_r) = \frac{1}{2}\left\{ \frac{\wp'(z) + \wp'(\beta_r)}{\wp(z) - \wp(\beta_r)} \right\},$$

from which we obtain

$$\sum a_{r,2}(\zeta(z - \beta_r) - \zeta(z)) = \sum a_{r,2}Q_r\{\wp(z), \wp'(z)\}, \tag{8.49}$$

where $Q_r\{\wp(z), \wp'(z)\}$ denotes a rational function of $\wp(z)$ and $\wp'(z)$.
  Again, since $\sum\limits_r a_{r,1} = 0$,

$$\sum_r a_{r,1} \log \sigma(z - \beta_r) = \sum_r a_{r,1} \log \frac{\sigma(z - \beta_r)}{\sigma(z)}. \tag{8.50}$$

Finally,

$$\log \frac{\sigma(z - \beta_r)}{\sigma(z)} = \frac{1}{2}\int \left\{ \frac{\wp'(z) + \wp'(\beta_r)}{\wp(z) - \wp(\beta_r)} \right\} dz - \zeta(\beta_r)z + C. \tag{8.51}$$

On combining the results from (8.47) to (8.51), we obtain

$$\int R(x, y)dx = \int g(z)dz$$

$$= \frac{1}{2}\sum_r a_{r,1}\int \frac{\wp'(z) + \wp'(\beta_r)}{\wp(z) - \wp(\beta_r)}dz + A + Bz + C\zeta(z) + Q(x, y). \tag{8.52}$$

We have

$$z = \int dz = \int \frac{d\{\wp(z)\}}{\wp'(z)} = \int \frac{dw}{v},$$

where $w = \wp(z) = Q_1(x, y)$, $v = \wp'(z) = Q_2(x, y)$.
  For the terms $\zeta(z)$ and $\dfrac{\sigma(z - \beta_r)}{\sigma(z)}$, we have

$$\zeta(z) = -\int \wp(z)dz = -\int \frac{w}{v}dw \tag{8.53}$$

and

$$\log \frac{\sigma(z - \beta_r)}{\sigma(z)} = \frac{1}{2}\int \frac{v + v_r}{w - w_r}\frac{dw}{v} - \zeta(\beta_r)\int \frac{dw}{v} + C, \tag{8.54}$$

where $w_r = \wp(\beta_r)$ and $v_r = \wp'(\beta_r)$.

Putting all that together, we obtain

$$\int R(x, y)dx = A + B \int \frac{dw}{v} - C \int \frac{wdw}{v}$$

$$+ \frac{1}{2} \sum a_{r,1} \int \frac{v + v_r}{w - w_r} \frac{dw}{v} + Q(w, v), \qquad (8.55)$$

where w and v are rational functions of x and y, $Q(w, v)$ is a rational function of w and v and A, B, C are constants.

So we have finally obtained the three kinds of Weierstrass normal integrals:

$$\int \frac{dw}{v} \qquad \text{(first kind)}, \qquad (8.56)$$

$$\int \frac{wdw}{v} \qquad \text{(second kind)}, \qquad (8.57)$$

$$\int \frac{v + v_r}{w - w_r} \frac{dw}{v} \qquad \text{(third kind)}. \qquad (8.58)$$

9

Applications of elliptic functions in geometry

9.1 Introduction

We already saw in Chapter 1 that the theory of elliptic functions arose from the study of two geometrical problems: the measurement of the arc-length of an ellipse and the problem of the rectification of a lemniscate and of the division of an arc into equal parts. The latter problem, in particular, is associated with the name of Count Fagnano (1682–1764), who coined the term 'elliptic integrals', and with its subsequent development by Gauss and Abel (see Prasolov and Solovyev, 1997 for further details and historical notes and Chapter 12 for the connection with the arithmetic-geometric mean of Gauss).

This chapter is concerned with a selection of examples of applications of the theory so far developed in this book to problems in geometry, though our treatment is by no means exhaustive, for the topics are so many and so richly varied and the literature correspondingly extensive (see the books by Cassels, 1991, McKean & Moll, 1997 and Prasolov & Solovyev, 1997 for more of the theory, and for further examples see the books by Bowman, 1961, Dutta & Debnath 1965, Greenhill, 1892, Halphen, 1886–91 and Lawden, 1989).

We begin (see Section 9.2) with another result due to Fagnano, which uses the properties of the integrals $F(u)$ and $E(u)$, of the first and second kinds and then, in Section 9.3, we look at the problem of finding the surface area of an ellipsoid. In Section 9.4 we shall look at some properties of space curves and then, in Sections 9.5 and 9.6 results concerning Poncelet's 'poristic polygons', spherical trigonometry and Seiffert's spiral. In Section 9.7 we present a new and very surprising result in elementary geometry, whose only known proof originally used elliptic functions but for which an elementary proof was given by Rigby (1981);[1] the reader is invited to find other elementary proofs.

---

[1] My attention was drawn to this paper by J. R. Snape.

232

Finally, in this chapter, we turn our attention to the theory of elliptic curves and the geometrical interpretation of the addition theorem for the Weierstrass function, $\wp(z)$, which presents an elliptic curve as an additive group. If the underlying field is changed from the complex numbers, $\mathbb{C}$, to a finite field, $\mathbb{F}_q$, of q elements, then that group is reminiscent of the finite group of residue classes modulo a prime, p, and Fermat's 'Little Theorem', and has a totally unexpected application in cryptography, to which we shall briefly refer (see Section 9.8).

We shall mention other problems relating to arithmetical questions in geometry, for example problems concerning the existence of points with rational coordinates on elliptic curves, and finally we shall touch very briefly upon the connection between elliptic curves and modular forms and the proof of Fermat's Last Theorem (see Section 9.9).

9.2 Fagnano's Theorem

Consider the ellipse, whose equation in Cartesian coordinates is:

$$\frac{x^2}{a^2} + \frac{y^2}{b^2} = 1. \tag{9.1}$$

The *eccentricity, e*, is given by

$$e^2 = 1 - \frac{b^2}{a^2},$$

and we note that we take the positive square root to find e and that $0 < e < 1$. We may express the coordinates of a point, P, on the ellipse by using the *eccentric angle, ϕ*, measured from the minor axis OB (see Figure 9.1); so that

$$x = a \sin\phi, \quad y = b\cos\phi. \tag{9.2}$$

(The semi-major and semi-minor axes of the ellipse are OA, OB, respectively, and the circle with radius $OA = OB'$ is the *auxiliary circle*.)

We may also express the coordinates of P in parametric form by writing

$$x = a\, sn(u, k), \quad y = b\, cn(u, k), \tag{9.3}$$

where the modulus, k, of the Jacobian elliptic functions is real and $0 < k < 1$. It is both natural and convenient to choose the modulus to be the *eccentricity*:

$$k = e = \sqrt{\left(1 - \frac{b^2}{a^2}\right)},$$
$$k' = e' = \sqrt{(1 - e^2)} = \frac{b}{a}. \tag{9.4}$$

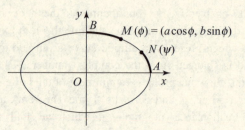

Figure 9.1 Division of the arc of an ellipse.

In what follows, we shall write simply $sn(u)$ etc., if there is no risk of confusion, where we assume that $k = e$.

We have already seen (Chapter 5, Theorem 5.12) that the elliptic integral, $E(u)$, of the second kind,

$$E(u) = \int_0^u dn^2 u \, du = \int_0^\phi \sqrt{(1 - k^2 \sin^2 \phi)} d\phi, \tag{9.5}$$

possesses an addition theorem

$$E(u + v) = E(u) + E(v) - k^2 sn\,(u)\,sn\,(v)\,sn(u + v). \tag{9.6}$$

Now if s denotes the length of the arc of the ellipse from B to the point P ($M(\phi)$, say) (see Figure 9.1), then

$$\frac{s}{a} = \int_0^u \sqrt{(cn^2 u \, dn^2 u + e'^2 sn^2 u \, dn^2 u)} \, du$$

$$= \int_0^u dn\,(u)\sqrt{(cn^2 u + e'^2 sn^2 u)} \, du$$

$$= \int_0^u dn^2 u \, du = E(u) \tag{9.7}$$

(and we would arrive at the same answer by using the parametric form in (9.2) and remembering that $a \, sn\,(u) = x = a \sin \phi$, which gives the second integral in (9.5)).

We now choose a second point, $Q, = N(\psi)$, say, on the ellipse, with parameter v, and then for the arc length s from B to Q we obtain

$$\frac{s}{a} = E(v).$$

It follows from the addition theorem (9.6) that the arc length from B to the point with parameter $u + v$ is given by $aE(u + v)$ where

$$aE(u) + aE(v) - aE(u + v) = e^2 sn\,(u)\,sn\,(v)\,sn(u + v). \tag{9.8}$$

Suppose that the parameter $u + v$ defines the point A; so that $u + v = K$ (recall that $sn(K) = 1$) and the corresponding value of the eccentric angle ϕ is $\pi/2$. It follows from (9.8) that if P, Q are points with parameters u, v, respectively, such that $u + v = K$, then

$$E(u) + E(v) - E = e^2 sn(u) sn(v),$$

where

$$E = E(K).$$

That result implies that

$$\text{arc } BP + \text{arc } BQ - \text{arc } BA = e^2 sn(u) sn(v)$$
$$= e^2 xx'/a, \qquad (9.9)$$

where P, Q are the points (x, y), (x', y'), respectively.

We shall give the result a more geometrical form by appealing to the eccentric angles ϕ and ϕ' corresponding to P and $P' = Q$.

We recall the properties (see the summary at the end of Chapter 2)

$$sn(u + K, k) = cd(u, k)$$
$$cn(u + K, k) = -k'sd(u, k) \qquad (9.10)$$
$$dn(u + K, k) = k'nd(u, k).$$

Hence $u + v = K$ implies

$$sn(v) = cn(u)/dn(u), \quad cn(v) = e'sn(u)/dn(u)$$

and so

$$\frac{cn(u)}{dn(u)}cn(v) = sn(v) \cdot e'\frac{sn(u)}{dn(u)},$$

or

$$cn(u)cn(v) = e'sn(u)sn(v). \qquad (9.11)$$

But (writing ϕ' for ψ)

$$cn(u) = \cos\phi, \quad sn(u) = \sin\phi,$$
$$cn(v) = \cos\phi', \quad sn(v) = \sin\phi',$$

and so (9.11) implies

$$\tan\phi \cdot \tan\phi' = \frac{1}{e'} = \frac{a}{b}. \qquad (9.12)$$

In other words, if $u + v$ is the parameter of the point A, then $\tan\phi \cdot \tan\phi' = a/b$, and conversely.

Accordingly, we have proved the following:

Theorem 9.1 (**Fagnano's theorem**) *Let $P(x, y)$ and $P'(x', y')$ be two points on the ellipse (9.1) whose eccentric angles ϕ, ϕ' are such that[2] $\tan \phi \tan \phi' = a/b$, then*

$$arc\, BP + arc\, BP' - arc\, BA = e^2 xx'/a. \tag{9.13}$$

If P and P' coincide then the point $F = P = P'$ is called Fagnano's point and one can prove the:

Corollary 9.1 *For Fagnano's point*

$$arc\, BF - arc\, AF = a - b. \tag{9.14}$$

Proof For the points P, P' the eccentric angles ϕ, ϕ' satisfy (as we saw in the proof of the theorem)

$$\tan^2 \phi \cdot \tan^2 \phi' = \frac{a^2}{b^2}.$$

Now $x = a \sin \phi$, $y = b \cos \phi$, $x' = a \sin \phi'$, $y' = b \cos \phi'$ and so

$$\tan^2 \phi \tan^2 \phi' = \frac{(x^2 x'^2)/a^4}{(1 - x^2/a^2)(1 - x'^2/a^2)}$$

$$= \frac{x^2 x'^2}{(a^2 - x^2)(a^2 - x'^2)}.$$

Whence

$$b^2 x^2 x'^2 = a^2 (a^2 - x^2)(a^2 - x'^2),$$

that is

$$\left(\frac{a^2}{b^2} - 1 \right) x^2 x'^2 - \frac{a^4}{b^2}(x^2 + x'^2) + \frac{a^6}{b^2} = 0,$$

or

$$e^2 x^2 x'^2 - a^2 (x^2 + x'^2) + a^4 = 0. \tag{9.15}$$

For Fagnano's point $F = P = P'$ and so $x = x'$, and then (9.15) gives

$$e^2 x^4 - 2a^2 x^2 + a^4 = 0.$$

---

[2] If the eccentric angles were measured from the major axis then this condition would read $\tan \phi$ $\tan \phi' = b/a$; see Bowman (1961), p.27.

Solving for x^2 we find

$$x^2 = \frac{a^2(1 \pm \sqrt{1 - e^2})}{e^2}. \tag{9.16}$$

Take $P = P' = F$ and $x = x'$ in (9.13) to obtain

$$\text{arc } BP - \text{arc } AP = e^2 x^2 / a$$
$$= a \left(1 - \frac{b}{a} \right)$$
$$= a - b,$$

on taking the minus sign in (9.16).

For further results relating to Fagnano's theorem, see the books by Bowman (1961), Greenhill (1892), Lawden (1989) and Prasolov & Solovyev (1997). A significant extension is to be found in that last, where the following question is (essentially) raised. Suppose $u + v$ is replaced by w and the eccentric angle corresponding to $u + v$ is then replaced by μ. Denote by ϕ, ψ, respectively, the eccentric angles corresponding to u and v. Then the formula (9.8) suggests that there might be a result of the form

$$aE(u) + aE(v) - aE(w) = e^2 sn\,(u)\,sn\,(v)\,sn\,(w), \tag{9.17}$$

which would reduce to the addition theorem for $E(u)$ if w were taken to be $u + v$. In terms of the eccentric angles ϕ, ψ, μ the result in (9.17) would then read

$$\int_0^\phi \sqrt{(1 - e^2 \sin^2 \phi)} \cdot d\phi + \int_0^\psi \sqrt{(1 - e^2 \sin^2 \psi)} \cdot d\psi$$
$$- \int_0^\mu \sqrt{(1 - e^2 \sin^2 \mu)} \cdot d\mu = a^{-1} e^2 \sin \phi \sin \psi \sin \mu. \tag{9.18}$$

That suggests that there might be a similar result connecting the integrals of the first kind, namely: can one express the sum

$$\int_0^\phi \frac{d\phi}{\sqrt{(1 - e^2 \sin^2 \phi)}} + \int_0^\psi \frac{d\psi}{\sqrt{(1 - e^2 \sin^2 \psi)}} - \int_0^\mu \frac{d\mu}{\sqrt{(1 - e^2 \sin^2 \mu)}}$$

$$\tag{9.19}$$

in terms of ϕ, ψ, μ? Moreover, can one deduce (9.18) from the appropriate form of (9.19)? The answer is 'yes', and Prasolov & Solovyev (1997) use the connection to give another proof of Fagnano's Theorem and of the addition theorem for the integrals of the second kind. We shall outline the argument and leave the details to the reader (see Exercises 9.2); see Prasolov & Solovyev (1997), Sections 2.7 and 3.2.

We use the notation

$$F(\phi) = \int_0^\phi \frac{d\phi}{\Delta(\phi)}, \quad E(\phi) = \int_0^\phi \Delta(\phi)\, d\phi,$$

where $\Delta(\phi) = \sqrt{(1 - k^2 \sin^2 \phi)}$ and k is the modulus of the Jacobi function $dn(u, k)$ for which $E(u) = \int_0^u dn^2(u)du$, (see (9.5)). We begin by showing that if

$$F(\phi) + F(\psi) = F(\mu),$$

then $\sin \mu$ may be expressed algebraically in terms of $\sin \phi$ and $\sin \psi$; in particular

$$\sin \mu = \frac{\sin \phi \cos \psi \, \Delta(\psi) + \sin \psi \cos \phi \, \Delta(\phi)}{1 - k^2 \sin^2 \phi \sin^2 \psi}, \tag{9.20}$$

which is the addition theorem for $sn(u + v)$ if we choose μ to be related to $u + v$ and write $\sin \mu = sn(u + v)$, $\sin \phi = sn(u)$, $\sin \psi = sn(v)$, etc. We outline the proof, as follows.

Consider the differential equation

$$\frac{d\phi}{\Delta(\phi)} + \frac{d\psi}{\Delta(\psi)} = 0, \tag{9.21}$$

whose integral is

$$F(\phi) + F(\psi) - F(\mu) = 0,$$

where μ is a constant.

One then shows that the integral of the differential equation (9.21) satisfies

$$\cos \phi \cos \psi - \sin \phi \sin \psi \sqrt{1 - k^2 \sin^2 \mu} = \cos \mu, \tag{9.22}$$

and two similar relations that are obtained by symmetry with respect to ϕ, ψ and $-\mu$. On squaring (9.22), we also obtain

$$\cos^2 \phi + \cos^2 \psi + \cos^2 \mu - 2 \cos \phi \cos \psi \cos \mu$$
$$+ k^2 \sin^2 \phi \sin^2 \psi \sin^2 \mu = 1, \tag{9.23}$$

which exhibits the underlying symmetry clearly.

In order to justify the claim made about the integral of (9.21), we divide both sides of (9.21) by $\sin\phi\sin\psi$ and then calculate $\frac{d\phi}{d\psi}$ to deduce

$$d\phi\left(\frac{\cos\psi - \cos\mu\cos\phi}{\sin\phi}\right) + d\psi\left(\frac{\cos\phi - \cos\mu\cos\psi}{\sin\psi}\right) = 0.$$

Using the formulae like (9.22) we then deduce that

$$\frac{d\phi}{\Delta(\phi)} + \frac{d\psi}{\Delta(\psi)} = 0.$$

So (9.22) is an integral of (9.21), and then an appeal to uniqueness shows that $F(\phi) + F(\psi) = F(\mu)$ implies (9.22). By writing $x = \cos\mu$ and $1 - x^2 = \sin^2\mu$, substituting in (9.22) and solving the resulting quadratic in x, we obtain (9.20) and

$$\cos\mu = \frac{\cos\phi\cos\psi - \sin\phi\sin\psi\Delta(\phi)\Delta(\psi)}{1 - k^2\sin^2\phi\sin^2\psi},$$

$$\Delta(\mu) = \frac{\Delta(\phi)\Delta(\psi) - k^2\sin\phi\sin\psi\cos\phi\cos\psi}{1 - k^2\sin^2\phi\sin^2\psi}.$$

We can now prove that $F(\phi) + F(\psi) = F(\mu)$ implies

$$E(\phi) + E(\psi) - E(\mu) = k^2\sin\phi\sin\psi\sin\mu,$$

where, by abuse of our earlier notation (see above)

$$E(\phi) = \int_0^\phi \frac{d\phi}{\sqrt{(1 - k^2\sin^2\phi)}} = E(u) = \int_0^u dn^2u\,du.$$

The idea of the proof is to show that if $F(\phi) + F(\psi) = F(\mu)$ and if $E(\phi) + E(\psi) - E(\mu)$ is a function $P(\phi, \psi, \mu)$, then

$$P(\phi, \psi, \mu) = k^2\sin\phi\sin\psi\sin\mu.$$

To do that we differentiate

$$E(\phi) + E(\psi) - E(\mu) = P(\phi, \psi, \mu),$$

on the assumption that the μ is a constant and then

$$\Delta(\phi)d\phi + \Delta(\psi)d\psi = dP.$$

Using our earlier results (see (9.22)) we find

$$dP = \frac{d(\sin^2\phi + \sin^2\psi + 2\cos\phi\cos\psi\cos\mu)}{2\sin\phi\sin\psi\sin\mu},$$

and then (9.23) implies

$$dP = \frac{d(k \sin \phi \sin \psi \sin \mu)^2}{2 \sin \phi \sin \psi \sin \mu} = k^2 d(\sin \phi \sin \psi \sin \mu).$$

By looking at the values when $\phi = 0$ we deduce $P = k^2 \sin \phi \sin \psi \sin \mu$, as required. (See Exercises 9.2 and Prasolov and Solovyev, 1997, 2.7.)

Exercises 9.2

9.2.1 Show that the coordinates (x, y) of the Fagnano point F are:

$$\frac{a}{\sqrt{(a+b)}}, \quad \frac{b}{\sqrt{(a+b)}}.$$

9.2.2 Prove that the tangents at the points P, Q on the ellipse (9.1) intersect in the point R on the confocal hyperbola whose equation is

$$\frac{x^2}{a} - \frac{y^2}{b} = a - b.$$

9.2.3 In the notation of 9.2.2, prove that

$$PR - \text{arc } PF = QR - \text{arc } QF.$$

The remaining questions are a development of the ideas outlined at the end of Section 9.2.

9.2.4 Show that one of the 'symmetric' identities similar to (9.22) is

$$\cos \mu \cos \phi + \sin \mu \sin \phi \Delta(\psi) = \cos \psi,$$

and write down the other one.

9.2.5 Using the identities like (9.20), prove that

$$\tan \mu = \frac{\tan \phi \Delta(\phi) + \tan \psi \Delta(\psi)}{1 - \tan \phi \tan \psi \Delta(\phi) \Delta(\psi)}.$$

Deduce that if ϕ', ψ' are angles such that

$$\tan \phi' = \tan \phi \Delta(\phi), \tan \psi' = \tan \psi \Delta(\psi),$$

then

$$\mu = \phi' + \psi'.$$

9.2.6 Take $\mu = \pi/2$. Show that $\sin \phi = \cos \psi' \Delta(\psi)$, $\sin \psi = \cos \phi/\Delta(\phi)$ and then, in the notation of the ellipse, show that $a \cos \phi \cos \psi = b \sin \phi \sin \psi$, whence $\tan \phi \tan \psi = a/b$. (Note that Prasolov and Solovyev take $a = 1$ so that $b = \sqrt{1 - k^2}$, whereas in the notation used here $b = \sqrt{1 - k^2} a = \sqrt{1 - e^2} a$.)

9.2.7 Prove that

$$\cos \phi \cos \psi = \Delta(\phi)\Delta(\psi) \sin \phi \sin \psi.$$

9.2.8 Complete the details of the proof of the theorem that $F(\phi) + F(\psi) - F(\mu) = 0$ implies

$$E(\phi) + E(\psi) - E(\mu) = k^2 \sin \phi \sin \psi \sin \mu.$$

9.2.9 Give a variation of the proof of Fagnano's Theorem, starting from the result in 9.2.8.

9.3 Area of the surface of an ellipsoid

See the books by Bowman (1961) and Lawden (1989).

We consider the ellipsoid

$$\frac{x^2}{a^2} + \frac{y^2}{b^2} + \frac{z^2}{c^2} = 1, \tag{9.24}$$

where a, b, c are not all equal, and for simplicity we shall suppose $a > b > c$. We denote by p the perpendicular from the centre of the ellipsoid on the tangent plane at the point (x, y, z), and then if $\cos \alpha$, $\cos \beta$ and $\cos \gamma$ denote the direction cosines of the normal at that point, we have

$$\frac{1}{p^2} = \frac{x^2}{a^4} + \frac{y^2}{b^4} + \frac{z^2}{c^4} \tag{9.25}$$

and

$$\cos \alpha = \frac{px}{a^2}, \quad \cos \beta = \frac{py}{b^2}, \quad \cos \gamma = \frac{pz}{c^2}, \tag{9.26}$$

from which we obtain in particular

$$\cos \gamma = 1 \bigg/ \sqrt{\left(1 + \frac{c^4 x^2}{a^4 z^2} + \frac{c^4 y^2}{b^4 z^2}\right)}. \tag{9.27}$$

It follows that the points at which the normals to the ellipsoid make a given, constant angle γ with the z-axis lie on the cone whose equation is

$$\left(\frac{x^2}{a^4} + \frac{y^2}{b^4} + \frac{z^2}{c^4}\right) \cos^2 \gamma = \frac{z^2}{c^4}. \tag{9.28}$$

From (9.24) and (9.25) we can substitute for z^2 to obtain the following equation:

$$\left(\frac{\cos^2 \gamma}{a^2} + \frac{\sin^2 \gamma}{c^2}\right) \frac{x^2}{a^2} + \left(\frac{\cos^2 \gamma}{b^2} + \frac{\sin^2 \gamma}{c^2}\right) \frac{y^2}{b^2} = \frac{\sin^2 \gamma}{c^2}, \tag{9.29}$$

which is the equation of a cylinder whose cross-sections are ellipses.

We consider the intersection of (9.29) with the ellipsoid (9.24) for $z > 0$, and if we denote by A the area of the cross-section of that cylinder and by S the area of that part of the surface of the ellipsoid so obtained, it follows that

$$dS = dA \cdot \sec \gamma. \qquad (9.30)$$

Now from (9.29) we obtain for the area A

$$A = \frac{\pi a^2 b^2 \sin^2 \gamma}{\sqrt{(c^2 \cos^2 \gamma + a^2 \sin^2 \gamma)}\sqrt{(c^2 \cos^2 \gamma + b^2 \sin^2 \gamma)}} \qquad (9.31)$$

$$= \frac{\pi a b \sin^2 \gamma}{\sqrt{(1 - e_1^2 \cos^2 \gamma)}\sqrt{(1 - e_2^2 \cos^2 \gamma)}},$$

where

$$e_1^2 = \frac{a^2 - c^2}{a^2}, \quad e_2^2 = \frac{b^2 - c^2}{b^2}, \qquad (9.32)$$

and where the assumption $a^2 > b^2 > c^2$ implies $e_1^2 > e_2^2$. Accordingly, we write

$$t = e_1 \cos \gamma, \quad k^2 = \frac{e_2^2}{e_1^2}, \qquad (9.33)$$

and then we find, using (9.31), that

$$A = \frac{\pi a b (e_1^2 - t^2)}{e_1^2 \sqrt{(1 - t^2)}\sqrt{(1 - k^2 t^2)}},$$

which reminds us of an elliptic integral. To make further progress we set

$$t = sn(u, k), \quad e_1 = sn(\theta, k),$$

whence

$$\sec \gamma = \frac{sn(\theta, k)}{sn(u, k)} = \frac{sn(\theta)}{sn(u)}, \qquad (9.34)$$

$$A = \frac{\pi a b}{sn^2\theta} \frac{sn^2\theta - sn^2 u}{cn(u)\, dn(u)}.$$

If we differentiate (9.34) with respect to u we obtain

$$\frac{sn(\theta)}{\pi a b} dA \sec \gamma = -\left(\frac{dn^2\theta}{dn^2 u} + \frac{cn^2\theta}{cn^2 u}\right) du. \qquad (9.35)$$

As γ varies from 0 to $\pi/2$, t varies from e_1 to 0 and u varies from θ to 0. We return to (9.30) and observe that if S denotes the area of the surface of the whole ellipsoid, then that result, together with (9.35), implies

$$\frac{S\,sn\,(\theta)}{2\pi ab} = \int_0^\theta \left(\frac{dn^2\theta}{dn^2 u} + \frac{cn^2\theta}{cn^2 u} \right) du. \tag{9.36}$$

So it remains to evaluate the integral in (9.36).

Now by differentiation one can verify that

$$\int dn^2 u \, du = k'^2 u + dn\,(u)\,sc\,(u) - k'^2 \int \frac{du}{cn^2 u}, \tag{9.37}$$

and also

$$\int dn^2 u \, du = k^2 sn\,(u)\,cd(u) + k'^2 \int \frac{du}{dn^2 u}. \tag{9.38}$$

It follows from (9.37) that

$$\int_0^\theta \frac{du}{cn^2 u} = \frac{1}{k'^2}\{k'^2\theta + dn\,(\theta) \cdot sc\,(\theta) - E(\theta)\} \tag{9.39}$$

and (9.38) gives

$$\int_0^\theta \frac{du}{dn^2 u} = \frac{1}{k'^2} E(\theta) - \frac{k^2}{k'^2} \frac{sn\,(\theta)\,cn\,(\theta)}{dn\,(\theta)}, \tag{9.40}$$

where we have appealed to the elliptic integral of the second kind

$$E(\theta) = \int_0^\theta dn^2 u \cdot du.$$

On substituting from (9.39) and (9.40) in the right-hand side of (9.36), we obtain

$$\frac{S\,sn\,(\theta)}{2\pi ab} = dn^2\theta \left\{ \frac{1}{k'^2} E(\theta) - \frac{k^2}{k'^2} \frac{sn\,(\theta)\,cn\,(\theta)}{dn\,(\theta)} \right\}$$
$$+ cn^2\theta \left\{ \theta + \frac{1}{k'^2} \frac{dn\,(\theta)\,sn\,(\theta)}{cn\,(\theta)} - \frac{1}{k'^2} E(\theta) \right\}. \tag{9.41}$$

Now we recall $dn^2\theta - cn^2\theta = k'^2 sn^2\theta$ and so the right-hand side of (9.41) is

$$sn^2\theta\,E(\theta) - \frac{k^2}{k'^2} sn\,(\theta)\,cn\,(\theta)\,dn\,(\theta)$$
$$+ cn^2\theta + \frac{1}{k'^2} sn\,(\theta)\,cn\,(\theta)\,dn\,(\theta) \tag{9.42}$$
$$= sn^2\theta\,E(\theta) + sn\,(\theta)\,cn\,(\theta)\,dn\,(\theta) + cn^2\theta.$$

On substituting from (9.32) and (9.33) in (9.42) we obtain finally

$$\frac{S \, sn \, (\theta)}{2\pi \, ab} = \frac{a^2 - c^2}{a^2} E(\theta) + \frac{\sqrt{(a^2 - c^2)} \, c^2}{a^2 b} + \frac{c^2}{a^2} \theta,$$

that is (after some simplification)

$$S = 2\pi c^2 + \frac{2\pi b}{\sqrt{(a^2 - c^2)}} \{(a^2 - c^2)E(\theta) + c^2\theta\}, \qquad (9.43)$$

which is the required formula.

Exercises 9.3

(See Bowman, (1961) and Lawden, (1986).)

9.3.1 Complete the verification of the calculations leading to (9.37) and (9.38). By a similar method prove that

$$\int dn^2 u \, du = u + dn \, (u) \, sc \, (u) - \int dc^2 u \, du.$$

9.3.2 The axes of two circular cylinders of radii a, b, respectively $(a > b)$, intersect at right angles. Show that their common volume is

$$\frac{8}{3}a\{(a^2 + b^2)E - (a^2 - b^2)K\}, \quad k = b/a,$$

where

$$E = \int_0^{\pi/2} \sqrt{(1 - k^2 \sin^2 \phi)} d\phi = \int_0^K dn^2 u \, du$$

and

$$K = \int_0^{\pi/2} \frac{d\phi}{\sqrt{(1 - k^2 \sin^2 \phi)}}.$$

9.3.3 Prove that

$$\int_0^{\pi/2} \frac{\sin^2 \theta}{\sqrt{(1 - k^2 \sin^2 \theta)}} d\theta = \frac{1}{k^2}(K - E).$$

By differentiating $\sin \theta \cos \theta \sqrt{(1 - k^2 \sin^2 \theta)}$ and integrating the result between 0 and $\pi/2$, deduce that

$$\int_0^{\pi/2} \frac{\sin^4 \theta}{\sqrt{(1 - k^2 \sin^2 \theta)}} d\theta = \frac{1}{3k^4}((k^2 + 2)K - 2(k^2 + 1)E).$$

9.3.4 Show that the volume common to the two elliptic cylinders

$$\frac{x^2}{a^2} + \frac{z^2}{c^2} = 1, \quad \frac{y^2}{b^2} + \frac{z^2}{c'^2} = 1, \quad (c < c'),$$

is

$$\left(\frac{8ab}{3c}\right)((c'^2 + c^2)E - (c'^2 - c^2)K), \quad k = \frac{c}{c'}.$$

(Hint: you may find question 9.3.3 helpful.)

9.3.5 Show that the surface area of the elliptic paraboloid

$$\frac{x^2}{a^2} + \frac{y^2}{b^2} = 2z, \quad a > b,$$

lying between the planes $z = 0$ and $z = h$ is

$$\frac{4}{3}\left\{ \frac{2ab^2h}{\sqrt{(b^2 + 2h)}}K + 2ah\sqrt{(b^2 + 2h)}E + a^2b^2((E - K)F(\phi, k'))\right.$$

$$\left. + KE(\phi, k') - \frac{1}{2}\pi\right\},$$

where

$$k^2 = \frac{2h(a^2 - b^2)}{a^2(b^2 + 2h)}, \quad \sin\phi = \frac{a}{\sqrt{(a^2 + 2h)}}, \quad E(\phi, k') = \int_0^\phi \sqrt{\Delta(\theta)}d\theta,$$

and K and E belong to the modulus k.

9.4   Some properties of space curves

This section is based on Dutta & Debnath (1965), Chapter 8.

Let t be a real parameter and let the Jacobi functions $sn\,(t)$, $cn\,(t)$, $dn\,(t)$ be defined with respect to the real modulus, $k, 0 < k < 1$. A *space curve* in general consists of points P whose coordinates depend on a real parameter, and such a curve is evidently given by

$$x = sn\,(t), \quad y = cn\,(t), \quad z = dn\,(t). \tag{9.44}$$

For such a curve we shall prove:

Theorem 9.2 *Let C be the space curve defined by* (9.44). *Then*

(a) *every plane $x = $ constant intersects the curve in exactly four points;*
(b) *the curve C is the intersection of the cylinders*

$$x^2 + y^2 = 1, \quad k^2x^2 + z^2 = 1;$$

(c)   *a plane intersects the curve, C, in four points; P_i, whose parameters
$t_i, i = 1, 2, 3, 4,$ satisfy*

$$t_1 + t_2 + t_3 + t_4 \equiv 0 (\text{mod } 4K, 4\mathrm{i}K'),$$

where, as usual, $4K, 4\mathrm{i}K'$ generate the period lattice for the three Jacobi functions.

Proof (a) Let x be given and consider the function $sn(t) - x$, which is an elliptic function having four poles in the parallelogram $4K, 4\mathrm{i}K'$. It must have four zeros and we may suppose that they are at the points with parameters t_1, t_2, t_3 and t_4, congruent modulo the periods $4K, 4\mathrm{i}K'$ to $t_1, 2K - t_1, 2\mathrm{i}K + t_1, 2K + 2\mathrm{i}K' - t_1$.

Suppose that t_1 corresponds to the point $P_1(x_1, y_1, z_1)$. Then t_2, t_3, t_4 will correspond respectively to the points $P_2(x_1, -y_1, z_1)$, $P_3(x_1, -y_1, -z_1)$ and $P_4(x_1, y_1, -z_1)$. Those four points are uniquely determined by t_1, t_2, t_3 and t_4 which are in turn determined by x. So (a) is proved.

(b) The Cartesian equation of the space curve is given by:

$$x^2 + y^2 = 1, \quad k^2x^2 + z^2 = 1. \tag{9.45}$$

In the first z is any real number, in the second y is any real number and so we obtain two cylinders; namely a circular cylinder and an elliptic cylinder, whose intersection is the given curve.

(c) Let $ax + by + cz + d = 0$ be the equation of a plane. Then the intersection of that plane and the curve defined by (9.44) is given by the points satisfying

$$a\, sn\,(t) + b\, cn\,(t) + c\, dn\,(t) + d = 0.$$

Now the function $g(t) = a\, sn\,(t) + b\, cn\,(t) + c\, dn\,(t) + d$ is an elliptic function having four poles in the fundamental parallelogram defined by $4K, 4\mathrm{i}K'$. So the function $g(t)$ has four zeros at the points t_1, t_2, t_3, t_4 in the fundamental parallelogram such that

$$t_1 + t_2 + t_3 + t_4 \equiv 0 (\text{mod } 4K, 4\mathrm{i}K'). \tag{9.46}$$

It follows that the plane intersects the curve in four points, P_i, whose parameters, t_i, satisfy (9.46).

Exercises 9.4

(See Dutta & Debnath, 1965.)

9.4.1 Given a curve in space, an *osculating plane* at a point P is a plane that meets the curve in three coincident points at t.

How many osculating planes can be drawn to the curve

$$x = sn\,(t), \quad y = cn\,(t), \quad z = dn\,(t),$$

for any point on it?

9.4.2 At how many points may a plane have a contact of the third order with a curve

$$x = sn\,(t), \quad y = cn\,(t), \quad z = dn\,(t)?$$

9.5 Poncelet's poristic polygons

Given two circles X and Y, with Y interior to X (or, more generally, two conics), is it possible to construct a closed polygon inscribed to X and circumscribed to Y? In general the answer is 'no', but if the circles are chosen appropriately, then there are such polygons, and the remarkable answer to the question is that if there is one such polygon there are infinitely many, having the same number of edges, inscribed to X and circumscribed to Y.

The books by Halphen (1886–1891) and Greenhill (1892) offer a more complete account than we have space for and, in particular, give extensive discussions of the cases when, for example, the polygons are pentagons or heptagons and also of the general cases when X and Y are conics. Halphen's treatment is based on Jacobi's geometrical construction (as given in the *Fundamenta Nova*, Jacobi, (1829), for the addition of the arguments in terms of two circles. To each point of one of the circles there corresponds an elliptic argument, and the chord that joins the points, in which the difference of the arguments is a constant, envelopes a second circle.

We shall give a geometrical interpretation similar to Jacobi's but based on the properties of the simple pendulum, as proved in Chapter 1, and derived from the presentation in Greenhill[3] of ideas due originally to Legendre, Abel and Jacobi.

We begin by recalling our work on the simple pendulum and then use it to motivate a geometrical argument for the addition of amplitudes.

---

[3] In his review of Ramanujan's *Collected Papers* (*Mathematical Gazette*, 14 (1928–29), pp. 425ff), Littlewood observes that: 'he (Ramanujan) was totally ignorant of Cauchy's Theorem and complex function theory. (This may seem difficult to reconcile with his complete knowledge of elliptic functions. A sufficient, and I think a necessary, explanation would be that Greenhill's very odd and individual *Elliptic Functions* was his text book.)' Greenhill's treatment of Jacobi's geometrical ideas is certainly unusual, but, based as it is on the simple pendulum, it fits very well with our treatment here.

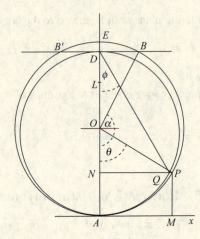

Figure 9.2 The simple pendulum: oscillatory motion.

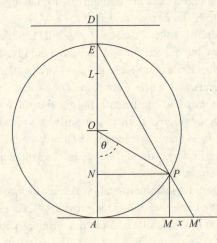

Figure 9.3 The simple pendulum: complete revolution.

Figures 9.2 and 9.3 illustrate the motion of the simple pendulum in the cases when the pendulum bob, P, oscillates between B and B (Figure 9.2) or makes complete revolutions (Figure 9.3); (cf. Figure 1.1 in Chapter 1).

We recall briefly the essential ideas from Chapter 1 in relation to the two diagrams.

To facilitate comparison with Greenhill's (and Halphen's) accounts, we recall briefly the results in Chapter 1 concerning the pendulum motion and the introduction of the Jacobi functions, and we draw attention to Greenhill's notation where it differs significantly from ours.

We consider first the case in which the pendulum oscillates in the way one expects a pendulum to behave; see Figure 9.2.

Let $l = OP$ and define $n = \sqrt{(g/l)}$. Then if

$$x = \sqrt{\frac{g}{l}} t = nt = u \qquad (9.47)$$

(that last being Greenhill's notation), we have (see (1.10))

$$\left(\frac{d\theta}{dx}\right)^2 = 4\left(\sin^2\frac{\alpha}{2} - \sin^2\frac{\theta}{2}\right) = 4\left(k^2 - \sin^2\frac{\theta}{2}\right),$$

where 2α denotes the maximum angle of swing; so that, in Figure 9.2, α is the angle AOB, and where $k = \sin \alpha/2$. If we write $\phi = \arcsin\left(k^{-1}\sin\theta/2\right)$ (see Section 1.4 and (1.15)) we obtain

$$\left(\frac{d\phi}{dx}\right)^2 = 1 - k^2\sin^2\phi,$$

and so (see (1.19))

$$x = \int_0^\phi \frac{d\phi}{\sqrt{(1 - k^2\sin^2\phi)}} = F(\phi, k). \qquad (9.48)$$

The angle ϕ is the amplitude, and in Jacobi's notation $\phi = am(x) = am(u)$. So (see Section 1.7) $\sin\phi = \sin am(x) = sn(x) = sn(u)$, (cf. (1.28) where we used ψ in place of ϕ). We also have

$$\cos\phi = \cos am(x) = cn(x) = cn(u);$$

$$\Delta\phi = \sqrt{1 - k^2\sin^2\phi} = \Delta am(x) = dn(x) = dn(u).$$

In the notation of Figure 9.2,

$$k = \sin\frac{1}{2}\alpha = \frac{AD}{AB} = \frac{AB}{AE}, \quad k^2 = \frac{AD}{AE}.$$

Also

$$AD = AE\sin\frac{1}{2}\theta = AB\, sn(nt),$$

$$PE = AE\cos\frac{1}{2}\theta = AE\, dn(nt)$$

and we recall the definition of the quarter-period

$$K = \int_0^{\pi/2} (1 - k^2\sin^\phi)^{-1/2} d\phi \qquad (9.49)$$

and the definition of the complementary modulus $k' = \cos \alpha/2$.

In Figure 9.2, the point Q moves according to the law

$$\phi = am(nt)$$

and so Q moves in a circle, centre C. In Figure 9.3, the pendulum makes complete revolutions and so the point P describes a circle, centre O.

In the case illustrated in Figure 9.2, the circle centre E and radius EB meets AE in a point L, called *Landen's point* (Landen of the transformation we met earlier). The velocity of the point Q is

$$n\frac{LQ}{1+k'}. \tag{9.50}$$

In Figure 9.3, when P makes complete revolutions, the velocity of P is

$$n\frac{LP(1+k')}{k},$$

where now the Landen point, L, is obtained by drawing a circle, centre D, cutting the line AE orthogonally and the vertical AD in L (where D is the point such that

$$\frac{1}{2}\frac{v^2}{g} + AN = AD = 2R, \tag{9.51}$$

R being a constant. If the angle $AEP = \phi, \phi = \theta/2$, then

$$\frac{d\phi}{dt} = \frac{n}{k}\sqrt{1 - k^2 \sin^2 \phi}$$

and so

$$\frac{nt}{k} = \int \frac{d\phi}{\Delta\phi} = F(\phi, k)$$

and

$$\frac{1}{2}\theta = \phi = am\left(\frac{nt}{k}, k\right). \tag{9.52}$$

In order to deal with the problem of the poristic polygons, we now consider further geometrical aspects of the pendulum motion in the cases of Figures 9.2, 9.3 as illustrated, respectively, in Figures 9.4 and 9.5.

Suppose first that P' would be the position of P (in Figure 9.2, cf. Figure 9.4) at time t if it started t_0 seconds later and then put $t' = t - t_0$. Then, using our earlier notation,

$$AN' = AD\, sn^2(nt'), \quad N'D = AD\, cn^2(nt'), \quad N'E = AE\, dn^2(nt').$$

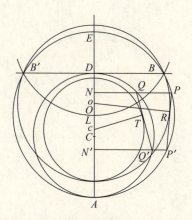

Figure 9.4 The simple pendulum: oscillatory motion.

The idea now is to prove that PP' touches a fixed circle through B and B' throughout the motion (Figure 9.4) – and that will lead to the construction of the poristic polygons.

To that end, suppose that in the time interval δt, P has moved to the point P_δ and P' to P'_δ and let PP', $P_\delta P'_\delta$ intersect at R; so that as $\delta t \to 0$, R is the point of contact on the envelope of the line PP'.

It follows that PP' cuts the circle $AP'P$ at equal angles at P and P' and

$$\frac{PR}{RP'} = \lim_{\delta t \to 0} \frac{PP_\delta}{P'P'_\delta} = \frac{v(P)}{v(P')} = \sqrt{\frac{ND}{N'D}},$$

where $v(P)$ denotes the velocity of P.

Now introduce a circle with centre o on AE, passing through B and B' and touching PP' at a point R'. Then, by appealing to the elementary geometry of the circle, we see that

$$PR'^2 = Po^2 - oR'^2 = 2 \, 0 \, oND$$

and similarly

$$R'P'^2 = 2 \, 0 \, oN'D.$$

It follows that

$$\frac{PR'}{R'P'} = \sqrt{\frac{ND}{N'D}} = \frac{PR}{RP'} \tag{9.53}$$

(see Exercise 9.5.4) and therefore R and R' coincide. So PP' touches at R the circle with centre o and radius oR. (Greenhill now uses that geometrical

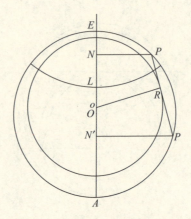

Figure 9.5 The simple pendulum: complete rotations.

construction to obtain one form of the addition theorem for the Jacobi functions).

In the foregoing, we supposed that the points P and P' oscillate (as a pendulum should) on a circle with a velocity related to the level of the horizontal line BB' (Figures 9.2 and 9.4). If, however, they perform complete revolutions (see Chapter 1 and Figures 9.3 and 9.5) with a velocity related to the level of a horizontal line BB' through D, not cutting the circle but lying above it, then a similar argument to that used above shows that PP' touches a fixed circle having a common radical axis, BB', with the circle on which P, P' move and those two circles do not intersect. The Landen point L is a limiting point of those two circles. The motion of P and P' in that case is imitated by the motion of Q and Q' on the circle AQ (Q a variable point) in Figure 9.4. So QQ' touches at T a fixed circle, centre c, and the horizontal line through E is the common radical axis of that circle and the circle CQ, and the Landen point is a limiting point.

9.5.1   Poncelet's poristic polygons

Either starting from the point A in Figure 9.4 and drawing the tangents $AQ_1, Q_1Q_2, Q_2Q_3, \ldots$ to the inner circle, centre c, from the points $Q_1, Q_2,$ Q_3 on the circle CQ; or starting from the point A in Figure 9.5 and drawing the tangents $AP_1, P_1P_2, P_2P_3, \ldots$ to the inner circle, centre o, from $P_1, P_2, P_3, \ldots$ on the circle OP, we can construct poristic polygons, as follows.

Denote the initial angle ADQ_1 or AEP_1, by $am(w)$ and then it follows from the construction that $ADQ_2 = AEP_2 = am(2w)$, $ADQ_3 = AEP_3 =$

am (3*w*), and we have obtained a geometrical construction (see Halphen, 1886-91) for the elliptic functions in which the argument is multiplied by a positive integer.

If $w = 2K/n$, where $2K$ denotes the half-period, then, after performing the construction n times we obtain a closed polygon $AQ_1 Q_2 \ldots Q_n$ or $AP_1 P_2 \ldots P_n$. During the subsequent motion of those points the polygon continues to be a closed polygon, inscribed in the circle CQ and circumscribed to the circle cT, or inscribed in the circle UP and circumscribed to the circle υR. So the theory of the simple pendulum affords a mechanical proof of the construction of Poncelet's poristic polygons.

For many detailed explorations of the particular cases when $n = 3, 5, 7$ and other results, see the books by Greenhill (1892) and Halphen (1886–91) .

Exercises 9.5

9.5.1 In the notation derived from Figure 9.2, prove that

$$AN = AD \, sn^2(nt), \quad ND = AD \, cn^2(nt), \quad NE = AE \, dn^2(nt).$$

Prove also that

$$NQ = (AN \, ND)^{1/2} = AD \, sn(nt)cn(nt), \quad NP = AB \, sn(nt)dn(nt).$$

9.5.2 (Landen's point and the velocity of Q; see Greenhill, 1892, p.23, Section 28.)

   In the notation of Figure 9.2, prove that

$$EB^2 = ED \, EA = EC^2 - CA^2$$

and deduce that the circle centre E and radius EB cuts the circle AQD, centre C, at right angles. Hence show that

$$LC^2 + CQ^2 = LC^2 + EC^2 - EL^2 = 2LC \, EC,$$

and

$$EL = EB = 2l \, K', \quad EC = l(1 + K'^2), \quad LC = l(1 - K'^2)$$

and deduce that

$$LQ^2 = LC^2 + CQ^2 + 2LC \, CN = 2LC \, EN = 2l(1 - K')^2 EN.$$

   Now use the fact that the velocity of Q is

$$(2g K^4 \, EN)^{1/2} = nK^2(2l \, EN)^{1/2} = n \, LQ(1 + K')$$

to prove the result in (9.49).

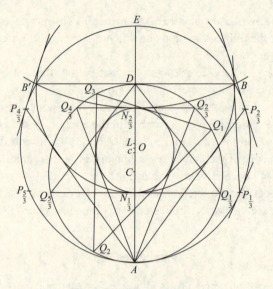

Figure 9.6 Poristic triangles.

9.5.3 According to the work in Chapter 1, the pendulum rod makes complete revolutions if the velocity at the lowest point, v_0, is such that $v_0^2 > 4gl$. We write

$$k^2 = 4gl/v_0^2 < 1$$

and it follows from

$$\left(\frac{d\theta}{dt}\right)^2 = 4\frac{g}{l}\left(\frac{v_0^2}{4gl} - \sin^2\frac{1}{2}\theta\right)$$

that

$$\sin\left(\frac{1}{2}\theta\right) = sn\left(\sqrt{\frac{g}{l}}\frac{t}{k}\right), \quad \cos\left(\frac{1}{2}\theta\right) = cn\left(\sqrt{\frac{g}{l}}\frac{t}{k}\right)$$

and

$$\frac{d\theta}{dt} = \frac{2}{k}\sqrt{\frac{g}{l}}dn\left(\sqrt{\frac{g}{l}}\frac{t}{k}\right), \quad \theta = 2\sin^{-1}\left(sn\frac{2gt}{v_0}\right).$$

Interpret (9.50) in the light of the foregoing.

9.5.4 Prove the result (9.52) using the recipe given in the text.

9.5.5 In the case of poristic triangles ($n = 3$, $w = 2K/3$ or $4K/3$), Figure 9.4 looks like Figure 9.6. Work out the details (see Greenhill, [1892], pp. 124–5, where the cases $n = 5$, $n = 7$ are also dealt with).

9.6 Spherical trigonometry

(See Bowman, 1961, Chapter III, and Lawden, 1989, pp. 103–105.)

9.6.1 Basic idea of spherical trigonometry

Trigonometry arose essentially from the need for accurate surveying of the Earth's surface. To begin with, of course, a process of triangulation was used that approximated the Earth by a plane surface, but (although it remains an approximation) it is better to treat the Earth as a perfect sphere and to replace ordinary triangles by spherical triangles, which are bounded by arcs of three great circles. Thus, in Figure 9.6, O is the centre of a sphere, A, B, C are three points on its surface and the arcs AB, BC and CA are arcs of three great circles and we call ABC a *spherical triangle*. If r denotes the radius of the sphere, then the lengths of the arcs BC, CA and AB are given, respectively, by

$$r \angle BOC, \quad r \angle COA, \quad r \angle AOB,$$

where $\angle BOC$ etc. denotes the angle BOC. We follow the custom in plane trigonometry and denote angles $\angle BOC, \angle COA, \angle AOB$ by a, b, c respectively; so for a sphere of unit radius those angles are the actual lengths of a, b, c measured along the surface of the sphere.

We consider a spherical triangle of vertices A, B, C and sides a, b, c on a sphere, centre O, of unit radius. By convention, the angles are all less than π. We denote the vectors $\overrightarrow{OA}$, $\overrightarrow{OB}$, $\overrightarrow{OC}$ by α, β, γ, respectively, and then

$$\cos a = \beta \cdot \gamma, \qquad \cos b = \gamma \cdot \alpha, \qquad \cos c = \alpha \cdot \beta,$$
$$\sin a = |\beta \times \gamma|, \quad \sin b = |\gamma \times \alpha|, \quad \sin c = |\alpha \times \beta|,$$

where $| \, |$ denotes the absolute value of the vector, $\beta \cdot \gamma$ denotes the scalar product and $\beta \times \gamma$ denotes the vector product. One readily verifies that

$$\cos A = \frac{(\alpha \times \beta) \cdot (\alpha \times \gamma)}{|\alpha \times \beta| |\alpha \times \gamma|}, \quad \sin A = \frac{|(\alpha \times \beta) \times (\alpha \times \gamma)|}{|\alpha \times \beta| |\gamma \times \alpha|}, \tag{9.54}$$

from which we obtain

$$\frac{\sin A}{\sin a} = \frac{\sin B}{\sin b} = \frac{\sin C}{\sin c}, \tag{9.55}$$

which is the sine rule for the spherical triangle.

We also find

$$\cos A = \frac{\beta \cdot \gamma - (\alpha \cdot \gamma)(\beta \cdot \alpha)}{|\alpha \times \beta| |\alpha \times \gamma|}$$
$$= \frac{\cos a - \cos b \cos c}{\sin c \sin b};$$

so that

$$\cos a = \cos b \cos c + \sin c \sin b \cos A, \qquad (9.56)$$

which is a form of the cosine rule, and one obtains similar expressions by cyclic interchange of a, b, c and A, B, C.

We may obtain a further set of formulae by interchanging the roles of angles and sides and considering the *polar triangle* $A'B'C'$ whose vertices are given by vectors α', β', γ' defined by

$$\alpha' = \frac{(\beta \times \gamma)}{|\beta \times \gamma|}, \quad \beta' = \frac{(\gamma \times \alpha)}{|\gamma \times \alpha|}, \quad \gamma' = \frac{(\alpha \times \beta)}{|\alpha \times \beta|}.$$

Those vectors are parallel to the normals to the planes BOC, COA and AOB, respectively, and if a', b', c' are the sides of $A'B'C'$, then

$$\cos a' = \frac{(\gamma \times \alpha) \cdot (\alpha \times \beta)}{|\gamma \times \alpha| \, |\alpha \times \beta|} = -\cos A,$$

and so on. One can now show (using properties of vector products) that

$$\cos A' = -\cos a, \quad \sin A' = \sin a$$

and

$$a' = \pi - A, \quad b' = \pi - B, \quad c' = \pi - C,$$
$$A' = \pi - a, \quad B' = \pi - b, \quad C' = \pi - c.$$

From the properties of the polar triangle, it follows that any identity involving the sides and angles of a spherical triangle may be replaced by another in which a is replaced by $\pi - A$, A by $\pi - a$ etc. The underlying principle is called the *duality principle*. In particular (for example) (9.56) may be replaced by

$$\cos A = -\cos B \cos C + \sin B \sin C \cos a, \qquad (9.57)$$

and similar expressions are obtained by cyclic interchange.

9.6.2   Elliptic functions and spherical trigonometry

We write (9.55) in the form

$$\frac{\sin a}{\sin A} = \frac{\sin b}{\sin B} = \frac{\sin c}{\sin C} = k < 1. \qquad (9.58)$$

Now introduce the Jacobi functions with modulus k, as follows:

$$sn\,(u) = \sin A, \quad sn\,(v) = \sin B, \quad sn\,(w) = \sin C. \qquad (9.59)$$

We can assume that the numbers u, v, w lie in the interval $(0, K)$ and, in Jacobi's notation,

$$A = am(u), \quad B = am(v), \quad C = am(w)$$

and so

$$cn(u) = \cos A, \quad cn(v) = \cos B, \quad cn(w) = \cos C. \tag{9.60}$$

The equations in (9.58) imply

$$\sin a = k \, sn(u), \quad \sin b = k \, sn(v), \quad \sin c = k \, sn(w), \tag{9.61}$$

whence

$$\cos a = dn(u), \quad \cos b = dn(v), \quad \cos c = dn(w). \tag{9.62}$$

On substituting from those results in (9.56) and (9.57), we obtain

$$dn(w) = dn(u) \, dn(v) + k^2 \, sn(u) \, sn(v) \, cn(w) \tag{9.63}$$

and

$$cn(w) = -cn(u) \, cn(v) + sn(u) \, sn(v) \, dn(w). \tag{9.64}$$

Solving those equations for $cn(w)$ and $dn(w)$ and then using the addition theorem for Jacobi functions, we find that

$$cn(w) = -cn(u+v), \quad dn(w) = dn(u+v). \tag{9.65}$$

Since we have $0 < u < K$, those equations have a unique solution $w = 2K - (u+v)$ and so we have shown that

$$u + v + w = 2K, \tag{9.66}$$

or *the sum of the angles of a spherical triangle is $2K$*, which reduces to the familiar theorem for plane triangles if we let k tend to 0.

If we substitute in the equation given in Exercise 9.6.1, we obtain

$$sn(u) \, dn(v) = cn(v) \, sn(w) + sn(v) \, cn(w) \, dn(u),$$

that is

$$cn(v) \, sn(u+v) = sn(u) \, dn(v) + dn(u) \, sn(v) \, cn(u+v). \tag{9.67}$$

We used the addition theorem to prove (9.65), but one can establish the result in (9.66) by a different method (using properties of the spherical triangles) and then we may derive the addition theorem for the elliptic functions (see Exercises 9.6.3 and 9.6.4, which are based on Lawden, 1989, p.105; the use of the formulae

(9.63) and (9.64) is reminiscent of one of the standard proofs of the addition theorems – see Hurwitz & Courant, (1964), pp. 218–219.

Exercises 9.6

See Lawden 65 (1989), Section 4.4.

9.6.1 Prove that, in the notation for spherical triangles,

$$\sin A \cos b = \cos B \sin C + \sin B \cos C \cos a.$$

9.6.2 Prove that

$$\cos C \cos b = \sin b \cot a - \sin C \cot A.$$

9.6.3 Suppose that in Equations (9.58) to (9.62), c and C are kept constant, but a, b, A, B are variables, subject to (9.58). By differentiating the result obtained from (9.54) by cyclic interchange (namely the polar formula

$$\cos C = -\cos A \cos B + \sin A \sin B \cos c),$$

show that

$$(\sin A \cos B + \cos A \sin B \cos c)\mathrm{d}A$$
$$+ (\cos A \sin B + \sin A \cos B \cos c)\mathrm{d}B = 0,$$

and then using the formula in 9.6.1 show that that reduces to

$$\mathrm{d}A \cos B + \mathrm{d}B \cos a = 0.$$

Differentiate (9.59) to obtain

$$cn\,(u)\,dn\,(u)\,\mathrm{d}u = \cos A\,\mathrm{d}A = cn\,(u)\,\mathrm{d}A,$$

that is $\mathrm{d}A = dn\,u\,\mathrm{d}u = \mathrm{d}u \cos a$ and, similarly, $\mathrm{d}B = \mathrm{d}v \cos b$. Conclude that

$$\mathrm{d}u + \mathrm{d}v = 0$$

and so

$$u + v = \text{constant.}$$

9.6.4 By taking $a = 0, b = c$ and then $A = 0, B = \pi - C$ in the result at the end of question 9.6.3, show that the constant in $u + v = \text{constant}$ must be $2K - w$ and conclude that $u + v + w = 2K$ (cf. (9.66)). Now obtain Equations (9.63) and (9.64), as before and, putting $w = 2K - (u + v)$

in those equations and solving for $cn(u + v)$ and $dn(u + v)$ obtain the addition theorems for those functions.

9.6.5 (See Bowman, 1961, p. 34, and also Lawden, 1989, pp. 105–107 for more details. For a fascinating account of ideas relating to *Seiffert's spiral* and properties of the Jacobi elliptic functions derived from the spiral see the paper by Erdos, [2000],[4] where the presentation of the subject is characteristically exciting and repays careful study.) In the notation for spherical polar coordinates and spherical triangles on a sphere of radius r whose equation is $x^2 + y^2 + z^2 = r^2$, let (r, θ, ϕ) be the usual spherical polar coordinates and write $\rho = r \sin \theta$. Denote by ds the element of *arc length* on a curve drawn on the sphere and show that

$$(ds)^2 = (\rho d\phi)^2 + (r d\rho)^2 / (r^2 - \rho^2).$$

Seiffert's spherical spiral is the curve on the sphere defined by the equation

$$r\phi = ks, \quad 0 < k < 1,$$

where s is measured from the pole of the sphere. Show that

$$\rho = r \, sn\left(\frac{s}{r}\right), \quad z = r \, cn\left(\frac{s}{r}\right)$$

and that $dn(s/r)$ is the cosine of the angle at which the curve cuts the meridian. (Hint: for simplicity, take $r = 1$ and then show that

$$ds^2 = \sin^2 \theta d\phi^2 + d\theta^2.$$

Substitute $d\phi = k \, ds$ to obtain

$$\frac{ds}{d\theta} = (1 - k^2 \sin^2 \theta)^{-1/2}.$$

Since $0 < k < 1$, we obtain by integration that

$$s = sn^{-1}(\sin \theta, k) + \text{constant};$$

that is, on choosing to measure s from the point where $\theta = 0$,

$$\sin \theta = sn(s, k).$$

In Jacobi's notation that reads

$$\theta = am(s, k).$$

The rest now follows.)

---

[4] I am indebted to J. R. Snape for drawing my attention to this paper.

9.7   The Nine Circles Theorem

9.7.1   Introduction

It is popularly supposed (not least by some mathematical educators) that there are no new theorems in elementary geometry, but the theorem described here is a fascinating counter-example, a theorem whose origins illustrate the pleasure to be derived from careful drawing, illuminated by insight, and the fact that there may still be interesting and significant theorems of elementary geometry awaiting discovery.[5] Its fascination is enhanced by the fact that there appears to be no known genuinely elementary proof, though for us it affords an example of an unexpected application of the theory of elliptic functions and in particular of the addition theorem for the Weierstrass function $\wp(z)$. For a very interesting alternative approach see the paper by J. F. Rigby (1981).

Let A, B and C be three circles of general position (the meaning of that phrase is intuitively obvious, but it will be made precise, if less obvious, in what follows) in the plane and let S_1 be any circle touching A and B. Consider the following chain of circles: S_2 is a circle touching B, C and S_1; S_3 touches C, A and S_2; S_4 touches A, B and S_3; S_5 touches B, C and S_4; S_6 touches C, A and S_5; and S_7 touches A, B and S_6. There are a finite number of choices at each stage for each successive circle $S_2, \ldots, S_7$ (see below), and the remarkable result, which we shall state formally in Theorem 9.3 and then prove, is that if the choices of circles at each stage are appropriately made (in a manner to be elucidated in what follows), then the last circle, S_7 coincides with the first circle S_1 – in other words the chain closes up to give a configuration of nine circles. (See Figure 9.7.)

It should be noted that the system of nine circles $A, B, C, S_1, S_2, S_3, S_4, S_5, S_6$ so obtained is a symmetrical system, each circle being touched by four others. The symmetry is illustrated by considering the array

$$
\begin{array}{ccc}
A & B & C \\
S_5 & S_3 & S_1 \\
S_2 & S_6 & S_4
\end{array}
$$

in which we see that two of the nine circles touch if (and in general only if) the corresponding letters are not in the same row or column of the array.

---

[5] For a very attractive account of similar theorems in elementary geometry, see the book by Evelyn, Money-Coutts & Tyrrell (1974). The proof given here is based partly on Section 3 of that book, but primarily on the paper by Tyrrell & Powell (1971).

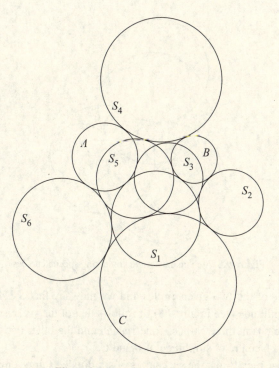

Figure 9.7 The nine circles theorem.

9.7.2 Formulation of Theorem 9.3

In order to formulate our theorem precisely, we have to explain how the 'appropriate choices' are to be made. To that end we begin by recalling that, in general, given three circles in the plane there are eight circles touching all three. If two of the circles touch (as is the case at each stage of the construction of our chain), then that number is reduced to six (two of the circles counting doubly and four simply.). For example, there are two circles in the contact coaxal[6] family determined by B and S_1, which also touch C and those are two of the six possible choices for S_2 in the chain. We call those the *special choices* for S_2 (they are the two that count double) and the other four possible choices for S_2 will be called the *general* or *non-special choices*. A similar distinction between special and non-special choices arises at each step in the construction of the chain.

Next we note that if a circle S is drawn to touch two given circles, then the line joining the two points of contact passes through one or other of the two centres

---

[6] For example, all circles that pass through two fixed points A, B form a *coaxal system* with *radical axis AB*. For further details about coaxal circles and the other results of circle geometry appealed to here, see books by Durell (1948) or Coxeter (1961).

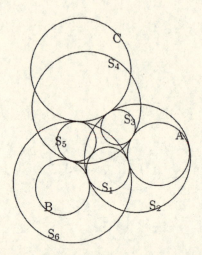

Figure 9.8 A chain constructed by taking special choices.

of similitude of the two given circles and we may say that S *belongs* to that centre of similitude (see Figure 9.8). It follows that of the six possible choices for S_2 in the chain, three choices (one *special* and the other two *non-special*) belong to each centre of similitude of B and C.

Finally, we recall[7] that the six centres of similitude of three circles, taken in pairs, lie in threes on four lines and so are the vertices of a *complete quadrilateral*.

We are now able to formulate our main result precisely, as follows.

Theorem 9.3 *Let A, B and C be three circles in the plane and select a set of three collinear centres of similitude (one for each pair of A, B and C). Suppose that a chain of six circles $S_1, \ldots, S_6$ is constructed, as above, touching pairs of the circles A, B and C in cyclic order and 'belonging' to the chosen centres of similitude. Then:*

(a) *if the special choice for S_i is made at each stage, the chain will close up (that is, $S_7 = S_1$); and*

(b) *if a non-special (general) choice is made at each stage, then, starting from S_1, the circles S_2, S_3 and S_4 can be chosen arbitrarily (two choices for each) and it is always possible then to choose S_5 and S_6 so that the chain closes up.*

It is straightforward (and a useful exercise in understanding the construction of the chain) to check that, starting from an assigned position of S_1, there are eighteen chains which close up, two special chains and sixteen general ones.

---

[7] See Durell (1948), Theorem 63, p. 136 and Figure 9.8.

9.7.3   Proof of Theorem 9.3

The proof given here is a very slightly expanded account of the proof given in Tyrrell & Powell (1971); a simpler proof for the case in which the circles A, B and C have equal radii, but still using elliptic functions, is given in Evelyn, Money-Coutts & Tyrrell (1974).

The circles in a plane may be regarded as the projections of plane sections of a given quadric surface; the projective model, Q, of the complete linear system of circles in the plane.[8] The idea is essentially that of the stereographic projection of a sphere, whose 'south pole' stands on the origin of the complex plane and whose 'north pole', N, is the centre of the projection, in which a point P of the sphere is mapped to the point z where the line NP meets the plane. A more precise definition is as follows. Given a quadric surface, Q, and a point U of Q, there are two generators of Q that pass through U, and if ϖ is a plane, not containing U, then the generators through U meet ϖ in two points I, J. We take I, J to be the *circular points*; so that any plane cuts the quadric in a conic section and that conic is projected into a conic in ϖ through I, J, that is, into a circle.[9]

If two plane sections of Q touch, then the corresponding circles in the plane also touch. So if we replace the word 'circle' in the foregoing by the phrase 'plane section of Q', we obtain the statement of a theorem about chains of plane sections of Q, and, by proving that as a theorem of complex projective geometry, we prove the original theorem about circles.

From now on, we shall not refer to the original formulation in terms of circles, but use instead the formulation in terms of plane sections, and we shall use the same notation to refer to the plane sections as was originally used for the circles to which those sections correspond.

The rules about 'appropriate choices' may be interpreted in terms of sections of Q, as follows. First, the distinction between special and non-special choices remains unchanged. However, the rule about centres of similitude, needs to be re-interpreted. If A, B, C are the three plane sections corresponding to the circles A, B and C, then each pair of sections lie on two quadric cones, giving six such cones whose vertices are the vertices of a complete quadrilateral in the polar plane, with respect to Q, of the point of intersection of the planes A, B and C. Now any plane section of Q that touches A and B lies in a tangent plane to one or other of the two cones through A and B; and there are similar alternatives for sections which touch B and C, and C and A, respectively. The rule of choice, corresponding to the rule in the plane

---

[8] See Baker (1992), Vol. 4, Chapters I and II. The proof in the case when the three circles have equal radii does not use the projective model.
[9] See Baker (1992).

concerning three fixed collinear centres of similitude, is to select at the outset
three of the six cones *with collinear vertices* and at every stage in the construction of the chain to choose a section that touches the appropriate one of those
cones.[10]

Choose homogeneous coordinates (x, y, z, w) in the space in which Q lies
and in which the planes of A, B and C are given by $x = 0$, $y = 0$ and $z = 0$,
respectively. The fourth plane of reference, $w = 0$, is the polar plane, with
respect to Q, of the point of intersection of the planes A, B and C.

By a proper choice of the unit point, the equation of Q may be put in the
form

$$x^2 + y^2 + z^2 - 2yz \cos\alpha - 2zx \cos\beta - 2xy \cos\gamma - w^2 = 0, \qquad (9.68)$$

where α, β and γ are constants. (That is possible because, by our hypothesis of
'generality' we may suppose that the planes A, B, C are linearly independent
and that, for the choice of tetrahedron of reference, the coefficients of x^2, y^2
and z^2 *are* non-zero – indeed that is what the vague phrase of 'general position'
is to be taken to mean.)

The equations of the pairs of cones through (A, B), (B, C) and (C, A) are,
respectively,

$$x^2 + y^2 + z^2 - 2yz \cos\alpha - 2zx \cos\beta + 2xy \cos(\alpha \pm \beta) - w^2 = 0,$$
$$x^2 + y^2 + z^2 + 2yz \cos(\beta \pm \gamma) - 2zx \cos\beta - 2xy \cos\gamma - w^2 = 0,$$
$$x^2 + y^2 + z^2 - 2yz \cos\alpha + 2zx \cos(\gamma \pm \alpha) - 2xy \cos\gamma - w^2 = 0. \quad (9.69)$$

The vertices of those cones are collinear if all three of the sign alternatives
in (9.69) are taken to be negative, or if two are positive and one is negative.
Without loss of generality, we may take all three signs to be negative and we
shall suppose that done in what follows.

The sections A, B and C may be parameterized as

$$A : (0, \cos\theta, \cos(\theta - \alpha), \sin\alpha),$$
$$B : (\cos(\phi - \beta), 0, \cos\phi, \sin\beta),$$
$$C : (\cos\psi, \cos(\psi - \gamma), 0, \sin\gamma), \qquad (9.70)$$

and we identify the sections $S_1, S_2, \ldots, S_7$ by parameters $\theta, \phi, \psi, \theta', \phi', \psi', \theta''$
of the points at which they touch A, B, C, A, B, C, A, respectively.

---

[10] See Baker (1992), Chapter II, for the origin of the first part of the proof. Baker deals with a
problem of Steiner and Cayley, which was to find the possible positions of the first section S_1
for which the chain closes up after three steps (that is, $S_4 = S_1$).

Accordingly, the tangent plane to the cone (9.69) at the point (9.70), whose equation is

$$x \sin(\theta - \beta) + y \sin(\theta - \alpha) - z \sin\theta + w = 0 \tag{9.71}$$

is the plane of a variable section S_1, which touches A at the point (9.70) and B at the point with parameter $\phi = \beta - \theta$. Similarly, any section touching B and C is cut by a plane of the form

$$-x \sin\phi + y \sin(\phi - \gamma) + z \sin(\phi - \beta) + w = 0, \tag{9.72}$$

and that is the plane of a possible position for S_2, provided that we choose ϕ so that the sections (9.71) and (9.72) touch. The resulting condition on ϕ, which is obtained by expressing the condition that the line of intersection of (9.70) and (9.71) should belong to the quadric complex of tangent lines to Q, turns out to be reducible. One of the factors gives $\phi = \beta - \theta$ and corresponds to the special choice for S_2. The other factor gives

$$\cos(\theta + \phi - \beta) - 2(\cos\beta + \rho\sin\beta)\cos\theta\cos\phi = 1, \tag{9.73}$$

where $\rho = \tan(\alpha + \beta + \gamma)/2$ and, for a given θ, that yields two values of ϕ, which correspond to the non-special choices for S_2.

We first consider the easy case in which a special choice is made at each stage. By permuting the letters cyclically, we see that the parameter ψ of S_3 is given in terms of ϕ by $\psi = \gamma - \phi$ and the parameter θ' of S_4 is given by $\theta' = \alpha - \psi$. So we have $\theta' = \alpha - \gamma + \beta - \theta$ and, since the relation is involutory (that is, on repetition one recovers the original), a further three steps in the chain bring us to a position of S_7 that coincides with S_1. So our theorem is proved in the 'easy' case in which special choices are made throughout without an appeal to elliptic functions.

We turn at last to the general case, where non-special choices are made, and the part played by Weierstrass elliptic functions, which appears to be an essential part, but see Rigby (1981).

In this case, instead of the trigonometric parameters θ, ϕ and ψ used hitherto, it is convenient to use a rational (projective) parameter and, in particular, any bilinear function of $\tan\theta/2$ may be used as a rational parameter on A. For reasons which will become apparent when we introduce the elliptic functions, it is convenient to choose a rational parameter on A, related to θ and denoted by t, where

$$\tan\frac{\theta}{2} = \frac{t + 2\rho - 3}{t + 2\rho + 3}. \tag{9.74}$$

Similarly, we choose parameters u and v on B and C, respectively, given by

$$\tan \frac{\theta}{2} = \frac{u + 2\rho - 3}{u + 2\rho + 3}, \quad \tan \frac{\psi}{2} = \frac{v + 2\rho - 3}{v + 2\rho + 3}. \tag{9.75}$$

In terms of those two new parameters, the relation (9.73) may be written (after some calculations – see Exercise 9.7.1) in the form

$$(t + u + b)(4btu - g_3) = \left(tu + bt + bu + \frac{1}{4} g_2 \right)^2, \tag{9.76}$$

where

$$g_2 = 48\rho^2 + 36, \quad g_3 = 64\rho^3 + 72\rho, \quad b = 3\tan\frac{\beta}{2} - 2\rho. \tag{9.77}$$

Now Equation (9.76) is a form of the addition theorem for the Weierstrass elliptic function, $\wp(z)$, (see Exercise 9.7.2) and may therefore be written in the form

$$\wp^{-1}(u) \equiv \pm\wp^{-1}(t) \pm \wp^{-1}(b) (\bmod \Lambda), \tag{9.78}$$

where $\wp^{-1}$ denotes the function inverse to the Weierstrass function $\wp$ and Λ denotes the period lattice. (See Chapter 7 for the background.)

Since $\wp$ is an even function of order 2 (see Chapter 7), $\wp^{-1}$ has two values, equal apart from the signs, plus or minus. Hence, given t, the right hand side of (9.78) has four possible values (mod Λ), which are two equal and opposite pairs, whence u is determined as having two possible values in terms of t. So (9.78) gives the two possible values of the parameter u of the section S_2 corresponding to a given value of the parameter t for the section S_1 (recall that we are now considering non-special choices only).

The essential point about the reduction to the form (9.78) is that the constants g_2 and g_3, on which $\wp$ depends, are functions of ρ only and ρ is an expression that is symmetric in α, β and γ. By permuting the letters cyclically, we conclude that the parameters v and t' of the sections S_3 and S_4 are given, respectively, by

$$\wp^{-1}(v) \equiv \pm\wp^{-1}(u) \pm \wp^{-1}(c), \quad \wp^{-1}(t') \equiv \pm\wp^{-1}(v) \pm \wp^{-1}(a), \quad \bmod \Lambda,$$

where $\wp$ is the *same* elliptic function as before and where $c = 3\tan\gamma/2 - 2\rho$ and $a = 3\tan\alpha/2 - 2\rho$. So, from the foregoing discussion, we obtain

$$\wp^{-1}(t') \equiv \pm\wp^{-1}(t) \pm \wp^{-1}(a) \pm \wp^{-1}(b) \pm \wp^{-1}(c), \quad \bmod \Lambda, \tag{9.79}$$

from which we see that the eight equal and opposite pairs of values of the right hand side determine t' as having eight possible values in terms of t. Similarly, the parameter t'' of S_7 is given as having eight values in terms of the parameter t'

of S_4 (an older, more classical usage would have said an 'eight-valued function', though to make that precise in terms of 'single-valued' functions, one needs the idea of a Riemann surface; our present usage will suffice) by

$$\wp^{-1}(t'') \equiv \pm\wp^{-1}(t') \pm \wp^{-1}(a) \pm \wp^{-1}(b) \pm \wp^{-1}(c), \quad \mod \Lambda. \tag{9.80}$$

It is clear now that, if we take the sign alternatives arbitrarily in (9.79), then *it is possible to choose* the sign alternatives in (9.80) so as to obtain $\wp^{-1}(t'') \equiv \pm\wp^{-1}(t)$, mod Λ, that is, $t'' = t$. In other words, if we choose the sections S_2, S_3, S_4 arbitrarily at each stage, then it is possible to choose S_5, S_6, S_7 so that S_7 coincides with S_1. So we have proved the theorem as formulated in terms of the plane sections and therefore the original Theorem 9.2 for the chain of nine circles.[11]

Exercises 9.7

9.7.1 Derive Equation (9.76) from the condition (9.73) in terms of the rational parameters in (9.74) and (9.75). (Hint: use the formulae expressing *sine* and *cosine* in terms of the *tangent* of the half-angle.)

9.7.2 (For the background to this question, see the section on the addition formula for the Weierstrass $\wp$-function in Chapter 7, or Copson (1935), pp. 362–364.)
Consider the elliptic function

$$
\begin{aligned}
F(z) &= \wp'^2(z) - \{A\wp(z) + B\}^2 \\
&= 4\wp^3(z) - A^2\wp^2(z) - (2AB + g_3)\wp(z) - (g_3 + B^2),
\end{aligned}
$$

where the constants A, B satisfy

$$
\begin{aligned}
\wp'(u) + A\wp(u) + B &= 0, \\
\wp'(v) + A\wp(v) + B &= 0, \\
\wp'(-u - v) + A\wp(-u - v) + B &= 0.
\end{aligned}
$$

Show that $F(z)$ is an elliptic function of order 6, with zeros at the points $\pm u$, $\pm v$, $\pm(u + v)$ and deduce that the cubic equation

$$4p^3 - A^2p^2 - (2AB + g_2)p - (g_3 + B^2) = 0$$

---

[11] The reader is referred to the original paper (Tyrrell & Powell, 1971) for a discussion of the possibility that $\pm\wp^{-1}(a) \pm\wp^{-1}(b) \pm\wp^{-1}(c) \equiv 0 \pmod{\Lambda}$. It turns out that that possibility cannot arise.

has the three roots $\wp(u)$, $\wp(v)$ and $\wp(u+v)$.

Hence show that

$$\wp(u) + \wp(v) + \wp(u+v) = \frac{1}{4}A^2$$

and that

$$A = -\frac{\wp'(u) - \wp'(v)}{\wp(u) - \wp(v)}$$

and conclude that

$$\wp(u+v) = \frac{1}{4}\left(\frac{\wp'(u) - \wp'(v)}{\wp(u) - \wp(v)}\right)^2 - \wp(u) - \wp(v),$$

which is a form of the addition theorem.

Take $\wp(u) = p_1$, $\wp(v) = p_2$, $\wp(w) = p_3$, where $u + v + w = 0$ and use the relation between the symmetric functions of the roots and the coefficients of a polynomial to obtain

$$(p_1 + p_2 + p_3)(4p_1 p_2 p_3 - g_3) = \left(p_1 p_2 + p_2 p_3 + p_3 p_1 + \frac{1}{4}g_2\right)^2.$$

9.7.3 Work through the proof given by Rigby (1981).

9.8   Elliptic curves, the addition theorem and the group law

9.8.1   Introduction

See Cassels (1991); McKean & Moll (1997); Prasolov & Solovyev (1997); and Koblitz (1991) for further details.

In Chapter 7 we saw that the Weierstrass elliptic function, $\wp(z)$, satisfies the differential equation

$$\wp'(z)^2 = 4\wp(z)^3 - g_2 \wp(z) - g_3,$$

where the functions $g_2 = g_2(\tau)$, $g_3 = g_3(\tau)$ are modular functions defined in terms of Eisenstein series on the lattice with ordered basis $(1, \tau)$; see Theorem 7.2. The functions $\wp(z)$, $\wp'(z)$ accordingly are parametric equations for the curve

$$y^2 = 4x^3 - g_2 x - g_3, \tag{9.81}$$

with $x = \wp(z)$, $y = \wp'(z)$, and the curve is non-singular if the discriminant $\Delta = g_2^3 - 27g_3^2 \neq 0$. The addition theorem for the function $\wp(z)$ (see Chapter 7, Theorem 7.7, (7.46) and Exercise 9.7.2) may be interpreted geometrically in

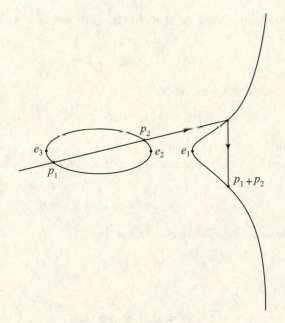

Figure 9.9 Addition on the cubic.

terms of the addition of points on the curve (9.81) in accordance with Figure 9.9. We shall look more closely into that.

We begin by observing that the curve in (9.81) is defined over the field of complex numbers, $\mathbb{C}$, and the properties of the Weierstrass functions used in relation to the addition theorem depend on complex variable theory. But many of the most interesting questions related to such curves are posed in terms that require the coefficients to be rational numbers or even elements of a finite field (for example, a curve whose coefficients are integers modulo a prime number, p). Indeed, those last seem to be the most abstract, yet they have important applications in cryptography (see Koblitz, 1991); so we begin by considering the case where the coefficients are in some field, K.

Let K be a field of characteristic not equal to 2 or 3 and let $x^3 + ax + b$, $a, b \in K$, be a cubic polynomial with no multiple roots and discriminant $4a^3 + 27b^2 \neq 0$. By analogy with (9.81) we say that an *elliptic curve* Γ, over K is the set of points (x, y) such that

$$\Gamma : y^2 = x^3 + ax + b, \quad 4a^3 + 27b^2 \neq 0, \tag{9.82}$$

together with a point, O, the 'point at infinity'. If the field K has characteristic 2, we replace (9.82) by

$$y^2 + y = x^3 + ax + b; \qquad (9.83)$$

if the characteristic is 3, we consider

$$y^2 = x^3 + ax^2 + bx + c, \qquad (9.84)$$

together with the point O at infinity.

The general form of the equation of an elliptic curve, that applies to any K, is

$$y^2 + a_1xy + a_3y = x^3 + a_2x^2 + a_4x + a_6,$$

which can be transformed when $char\ (K) \neq 2$ to

$$y^2 = x^3 + ax^2 + bx + c$$

and if $char\ (K) > 3$ to

$$y^2 = x^3 + ax + b.$$

The vague definition of O as the 'point of infinity' may be made more precise by introducing homogeneous coordinates (X, Y, Z) such that $x = X/Z$, $y = Y/Z$ and the 'point at infinity' is then the point for which $Z = 0$. The equation of the curve (9.82) is then:

$$Y^2Z = X^3 + aXZ^2 + bZ^3. \qquad (9.85)$$

9.8.2   The addition of points on an elliptic curve and the group law

We start with an irreducible, non-singular cubic curve as in (9.82); we propose to define an addition of points on the cubic, which, as we shall see, is related to the addition formula for the Weierstrass function, $\wp(z)$.

To that end, we take two points P, Q on the curve Γ defined by (9.82) and then there is a unique line, l, passing through P and Q. The line l meets the curve Γ in a third point $\overline{T}$ (if $P = Q$, then l is the tangent to Γ at P). We can now define a function $\phi : \Gamma \times \Gamma \to \Gamma$ by taking $\phi(P, Q) = T$, where T is the reflection of $\overline{T}$ in the x-axis. In order to make ϕ behave like addition, we define

$$P + Q = \phi(O, \phi(P, Q)), \qquad (9.86)$$

where O is the point at infinity (see Figure 9.9). In the case when l is the tangent to Γ at P and $P = Q$, then $\phi(P, Q) = P$. The fundamental result is the following one.

Theorem 9.4 *The definition (9.86) yields the structure of an Abelian group on the curve Γ.*

Proof It is easy to see that the addition (9.86) is commutative and also that O is the (additive) identity, since

$$O + P = P$$

for all P. Next we construct the inverse. Let the third intersection of the tangent at O be S, and denote by P' the third intersection of the line through P and S with Γ. Then, by definition,

$$P + P' = O;$$

so P' is the additive inverse to P.

The difficulty is to prove the associate law:

$$(P + Q) + R = P + (Q + R). \tag{9.87}$$

In order to prove (9.87) we appeal to the following result (whose proof we shall outline) from algebraic geometry. Suppose that Γ is an irreducible cubic and let Γ_1, Γ_2 be cubics. Suppose that Γ and Γ_1 intersect in nine simple points on Γ. Then if Γ and Γ_2 intersect in eight of those points; they also intersect in the ninth.

A cubic curve is defined by a cubic form $F(X_1, X_2, X_3)$ having ten coefficients. If (x_1, x_2, x_3) is a point on the curve then the equation $F(x_1, x_2, x_3) = 0$ imposes a linear condition on the coefficients, and so if the curve passes through eight points, eight linear conditions are imposed. It follows that if $F_1(X_1, X_2, X_3)$ and $F_2(X_1, X_2, X_3)$ are linearly independent forms through the eight points, then any other F may be expressed as

$$F(X_1, X_2, X_3) = \lambda F_1(X_1, X_2, X_3) + \mu F_2(X_1, X_2, X_3).$$

Now the curves $F_1 = 0$ and $F_2 = 0$ have nine points in common and the curve with equation $F = 0$ must pass through them all. Whence the result.

We return to the proof of (9.87). Let l_i, m_i be lines and let P, Q, R be points on the elliptic curve Γ.
Let

$$l_1 \cap \Gamma = \{P, Q, \overline{S}\}, \quad m_1 \cap \Gamma = \{O, \overline{S}, S\}, \quad l_2 \cap \Gamma = \{S, R, \overline{T}\}.$$

Let

$$m_2 \cap \Gamma = \{Q, R, \overline{U}\}, \quad l_3 \cap \Gamma = \{O, \overline{U}, U\}, \quad m_3 \cap \Gamma = \{P, U, T'\}.$$

Since

$$(P + Q) + R = O - \overline{T}, \quad P + (Q + R) = O - T',$$

it suffices to show that $T' = \overline{T}$.

Let Γ_1 be the cubic defined by the product $l_1 l_2 l_3$ of three lines and let Γ_2 be defined similarly by $m_1 m_2 m_3$. Then by the result proved above we have the desired result.

We indicate briefly how the foregoing may be related to the group law given by the picture in Figure 9.9.

Every cubic in the two dimensional projective space over $\mathbb{R}$ can be written in the form

$$\Gamma : Y^2 Z = X^3 + a X Z^2 + b Z^3.$$

We take the point at infinity, O, to be given by $(0, 1, 0)$ and then Γ is the affine curve

$$y^2 = x^3 + ax + b,$$

with the additional point at infinity.

As for the group law, if $S = P + Q, \overline{S} = \phi(P, Q)$ and if the coordinates of S are (x, y), we have $\overline{S} = (x, -y)$; that is, the map $S \mapsto \overline{S}$ defines a symmetry of the curve (reflection in the x-axis) and $\overline{S} = -S$.

The result proved in Theorem 9.4 is fundamental not only in the arithmetical applications of the theory to problems associated with finding points on elliptic curves with coordinates in fields of arithmetical interest, but also to practical problems in cryptography, where the group law on an elliptic curve over a finite field, $\mathbb{F}_q$, of $q = p^n$ elements may be used to encrypt messages using the group law on the curve to replace the group law in the finite field used in the RSA method in public key cryptography. (See the book by Koblitz, 1991, for a complete account.)

9.9   Concluding remarks and suggestions for further reading

We conclude with a brief survey of some of the most exciting connections between elliptic functions and algebraic geometry and arithmetic in the hope of encouraging the reader to turn to more authoritative sources; to use old-fashioned language a kind of 'U-trailer advertising an A-film'.

The book by Cassels (1991), affords an excellent introduction to the connection between elliptic functions and *Diophantine geometry*; that branch of algebraic geometry that connects the subject with the theory of Diophantine equations – the solution of algebraic equations in integers or rational numbers. In Diophantine geometry, as in Section 9.8, one replaces the complex numbers, $\mathbb{C}$, by the rational numbers, $\mathbb{Q}$, or by a finite field $\mathbb{F}_q$ of q elements, or by finite extensions of those, in order to investigate such questions as: how many solutions does the equation $y^2 = x^3 + ax + b$, where a, b are elements of the finite field $\mathbb{F}_q$, such that $4a^3 + 27b^2 \neq 0$, have? Thus if we denote by Γ the curve whose equation is

$$y^2 = x^3 + ax + b, \quad \Delta = -(4a^3 + 27b^2) \neq 0, \tag{9.88}$$

then, if N_q denotes the number of points of Γ defined over $\mathbb{F}_q$ (that is with coordinates in $\mathbb{F}_q$) then a famous theorem, due to Hasse and generalized by Weil and by Deligne, asserts that

$$|N_q - (q + 1)| \leq 2q^{1/2}.$$

That result is referred to as the 'Riemann Hypothesis for function fields' because of an analogy with Riemann's 'Riemann Hypothesis', related to the distribution of prime numbers.

To investigate the connection a little more closely, we denote by c_n the number of points of Γ defined over an extension of degree n of the underlying finite field, k, and then the function $Z(u)$ is defined by

$$Z'(u)/Z(u) = u^{-1} \sum_{n=1}^{\infty} c_n u^n. \tag{9.89}$$

If we write $u = q^{-s}$, $\operatorname{Re} s > 1$, we see that $Z(u)$ resembles the Riemann zeta function and if $Card(k) = q$, then we have the functional equation

$$Z\left(\frac{1}{qu}\right) = Z(u), \tag{9.90}$$

which resembles the functional equation of the Riemann zeta function, which we obtain in Chapter 12. Moreover,

$$(1 - u)(1 - qu)Z(u) = (1 - \lambda_1 u)(1 - \lambda_2 u), \tag{9.91}$$

where Hasse's Theorem implies $|\lambda_i| \leq q^{1/2}$, which resembles the Riemann Hypothesis, which asserts that the zeros of the Riemann zeta function, $\zeta(s)$, have real part 1/2. (It should be noted that the form of $Z(u)$ given in (9.91) and its functional equation, (9.90), assume that we are dealing with an *elliptic curve*, that is a curve of *genus one*. See Theorem 7.9 and the remark preceding the statement of it.) The ideas here generalize to curves of higher genus and to algebraic varieties of higher dimension. (See Chapter 25 of Cassels, 1991.)

We return to the ideas in Section 9.8. The addition theorem on an elliptic curve leads to the idea of a group law on the curve. The structure of that group is connected with the structure of the set of rational points on an elliptic curve and leads to the notion of the group, G, of rational points on an elliptic curve defined over $\mathbb{Q}$. It was proved by Mordell (see Cassels, 1991, Chapter 13) that the group G is finitely generated – that is one can find a finite basis in terms of which all the rational points may be expressed – and that theorem was generalized by Weil to the case of number-fields – the *Mordell–Weil Finite Basis theorem* (Cassels, 1991, and see also McKean & Moll, 1997, Chapter 7 and Prasolov & Solovyev, 1997, Chapter 5).

It follows from the finite basis theorem that if E denotes an elliptic curve defined over $\mathbb{Q}$, then the group $E(\mathbb{Q})$ of rational points is finitely generated and accordingly admits the decomposition

$$E(\mathbb{Q}) = \mathbb{Z}^{r_E} \times Tors\, E(\mathbb{Q})$$

where r_E denotes the *rank* of $E(\mathbb{Q})$ and *Tors* $E(\mathbb{Q})$ is the subgroup of elements of finite order, the *torsion group*.

A celebrated conjecture by Birch and Swinnerton-Dyer (1965), based on insight supported by massive calculations, relates the number r_E to the order of the zero at $s = 1$ of an analytic function of a complex variable, s, similar to the zeta function introduced above and to the Riemann zeta function and the associated L-functions. To formulate it we return to the elliptic curve

$$y^2 = x^3 + ax + b, \quad \Delta = -(4a^3 + 27b^2) \neq 0,$$

defined over $\mathbb{Q}$ and its reduction mod p,

$$y^2 = x^3 + \bar{a}x + \bar{b}, \quad \bar{a}, \bar{b} \in \mathbb{F}_p, \quad p \nmid \Delta.$$

If N_p denotes the number of points defined over $\mathbb{F}_p$ (c_p in our earlier notation, but we use N_p to facilitate comparison with our references) then, as we have seen, Hasse proved that $|p + 1 - N_p| \leq 2\sqrt{p}$ and one defines the L-function by

$$L(E, s) = \prod_{p|\Delta}(1 - a_p p^{-s})^{-1} \prod_{p \nmid \Delta}(1 - a_p p^{-s} + p^{1-2s})^{-1}, \tag{9.92}$$

where $a_p = p + 1 - N_p$. The conjecture of Birch and Swinnerton-Dyer is that the rank r_E is equal to the order of the zero of

$$\tilde{L}(E, s) = \prod_{p \nmid \Delta}(1 - a_p p^{-s} + p^{1-2s})^{-1} \qquad (9.93)$$

at $s = 1$. (See Prasolov & Solovyev, 1997, for details. See also the paper by Zagier, 1991 for a very clear introduction.)

Our final application, which alas, will fail to do justice to its theme, though we hope the reader will be persuaded to read more (for example in Cornell, Silverman and Steven, 1997)[12] is to the proof of Fermat's Last Theorem, by Taylor and Wiles. We recall that the assertion of this famous 'theorem' whose proof was announced by Wiles in 1993, with a gap that was not closed but circumvented by Taylor and Wiles in 1994, asserts that the equation $x^n + y^n = z^n$, $n \geq 3$, has no solution in non-zero integers x, y, z. Wiles' method associates with a solution of $a^l + b^l = c^l$, a, b, c relatively prime and $l \geq 3$, the elliptic curve

$$y^2 = x(x - a^l)(x - c^l), \qquad (9.94)$$

called the Frey curve. An elliptic curve $y^2 = f(x)$, where $f(x)$ is a cubic, is said to have *good reduction at* p if the zeros modulo p of f are distinct, and *bad reduction* otherwise. The Frey curve (9.94) has bad reduction exactly at the prime divisions of abc. The proof of Taylor and Wiles is indirect; one supposes that Fermat's theorem is false and then one shows that the curve (9.94) must have good reduction. But the curve has bad reduction, as already noted, and that contradiction establishes the truth of Fermat's Last Theorem.

---

[12] See also the expository article by G. Faltings (1995).

10

An application of elliptic functions in algebra – solution of the general quintic equation

Introduction

One of the earliest applications of the theory of elliptic functions, indeed one of its origins, is to be found in ideas arising from Gauss' criterion for the constructibility by ruler and compass of a regular polygon of n sides, namely that $n = 2^m p_1 p_2 \ldots p_r$, where the p_i are distinct Fermat primes, of the form $p_i = 2^{2^{k_i}} + 1$ (see Gauss, 1801, Section 7, or Hardy & Wright, 1979, Chapter V). That problem played a significant part in the development of Galois theory and also in the work of Abel, who proved that, for those values of n given by Gauss, it is possible to divide Bernoulli's lemniscate into n equal parts by ruler and compass, a problem that in turn played a part in the creation of the theory of elliptic functions (see Prasolov & Solovyev, 1997, for historical background).

In this Chapter and in Chapter 11 we shall be looking at some applications of the theory of elliptic functions to problems in algebra and arithmetic. We have already encountered applications to arithmetic in the references to Diophantine geometry in Chapter 9. Some of the most important applications of the theory, which, again, were significant influences in the development of the theory, are to be found in the solutions of polynomial equations – to the theory of the quintic as outlined here and in generalizations that require the theory of Abelian functions. We shall offer two approaches to the problem of the solution of the general quintic equation, one due to Hermite and Klein (see Hermite, 1861, Klein, 1884, and the exposition by McKean & Moll, 1997), the other to be found in Weber (1908) and the exposition by Prasolov & Solovyev (1997).

10.1 Revision of the quadratic, cubic and quartic equations

We begin with the familiar quadratic equation

$$ax^2 + bx + c = 0 \quad (a, b, c \in \mathbb{C}, a \neq 0),$$

or, equivalently,

$$x^2 + Bx + C = 0 \quad (B = b/a, \ C = c/a).$$

One solves that equation by writing $y = x + B/2$ (so as to 'complete the square') and so obtain

$$y^2 = \frac{B^2}{4} - C.$$

Whence, on substituting back to the original equation,

$$x = (-b \pm \sqrt{b^2 - 4ac})/2a. \tag{10.1}$$

Of course, all that is very familiar, but we begin with it in order to throw light on the solutions of the general cubic and quartic equations — and then to see how (and why) the general quintic is different.

Consider then the general cubic equation

$$ax^3 + bx^2 + cx + d = 0, \quad a, b, c, d \in \mathbb{C}, \quad a \neq 0,$$

and divide through by a and replace x by $y = x + b/(3a)$, the latter being an (unsuccessful) attempt to 'complete the cube' to obtain the equation

$$y^3 + py + q = 0, \tag{10.2}$$

where $p = c/a - b^2/(3a^2), q = d/a - bc/(3a^2) + 2b^3/(27a^3)$. The roots of that new equation differ from those of the original equation by $b/3a$.

Now make the substitution[1] $y = z - p/(3z)$ to yield

$$z^3 - \frac{p^3}{27z^3} + q = 0,$$

from which we obtain a quadratic in z^3 that gives the solutions

$$z^3 = -\frac{q}{2} \pm \sqrt{\frac{q^2}{4} + \frac{p^3}{27}}, \tag{10.3}$$

---

[1] This substitution is usually attributed to Vieta. The general solution of the cubic is associated with Cardan and Tartaglia (the latter having priority) and the solution of the quartic given here with Ferrari. We shall not go into the involved history of those solutions – see Stewart (2003). The solution of the quadratic goes back to the Babylonians, and to the Hindus and, in its geometrical form, was known to the Greeks.

whence *six* solutions to the original cubic. If we now write $y = z - p/(3z)$ we obtain three pairs of solutions for y, paired solutions being equal.

To solve the quartic, we start with

$$ax^4 + bx^3 + cx^2 + dx + e = 0, \quad a \neq 0,$$

and, after dividing through by a and replacing x by $z = x + b/(4a)$ (again an attempt to 'complete the quartic') we obtain

$$z^4 + pz^2 + qz + r = 0, \tag{10.4}$$

whose roots differ from those of the original equation by $b/4a$.

Now try to represent our quartic as the difference of two squares (we revert to using x in place of z)

$$x^4 + px^2 + qx + r = \left(x^2 + \frac{p}{2} + t\right)^2 - \left(2tx^2 - qx - r + \frac{p^2}{2}\right)$$

$$= \left(x^2 + \frac{p}{2} + t\right)^2 - \left(\left(\sqrt{2t}x - \frac{q\sqrt{2}}{4\sqrt{t}}\right)^2 - \frac{q^2}{8t}\right.$$

$$\left. + \left(t^2 + pt - r + \frac{p^2}{2}\right)\right),$$

where the idea is to choose t so that

$$q^2 - 8t\left(t^2 + pt - r + \frac{p^2}{2}\right) = 0. \tag{10.5}$$

Then we can solve our quartic by first solving the cubic equation (10.5) for t, and then, if t_0 is one of its roots, the problem of solving the quartic

$$x^4 + px^2 + qx + r = 0$$

amounts to solving the quadratic

$$x^2 + \frac{p}{2} + t_0 = \pm\sqrt{2t_0}\left(x - \frac{q}{4t_0}\right). \tag{10.6}$$

Thus the problem of solving the general quartic is equivalent to the problem of solving a cubic and a quadratic.

Accordingly, we can always solve the general quadratic, cubic or quartic in terms of square roots, cube roots or fourth roots of expressions already derived from the coefficients of the original equations; such solutions lead us to say that the equations are 'solvable in radicals'.

It is natural to ask whether or not the general quintic can be solved by similar arguments using square roots, cube roots, fourth roots and fifth roots. Many attempts were made to solve equations of degree greater than 4 in terms of 'radicals', but they were doomed to failure. It was Abel (whose ideas have influenced much of our development of the theory of elliptic functions) who first showed that the general quintic is *not* solvable by radicals; his argument is reproduced in Prasolov and Solovyev (1997). Another proof is due to Galois (see Stewart, 2003), who interpreted the problem of solving an equation by successive adjunction of radicals in terms of the solvability of Galois groups of equations. More precisely, a polynomial with coefficients in a field, F, which contains all the roots of unity is said to be solvable by radicals if there is a normal extension, K, of F and a sequence of intermediate fields, K_i, such that

$$F = K_0 \subset K_1 \subset K_2 \subset \ldots \subset K_l = K,$$

where $K_i = K_{i-1}(x_i)$, $x_i^{n_i} \in K_{i-1}$. If G denotes the Galois group of K/F, then G is said to be solvable if and only if there is a chain of subgroups that corresponds to a sequence of field extensions K_i, and in order to prove that an equation is not solvable in radicals it suffices to show that the corresponding group is not solvable. Since one can prove that the symmetric group on n letters is not solvable if $n > 4$, and in particular there exists a real quintic equation (whose coefficients are in $\mathbb{R}$) whose Galois group is the symmetric group on five letters, it follows that the general quintic is not solvable by radicals. (See Stewart, 2003 for details.)

But the general quintic can be solved in terms of elliptic functions (or theta functions) and we shall now give two proofs of that truly remarkable application of the theory of elliptic functions.

Before proceeding to the proof, however, we remind the reader that a cubic can also be solved using trigonometric or hyperbolic functions and we give an example, which may help to throw some light on our solution of the quintic.

We begin with our cubic equation in normal form

$$y^3 + py + q = 0, \tag{10.7}$$

and we shall suppose that $p > 0$ and $q^2/4 + p^3/27 > 0$ (two of the roots are imaginary).

Consider the equation

$$4 \sinh^3 u + 3 \sinh u = \sinh 3u, \tag{10.8}$$

which is derived from (10.7) by the substitution

$$y = a \sinh u,$$

where $p = 3a^2/4$, $q = -(a^3 \sinh 3u)/4$; whence

$$y^3 + \frac{3}{4}a^2 y - \frac{1}{4}a^3 \sinh 3u = 0. \tag{10.9}$$

Note that

$$\sinh 3u = -4 \left(\frac{27}{64} \frac{q^2}{p^3} \right)^{1/2}. \tag{10.10}$$

The roots of (10.8) are

$$\sinh u, \quad \sinh \left(u + \frac{2}{3} \pi i \right), \quad \sinh \left(u + \frac{4}{3} \pi i \right); \tag{10.11}$$

So the roots of (10.7) are

$$\sqrt{\frac{4}{3} p} \sinh u, \quad \sqrt{\frac{1}{3} p} (- \sinh u \pm i \sqrt{3} \cosh u). \tag{10.12}$$

From (10.10) we obtain the value of u and then we substitute that in (10.12) to obtain the roots of (10.7).

The reader may find it helpful to bear in mind how the hyperbolic functions are used when working through the more complicated details in the case of the quintic, when the hyperbolic functions are replaced by functions related to elliptic functions.

10.2   Reduction of the general quintic equation to normal form

Our earlier work on the cubic and quartic equations started from the observation that those equations can be reduced to equations in which the coefficient of the second highest power of the variable is 0 (see (10.2) and (10.4)). Our aim here is to show that the general quintic may be reduced to the form $x^5 + px + q$ (called Bring's form after the Swedish lawyer who discovered it) and to do that we begin by showing that the second, third and fourth terms may be removed from an equation of degree n. We shall follow the method of Tschirnhausen (or Tschirnhaus as perhaps it should be), as described in Burnside and Panton (1901); see also Prasolov and Solovyev (1997).

We begin by proving:

Proposition 10.1 *Denote by $V = V(x_1, \ldots, x_n)$ a homogeneous function of degree 2 in the n variables $x_1, \ldots, x_n$. Then V can be expressed as the sum of n squares.*

Proof We may write V in the form

$$V(x_1, \ldots, x_n) = P_1 x_1^2 + 2Q_1 x_1 + R_1,$$

where P_1 does not contain any of the variables $x_1, \ldots, x_n$ and where Q_1, R_1 are linear and quadratic functions, respectively, of $x_2, x_3, \ldots, x_n$. It follows that

$$V = \left(\sqrt{P_1}x_1 + \frac{Q_1}{\sqrt{P_1}}\right)^2 + R_1 - \frac{Q_1^2}{P_1}.$$

Now $R_1 - Q_1^2/P_1$ is a quadratic function of $x_2, \ldots, x_n$ and so we may repeat the argument to show that

$$V_1 = R_1 - \frac{Q_1^2}{P_1} = P_2 x_2^2 + 2Q_2 x_2 + R_2,$$

where P_2 is a constant and Q_2, R_2 do not contain x_1 and x_2. Whence

$$V_1 = \left(\sqrt{P_2}x_2 + \frac{Q_2}{\sqrt{P_2}}\right)^2 + R_2 - \frac{Q_2^2}{P_2},$$

and it follows that we may write

$$V = \left(\sqrt{P_1}x_1 + \frac{Q_1}{\sqrt{P_1}}\right)^2 + \left(\sqrt{P_2}x_2 + \frac{Q_2}{\sqrt{P_2}}\right)^2 + R_2 - \frac{Q_2^2}{P_2}.$$

We now repeat the argument, removing $x_3, x_4, \ldots$ in succession until we reach

$$R_{n-1} - \frac{Q_{n-1}^2}{P_n},$$

which is equal to $P_n x_n^2$. That proves the proposition.

Now consider the equation

$$p(x) = x^n + p_1 x^{n-1} + p_2 x^{n-2} + \cdots + p_n = 0 \tag{10.13}$$

and write

$$y = \alpha x^4 + \beta x^3 + \gamma x^2 + \delta x + \varepsilon, \tag{10.14}$$

to obtain the transformed equation

$$q(y) = y^n + q_1 y^{n-1} + q_2 y^{n-2} + \cdots + q_n = 0. \tag{10.15}$$

We show that the coefficients $q_1, q_2, \ldots, q_r, \ldots$ are homogeneous functions of degrees $1, 2, \ldots, r, \ldots$ (respectively) in $\alpha, \beta, \gamma, \delta, \varepsilon$, and to that end we adopt the ideas in Burnside and Panton (1901).

We have

$$y^n = (\alpha x^4 + \cdots + \varepsilon)^n,$$
$$y^{n-1} = (\alpha x^4 + \cdots + \varepsilon)^{n-1},$$

and so on, and we can reduce each polynomial y^r, $1 \leq r \leq n$, mod $p(x)$ to obtain polynomials of degrees $\leq n - 1$; thus

$$
\begin{aligned}
y &= \varepsilon + \delta x + \gamma x^2 + \beta x^3 + \alpha x^4 \\
&= a_{10} + a_{11}x + a_{12}x^2 + a_{13}x^3 + a_{14}x^4, \\
y^2 &= a_{20} + a_{21}x + \cdots \qquad\qquad\quad + a_{2,n-1}x^{n-1}, \\
y^n &= a_{n0} + a_{n1}x + \cdots \qquad\qquad\quad + a_{n,n-1}x^{n-1}.
\end{aligned}
\tag{10.16}
$$

Now denote by s_r the sum of the rth powers of the roots of $p(x) = 0$ and by σ_r the sum of the rth powers of the roots of $q(y) = 0$. Then

$$
\sigma_1 = n\varepsilon + \delta s_1 + \gamma s_2 + \beta s_3 + \alpha s_4,
$$
$$
\sigma_2 = na_{20} + a_{21}s_1 + a_{22}s_2 + \cdots
$$

and so on. If we factorize $p(x)$ as

$$
p(x) = (x - \xi_1)(x - \xi_2) \cdots (x - \xi_n),
\tag{10.17}
$$

then

$$
s_1 = \sum_{i=1}^{n} \xi_i, \quad s_2 = \sum \xi_i^2, \quad \ldots, \quad s_n = \sum \xi_i^n
$$

and so

$$
s_1 = -p_1, \quad s_2 = p_1^2 - 2p_2, \quad s_3 = -p_1^3 + 3p_1p_2 - 3p_3.
\tag{10.18}
$$

It follows that every rational symmetric function of the roots can be expressed rationally in terms of the coefficients. Evidently p_r involves no sums of powers beyond s_r.

One can also express the coefficients $p_1, p_2, \ldots$ in terms of the s_r, for example

$$
p_1 = -s_1, \quad p_2 = -\frac{1}{2}s_2 + \frac{1}{2}s_1^2, \quad p_3 = -\frac{1}{3}s_3 + \frac{1}{1 \cdot 2}s_1s_2 - \frac{1}{6}s_1^3.
$$

Denote by $\sigma_1, \sigma_2, \ldots$ the sums of the powers of the roots of the equation (10.15) (thus if

$$
q(y) = (y - \eta_1)(y - \eta_2) \cdots (y - \eta_n),
\tag{10.19}
$$

then

$$
\sigma_1 = \sum_{k=1}^{n} \eta_k, \quad \sigma_2 = \sum \eta_k^2, \quad \ldots).
\tag{10.20}
$$

By comparison with (10.16), we see that

$$\sigma_1 = n\varepsilon + \delta s_1 + \gamma s_2 + \beta s_3 + \alpha s_4,$$
$$\sigma_2 = na_{20} + a_{21}s_1 + a_{22}s_2 + \cdots,$$

$$\cdots$$

$$\sigma_n = na_{n0} + a_{n1}s_1 + a_{n2}s_2 + \cdots + a_{n,\,n-1}s_{n-1}. \qquad (10.21)$$

As we have seen, we can express the $s_1, s_2, \ldots$ in terms of the coefficients of the polynomial (10.12) and so we can express the $\sigma_1, \ldots, \sigma_n$ in terms of the coefficients in (10.12) Moreover, as already noted, we can express the coefficients of the equation whose roots are given by substituting $\xi_1, \ldots, \xi_n$ in (10.15) in terms of $\sigma_1, \ldots \sigma_n$ and so in terms of the coefficients in (10.15) and (10.12). That completes our assertion about the coefficients $q_1, \ldots, q_n$.

If we can now determine $\alpha, \beta, \gamma, \delta, \varepsilon$ so that

$$q_1 = 0, \quad q_2 = 0, \quad q_3 = 0,$$

the problem will be solved. To that end, eliminate ε from q_2 and q_3 by substituting its value derived from $q_1 = 0$ and so obtain two homogeneous equations

$$r_2 = 0, \quad r_3 = 0$$

of degrees 2 and 3 respectively in $\alpha, \beta, \gamma, \delta$. But by Proposition 10.1 we may write r_2 in the form

$$u^2 - v^2 + w^2 - t^2$$

and so $r_2 = 0$ if we take $u = v, w = t$. From those equations we obtain

$$\gamma = l_1\alpha + m_1\beta, \quad \delta = l_2\alpha + m_2\beta$$

and, substituting those values in $q_3 = 0$, we obtain a cubic equation to determine β/α.

It follows that if any of $\alpha, \beta, \gamma, \delta, \varepsilon$ is given a particular value, then the others are determined. So our original equation is reduced to the form

$$y^n + q_4y^{n-4} + q_5y^{n-5} + \cdots + q_n = 0.$$

We state our results as:

Proposition 10.2 *The equation*

$$x^n + p_1x^{n-1} + p_2x^{n-2} + \cdots + p_n = 0$$

may be reduced to the form

$$y^n + q_4y^{n-4} + q_5y^{n-5} + \cdots + q_n = 0$$

by a substitution

$$y = \alpha x^4 + \beta x^3 + \gamma x^2 + \delta x + \varepsilon.$$

Corollary 10.1 *The general quintic equation*

$$x^5 + p_1 x^4 + p_2 x^3 + p_3 x^2 + p_4 x + p_5 = 0 \tag{10.22}$$

may be reduced to the form (Bring's form)

$$x^5 + ax + b = 0. \tag{10.23}$$

10.3   Elliptic functions and solution of the quintic: outline of the proof

In this section we outline two approaches to the problem: one uses theta functions and the transformation theory developed in Chapter 6, especially that related to Dedekind's eta function; the other is related to the theory of the icosahedral group and properties of the modulus, k, and its representation by theta functions. In both approaches, but especially in the second, we content ourselves with the essentials of the argument. More details concerning the first method can be found in the books by Weber (1908) (though from a different point of view) and Prasolov and Solovyev (1997), on which our account is based, and for the second method the reader is referred to the books by Klein (1884), McKean & Moll (1997), and Briot & Bouqet (1875), for further details. Our summary is a somewhat expanded version of that in Dutta and Debnath (1965). Both methods rely on obtaining an equation of degree 6, the modular equation, and thence a quintic of the form (10.5), with $a = 5$ or $a = 1$, and in which b is expressible in terms of theta functions (or related functions). One then appeals to Corollary 10.1 to obtain the solution of the general quintic.

The first method begins by reducing the general quintic, (10.22), to the special form of (10.23) given by

$$x^5 + 5x + b = 0. \tag{10.24}$$

The idea then is to show that for every $b \in \mathbb{C}$, we can express b in terms of functions of τ, where τ is in the upper half-plane, H, and the expression involves theta functions or the functions introduced in Section 6.7. Therein lies the crux of the argument, and the connection is established by relating the quintic (10.24) to an equation of degree 6 in two variables u, v, called the *modular equation*, where $u = f(\tau)$, $v = f(5\tau)$ and f is the function

$$f(\tau) = q^{-\frac{1}{4}} \prod_{k=1}^{\infty} (1 + q^{2k-1}), \quad q = e^{\pi i \tau}, \quad \operatorname{Im} \tau > 0. \tag{10.25}$$

(See Section 6.7 and Exercises 6.7 for properties of the function $f(\tau)$ and its connection with theta functions.)

By using the transformation properties of $u = f(\tau)$ and $v = f(5\tau)$ (as established in Chapter 6 and see (10.37) below) it will be shown that u, v satisfy the equation

$$u^6 + v^6 - u^5 v^5 + 4uv = 0, \tag{10.26}$$

and that equation, one form of the modular equation, will be used to construct an equation of degree 5 of the form

$$y^5 + 5y - \frac{f_1^8 - f_2^8}{f^2} = 0, \tag{10.27}$$

where f, f_1 and f_2 are the functions defined in Section 6.7 and Exercises 6.7, and one notes that (10.27) is of Bring's form.

So to solve the general quintic equation one first reduces it to the form (10.24) and one then finds $\tau \in H$ (the upper half-plane) such that

$$\frac{f_1^8(\tau) - f_2^8(\tau)}{f^2(\tau)} = b \tag{10.28}$$

(which, as we shall see, is possible). That will solve the problem, since the roots of the quintic (10.27) are expressible in terms of functions involving τ; the details are given in Sections 10.4 to 10.8.

As already remarked, there is a variation on that theme, which we outline here. The basic idea is essentially the same: find a 'modular equation' in two variables u, v of degree 6 and then use it to obtain an equation of the form

$$x^5 - x - A = 0, \tag{10.29}$$

where

$$A = \frac{2}{5^{5/4}} \frac{1 + u^8}{u^2 (1 - u^8)^{1/2}} \tag{10.30}$$

and $u = (k(\tau))^{1/4}$, which may be expressed as a quotient of theta functions. Here $u^4 = k(\tau)$ is the modulus of the Jacobi function and we take $v^4 = k(5\tau)$. The modular equation in this case, of degree 6, in u and v, is obtained by comparing the integrals

$$\int \frac{dx}{\sqrt{(1-x^2)(1-k^2x^2)}}, \quad \int \frac{dy}{\sqrt{(1-y^2)(1-k_1^2y^2)}}, \tag{10.31}$$

where $k = k(\tau)$, $k_1 = k(5\tau)$. As we shall see, in that case the modular equation reads

$$u^6 - v^6 + 5u^2v^2(u^2 - v^2) + 4uv(1 - u^4v^4) = 0. \tag{10.32}$$

So, finally, one solves the general quintic by reducing it to the form (10.29), then one solves (10.30) for τ (as is possible) and then the solutions of the quintic are given in terms of the theta functions given by that value of τ.

The first method (Sections 10.4 to 10.8)

10.4   Transformation theory and preparation for application to the modular equation for the quintic

As already outlined, the first method for the solution of the quintic, (10.12), begins by reducing it to the special form (Bring's form)

$$x^5 + 5x + b = 0 \tag{10.33}$$

and then we write that in the form

$$y^5 + 5y = \frac{f_1^8 - f_2^8}{f^2}, \tag{10.34}$$

where $f = f(\tau)$, etc., and the functions $f(\tau)$, $f_1(\tau)$, $f_2(\tau)$ are defined in Chapter 6, Exercises 6.7. We recall their definitions, as follows.

Let Im $\tau > 0$, put $q = e^{\pi i \tau}$ and then

$$f(\tau) = q^{-1/24} \prod_{n=1}^{\infty} (1 + q^{2n-1}),$$

$$f_1(\tau) = q^{-1/24} \prod_{n=1}^{\infty} (1 - q^{2n-1}),$$

$$f_2(\tau) = \sqrt{2} q^{1/12} \prod_{n=1}^{\infty} (1 + q^{2n}). \tag{10.35}$$

Each of those functions is expressible in terms of the Dedekind eta function

$$\eta(\tau) = q^{1/12} \prod_{n=1}^{\infty} (1 - q^{2n}) \tag{10.36}$$

(see Exercise 6.7.4) and we shall make extensive use of the results in Exercises 6.7 throughout the following discussion.

In order to obtain the modular equation (10.32) for the first method, we introduce the variables $u = f(\tau)$ and we allow v to take the values

$$v_c = f\left(\frac{\tau + c}{5}\right), \quad c = 0, 1, 2, 3, 4; \quad v_\infty = f(5\tau), \tag{10.37}$$

and it will emerge that, for a given value of τ, those six values of v_c are the six corresponding values of v in the modular equation (10.26). In order to obtain (10.26), we study the behaviour of the functions u, v_c, as functions of τ, under the transformations of τ given by

$$\tau \mapsto \tau + 2, \quad \tau \mapsto -\tau^{-1}, \quad \tau \mapsto (\tau - 1)/(\tau + 1), \tag{10.38}$$

and we shall see that the v_c, considered as functions of c, also transform in a similar way.

We begin by recalling (see Chapter 6, Sections 6.2 and 6.3) that the transforms of τ (thought of as a column vector $\binom{\tau}{1}$) correspond to the matrices

$$\begin{pmatrix} 1 & 2 \\ 0 & 1 \end{pmatrix}, \quad \begin{pmatrix} 0 & -1 \\ 1 & 0 \end{pmatrix}, \quad \begin{pmatrix} 1 & -1 \\ 1 & 1 \end{pmatrix}, \tag{10.39}$$

respectively, and the formulae for the transformations of $\eta(\tau)$ under $\tau \mapsto \tau + 1$ and $\tau \mapsto -\tau^{-1}$ may be written in terms of the generators

$$T = \begin{pmatrix} 1 & 1 \\ 0 & 1 \end{pmatrix}, \quad S = \begin{pmatrix} 0 & -1 \\ 1 & 0 \end{pmatrix}$$

of the group $SL_2(\mathbb{Z})$ as

$$\eta(T \circ \tau) = \eta(T\tau) = e^{\pi i/12}\eta(\tau),$$
$$\eta(S \circ \tau) = \eta(S\tau) = \sqrt{-i\tau}\,\eta(\tau), \tag{10.40}$$

respectively. (We shall use the simpler $T\tau$ in place of $T \circ \tau$ in what follows, the former implies the 'vector' interpretation of τ, as above.) Thus:

$$T\tau = \begin{pmatrix} 1 & 1 \\ 0 & 1 \end{pmatrix}\begin{pmatrix} \tau \\ 1 \end{pmatrix} = \begin{pmatrix} \tau + 1 \\ 1 \end{pmatrix},$$

$$S\tau = \begin{pmatrix} 0 & -1 \\ 1 & 0 \end{pmatrix}\begin{pmatrix} \tau \\ 1 \end{pmatrix} = \begin{pmatrix} -1 \\ \tau \end{pmatrix},$$

(and the notation is meant to convey the idea of dividing the first component by the second.)

We may express the transformations of $f(\tau)$, $f_1(\tau)$ and $f_2(\tau)$ in a similar way.

The transformation properties are essential to our derivation of the modular equation (10.32) (in its first form) and thence to the construction of the quintic equation (10.34).

We conclude this section by interpreting the definitions (10.37) in matrix form.

We begin by observing that if $A \in SL_2(\mathbb{Z})$, then the transformation $\eta(A\tau)$ may be obtained from (10.40) by using the fact that S, T generate the group $SL_2(\mathbb{Z})$. Thus, for example, if $A = T^2S$ then

$$\eta(A\tau) = \eta(T^2 S\tau) = e^{\pi i/12}\eta(TS\tau) = e^{\pi i/6}\eta(S\tau) = e^{\pi i/6}\sqrt{-i\tau}\,\eta(\tau).$$

We may obtain formulae for $f(A\tau)$, $f_1(A\tau)$ and $f_2(A\tau)$ in a similar way, but the formulae for $f(A\tau)$ may involve f_1 and f_2 (see Exercise (10.4.1)).

It is clear that the transformations defining the $v_c(\tau)$ are of the type

$$\tau \mapsto \begin{pmatrix} 1 & 1 \\ 0 & 5 \end{pmatrix} \tau = \frac{\tau + 1}{5}, \qquad \tau \mapsto \begin{pmatrix} 5 & 0 \\ 0 & 1 \end{pmatrix} \tau = 5\tau$$

and we note that each of those has determinant 5. If $A \in SL_2(\mathbb{Z})$ and if $P = \binom{p\ q}{r\ s}$ is a matrix with integer elements and determinant 5, then the matrix AP also has integer elements and determinant 5 and $f(AP\tau)$ may be expressed in terms of $f(P\tau)$ by expressing A in terms of S, T. So we begin by looking for the simplest forms to which P can be reduced by left-multiplication by $A \in SL_2(\mathbb{Z})$ in order to find $f(P\tau)$, and we hope that our analysis will yield the forms in (10.37)!

So suppose that

$$P = \begin{pmatrix} p & q \\ r & s \end{pmatrix} \in GL_2\mathbb{Z}, \quad \det P = 5$$

and choose $c, d \in \mathbb{Z}$ so that $cp + dr = 0$ and g.c.d.$(c, d) = 1$. Then there exist $a, b \in \mathbb{Z}$ with $ad - bc = 1$ and if

$$A = \begin{pmatrix} a & b \\ c & d \end{pmatrix} \in SL_2(\mathbb{Z})$$

we obtain

$$AP = \begin{pmatrix} a & b \\ c & d \end{pmatrix}\begin{pmatrix} p & q \\ r & s \end{pmatrix} = \begin{pmatrix} ap + br & aq + bs \\ cp + dr & cq + ds \end{pmatrix} = \begin{pmatrix} p' & q' \\ 0 & s' \end{pmatrix} = P',$$

say. We must have $\det P' = 5$ and accordingly $p's' = 5$. Moreover, if

$$\begin{pmatrix} 1 & n \\ 0 & 1 \end{pmatrix} \in SL_2(\mathbb{Z}),$$

then

$$\begin{pmatrix} 1 & n \\ 0 & 1 \end{pmatrix} \begin{pmatrix} p' & q' \\ 0 & s' \end{pmatrix} = \begin{pmatrix} p' & q' + ns' \\ 0 & s' \end{pmatrix},$$

and by choosing n suitably we can ensure that $-s/2 \leq q' \leq s/2$. Evidently the possibilities for p' and s' are $p' = 1, s' = 5$ or $p' = 5, s' = 1$. Accordingly we obtain the six matrices

$$P_\infty = \begin{pmatrix} 5 & 0 \\ 0 & 1 \end{pmatrix}, \quad P_0 = \begin{pmatrix} 1 & 0 \\ 0 & 5 \end{pmatrix},$$

$$P_{\pm 1} = \begin{pmatrix} 1 & \pm 1 \\ 0 & 5 \end{pmatrix}, \quad P_{\pm 2} = \begin{pmatrix} 1 & \pm 2 \\ 0 & 5 \end{pmatrix}, \tag{10.41}$$

which, mod 5, yield the transformations in (10.37).

We shall be interested in finding out how the functions $v_c(\tau)$ behave as functions of c under the transformations (10.38) (or (10.39)) and that will occupy our attention in the following sections. To conclude this section we note that a change of parameter corresponds to the right multiplication of P_c by some matrix B. By multiplying $P_c B$ on the left by $A \in SL_2(\mathbb{Z})$ we can obtain the matrix P_d for the new parameter; so that $f(P_c B\tau)$ can be expressed in the form

$$v_d(\tau) = f(P_d \tau) = f(P_c B\tau).$$

Similar remarks apply to $f_1(\tau)$ and $f_2(\tau)$.

Exercise 10.4

10.4.1 Show that if $A = T^2 S$ then

$$\eta(A\tau) = e^{\pi i/6} \sqrt{-i\tau} \, \eta(\tau)$$

and find expressions for $f(A\tau)$, $f_1(A\tau)$ and $f_2(A\tau)$.

10.4 The transformation $\tau \mapsto \tau + 2$ and $v_c(\tau)$

From Exercises 6.7.5 and 6.7.6, we recall that:

$$f(\tau + 1) = e^{-\pi i/24} f_1(\tau),$$
$$f(\tau + 2) = e^{-\pi i/12} f(\tau),$$

and

$$f(\tau + 48) = f(\tau). \tag{10.42}$$

It will be noticed that the transformation $\tau \mapsto \tau + 2$ yields an answer involving $f(\tau)$, whereas $\tau \mapsto \tau + 1$ introduces $f_1(\tau)$ and so the former has obvious advantages. We note that, under $\tau \mapsto \tau + 2$,

$$f(5\tau) \mapsto f(5\tau + 10) = e^{-5\pi i/12} f(5\tau)$$

and

$$f\left(\frac{\tau}{5}\right) \mapsto f\left(\frac{\tau+2}{5}\right);$$

so that v_∞ appears to go to v_∞ and v_0 to v_2; one might conjecture that $v_c \mapsto v_{c+2}$ under $\tau \mapsto \tau + 2$. We shall try to introduce a notation in which the effect of $\tau \mapsto \tau + 2$ on v_c is more readily accessible.

We saw in (10.42) that $f(\tau)$ has period 48 and so there are 240 possible values in $f((\tau + c)/5)$. We can partition the additive group $\mathbb{Z}_{240}$ into 48 sets, each having five elements, and we can choose the number c from the same five-tuple by requiring that $c \equiv 0 (\mathrm{mod}\ 48)$. So we change our notation so that

$$v_c = f\left(\frac{\tau+c}{5}\right), \quad c \equiv 0(\mathrm{mod}\ 48), \tag{10.43}$$

and then

$$v_{48} = f\left(\frac{\tau+48}{5}\right), \quad v_{96} = f\left(\frac{\tau+96}{5}\right), \ldots \tag{10.44}$$

It follows that $v_c = v_d$ if $c \equiv d(\mathrm{mod}\ 5)$. That means, since, for example, $48 \equiv -2(\mathrm{mod}\ 5)$, that we have

$$v_{\pm 2} = f\left(\frac{\tau \mp 48}{5}\right), \quad v_{\pm 1} = f\left(\frac{\tau \pm 96}{5}\right). \tag{10.45}$$

Note that $v_{-2} = v_3$ and $v_{-1} = v_4$. Accordingly, we may use (10.45) to give us four expressions for v_c, and the other two are

$$v_0 = f\left(\frac{\tau}{5}\right), \quad v_\infty = f(5\tau). \tag{10.46}$$

In that notation, the change of parameter $\tau \mapsto \tau + 2$ produces the transformation on v_c given by

$$f\left(\frac{\tau+c}{5}\right) \mapsto f\left(\frac{\tau+2+c}{5}\right) = f\left(\frac{\tau+50-48+c}{5}\right)$$

$$= e^{-5\pi i/12} f\left(\frac{\tau+c'}{5}\right),$$

where $c' = c - 48 \equiv 0(\mathrm{mod}\ 48)$ and $c' \equiv c + 2(\mathrm{mod}\ 5)$.

So $v_c \mapsto e^{-5\pi i/12} v_{c+2}$ and that result holds also in the case when $c = \infty$. So the index c is transformed according to the rule $c \mapsto c + 2$, which is the same as for $\tau \mapsto \tau + 2$. In the next two sections we shall see that a similar rule holds for the transformations $\tau \mapsto -1/\tau$ and $\tau \mapsto (\tau - 1)/(\tau + 2)$; namely v_c is transformed into $v_{-1/c}$ in the first case and v_c is transformed into $-\sqrt{2}/v_d$, where $d = (c - 1)/(c + 1)$ in the second case.

Exercises 10.5

10.5.1 Verify that the group $\mathbb{Z}_{240}$ may be partitioned into 48 sets each with five elements that are congruent mod 48 (for example, one such set is $46, 94, 142, 190, 238 \in \mathbb{Z}_{240}$) and $46 \equiv 94 \equiv 142 \equiv 190 \equiv 238 (\mathrm{mod}\ 48)$. Note that $46 \equiv 1, 94 \equiv 4, 142 \equiv 2, 190 \equiv 0, 238 \equiv 3$, $(\mathrm{mod}\ 5)$.

10.5.2 Check the calculations for $v_c \mapsto e^{-5\pi i/12} v_{c+2}$ under $\tau \mapsto \tau + 2$ and note that that result holds also for $c = \infty$, which is the main reason for introducing the new notation.

10.6   The transformations $\tau \mapsto -1/\tau$ and $v_c(\tau)$

The change of parameter maps $v_0 \mapsto v_\infty$, $v_\infty \mapsto v_0$; for example

$$v_\infty = f(5\tau) \mapsto f\left(-\frac{\tau}{5}\right) = f\left(\frac{\tau}{5}\right) = v_0.$$

So certainly the transformation $\tau \mapsto -1/\tau$ induces the transformation $v_c \mapsto v_{-1/c}$ for $c = 0, \infty$. We shall see that that is also true for the other values of c, though the calculations are more complicated and use the methods introduced in Section 10.4.

We begin by using

$$v_{\pm 2} = f\left(\frac{\tau \mp 48}{5}\right) = f\left(\frac{\tau \pm 2}{5} \mp 10\right) = e^{\mp 5\pi i/12} f\left(\frac{\tau \pm 2}{5}\right),$$

and

$$v_{\pm 1} = f\left(\frac{\tau \mp 96}{5}\right) = f\left(\frac{\tau \mp 4}{5} \pm 20\right) = e^{\mp 5\pi i/6} f\left(\frac{\tau \mp 4}{5}\right),$$

the idea being to prefer matrices with smaller elements when performing calculations

Now we consider the change of parameter $\tau \mapsto -1/\tau$ on the expressions on the right-hand side and obtain

$$v_{\pm 2} \mapsto e^{\mp 5\pi i/12} f \left(\dfrac{-\dfrac{1}{\tau} \pm 2}{5} \right) = e^{\mp 5\pi i/12} f \left(\dfrac{\pm 2\tau - 1}{5\tau} \right), \qquad (10.47)$$

and

$$v_{\pm 1} \mapsto e^{\mp 5\pi i/6} f \left(\dfrac{-\dfrac{1}{\tau} \mp 4}{5} \right) = e^{\mp 5\pi i/6} f \left(\dfrac{\mp 4\tau - 1}{5\tau} \right). \qquad (10.48)$$

We shall now try to reduce each of those maps to one of the standard forms (10.41) of Section 10.4 by using the matrices S, T as generators of $SL_2(\mathbb{Z})$. We start with $v_{-2}(\tau) = f((\tau + 48)/5)$ and we see from (10.47) that the map $\tau \mapsto -1/\tau$ is represented in that case by the matrix

$$P = \begin{pmatrix} -2 & -1 \\ 5 & 0 \end{pmatrix}, \qquad (10.49)$$

and we seek to reduce P to one of the forms (10.41) by pre-multiplying by a product of matrices S, T. Now, from (10.41),

$$-P_{-2} = \begin{pmatrix} -1 & 2 \\ 0 & -5 \end{pmatrix}$$

and so we look for a matrix A such that

$$AP = -P_{-2}$$

and then we express A in terms of S, T. Let

$$A = \begin{pmatrix} a & b \\ c & d \end{pmatrix}, \qquad a, b, c, d \in \mathbb{Z}, \qquad ad - bc = 1,$$

then we require

$$\begin{pmatrix} a & b \\ c & d \end{pmatrix} \begin{pmatrix} -2 & -1 \\ 5 & 0 \end{pmatrix} = \begin{pmatrix} -1 & 2 \\ 0 & -5 \end{pmatrix},$$

so we look for $c, d \in \mathbb{Z}$, with greatest common divisor 1, and $-2c + 5d = 0$. Choose $c = 5$ and $d = 2$ and then select $a = -2$, $b = -1$ to obtain

$$A = \begin{pmatrix} -2 & -1 \\ 5 & 2 \end{pmatrix} \in SL_2(\mathbb{Z}),$$

which gives

$$AP = \begin{pmatrix} -2 & -1 \\ 5 & 2 \end{pmatrix} \begin{pmatrix} -2 & -1 \\ 5 & 0 \end{pmatrix} = \begin{pmatrix} -1 & 2 \\ 0 & -5 \end{pmatrix} = -P_{-2}.$$

We note that

$$A^{-1} = \begin{pmatrix} 2 & 1 \\ -5 & -2 \end{pmatrix}$$

and then check that

$$\begin{pmatrix} 2 & 1 \\ -5 & -2 \end{pmatrix} = ST^2 ST^{-2} S.$$

So

$$
\begin{aligned}
f\left(\frac{-2\tau - 1}{5\tau}\right) &= f\left(ST^2 ST^{-2} S\left(\frac{\tau - 2}{5}\right)\right) \\
&= f\left(T^2 ST^{-2} S\left(\frac{\tau - 2}{5}\right)\right) \\
&= e^{-\pi i/12} f\left(ST^{-2} S\left(\frac{\tau - 2}{5}\right)\right) \\
&= e^{-\pi i/12} f\left(T^{-2} S\left(\frac{\tau - 2}{5}\right)\right) \\
&= f\left(S\left(\frac{\tau - 2}{5}\right)\right) \\
&= f\left(\frac{\tau - 2}{5}\right)
\end{aligned}
$$

(on using $f(S\tau) = f(\tau)$ and $f(T^2\tau) = e^{-\pi i/12} f(\tau)$). Therefore, under the transformation $\tau \mapsto -1/\tau$ we obtain

$$
\begin{aligned}
v_{-2} \mapsto e^{-5\pi i/12} f\left(\frac{\tau - 2}{5}\right) &= e^{-5\pi i/12} f\left(\frac{\tau + 48 - 50}{5}\right) \\
&= e^{-5\pi i/12} f\left(\frac{\tau + 48}{5} - 10\right) \quad (10.50) \\
&= f\left(\frac{\tau + 48}{5}\right) = v_{-2}.
\end{aligned}
$$

A similar calculation shows that $v_2 \mapsto v_2$, and we conclude that under $\tau \mapsto -1/\tau$, $v_c \mapsto v_c$, $c = \pm 2$. But since we are considering values of $c \pmod 5$, we know that $c = -1/c$, $c = \pm 2$, and so we conclude that the transformation law for the suffix c is the same as that for $\tau \mapsto -1/\tau$, namely $c \mapsto -1/c$, if $c = \pm 2$.

For the functions $f(\frac{\tau \mp 96}{5})$, which give $v_{\pm 1}(\tau)$, similar calculations (outlined in Exercises 10.6.3 and 10.6.4) show that v_1 becomes v_{-1} and v_{-1} becomes v_1 and so, as before, the transformation $\tau \mapsto -1/\tau$ induces the transformation $v_{\pm 1} \mapsto v_{\mp 1}$ and in all cases $\tau \mapsto -1/\tau$ sends v_c into $v_{-1/c}$.

Exercises 10.6

10.6.1 Show that

$$\begin{pmatrix} 2 & 1 \\ -5 & -2 \end{pmatrix} = ST^2ST^{-2}S.$$

10.6.2 Carry out the details of the calculation, similar to that leading to (10.50), then show that $v_2 \mapsto v_2$ under $\tau \mapsto -1/\tau$.

10.6.3 (In this exercise we outline the details of the calculation that shows that $v_1 \mapsto v_{-1}$ under $\tau \mapsto -1/\tau$. For a complete account see Prasolov & Solovyev, 1997, p. 162.) Consider $v_1 = f((\tau + 96)/5)$. As in (10.45) and the preceding calculations, verify that v_1 is represented by the matrix $P'(\begin{smallmatrix} -4 & -1 \\ 5 & 0 \end{smallmatrix})$ and show that if $A' = (\begin{smallmatrix} -1 & -1 \\ 5 & 4 \end{smallmatrix})$, then

$$A'P' = \begin{pmatrix} -1 & 1 \\ 0 & -5 \end{pmatrix} = -P_{-1}.$$

Deduce that $(A')^{-1} = \begin{pmatrix} 4 & 1 \\ -5 & -1 \end{pmatrix}$ and then verify that

$$(A')^{-1} = STST^{-4}S.$$

Then show that

$$f\left(\frac{-4\tau - 1}{5\tau}\right) = f\left(STST^{-4}S\left(\frac{\tau - 1}{5}\right)\right) = e^{-\pi i/3} f\left(\frac{\tau + 4}{5}\right),$$

by repeated use of

$$f(S\tau) = f(\tau), \quad f(T\tau) = e^{-\pi i/24} f_1(\tau), \quad f_1(-1/\tau) = f_2(\tau),$$
$$f_2(\tau + 1) = e^{\pi i/12} f_2(\tau), \quad f_1(\tau + 1) = e^{-\pi i/24} f(\tau).$$

Hence show that under $\tau \mapsto -1/\tau$,

$$v_1 \mapsto e^{-5\pi i/6} e^{-\pi i/3} f\left(\frac{\tau + 4}{5}\right) = e^{-7\pi i/6} f\left(\frac{\tau - 96 + 100}{5}\right)$$

$$= f\left(\frac{\tau - 96}{5}\right)$$

and conclude that $v_1 \mapsto v_{-1}$.

10.6.4 By using the arguments outlined in 10.6.3, show that under the transformation $\tau \mapsto -1/\tau$, $v_{-1} \mapsto v_1$.

By using 10.6.3 and 10.6.4 and the argument given in the text, conclude that the transformation $\tau \mapsto -1/\tau$ sends v_c to $v_{-1/c}$.

10.7 The transformation $\tau \mapsto (\tau - 1)/(\tau + 1)$ and $v_c(\tau)$

Finally, we shall consider the effect of the transformation $\tau \mapsto (\tau - 1)/(\tau + 1)$ on v_c (the reasons why that and the transformations discussed in Sections 10.5 and 10.6 are important in the solution of the quintic will emerge in Section 10.8). Our object is to show that the transformation on τ affects v_c in the same way; so that $\tau \mapsto (\tau - 1)/(\tau + 1)$ maps v_c into $(-\sqrt{2})/(v_d)$, where $d = (c - 1)/(c + 1)$.

We shall appeal to ideas similar to those used in the preceding sections, but in order to deal with the calculations it will be convenient to use the relation

$$f(\tau)f\left(\frac{\tau - 1}{\tau + 1}\right) = \sqrt{2}, \tag{10.51}$$

which is obtained in Exercise 6.7.8.

Let $\begin{pmatrix} a & b \\ 0 & d \end{pmatrix}$ be the matrix representation of v_c (see (10.41)), where, for the time being, we prefer *not* to use the changed notation for v_c (as in Section 10.5), but rather to represent (for example) v_1 by $\begin{pmatrix} 1 & 1 \\ 0 & 5 \end{pmatrix}$. It follows that under the transformation $\tau \mapsto (\tau - 1)/(\tau + 1)$,

$$v_c \mapsto f\left(\begin{pmatrix} a & b \\ c & d \end{pmatrix}\begin{pmatrix} 1 & -1 \\ 1 & 1 \end{pmatrix}\tau\right), \tag{10.52}$$

and we see that leads to a term in τ in the denominator of the argument of f, and that does not readily lend itself to one of our standard representations for v_c. To achieve a more readily recognisable form we pre-multiply the matrix product in (10.52) by an element of $SL_2(\mathbb{Z})$ to obtain an expression of the form

$$\begin{pmatrix} \alpha & \beta \\ \gamma & \delta \end{pmatrix}\begin{pmatrix} a & b \\ 0 & d \end{pmatrix}\begin{pmatrix} 1 & -1 \\ 1 & 1 \end{pmatrix} = \begin{pmatrix} 1 & -1 \\ 1 & 1 \end{pmatrix}\begin{pmatrix} x & y \\ 0 & z \end{pmatrix}. \tag{10.53}$$

By taking determinants, we see that $xz = 5$ and so $\begin{pmatrix} x & y \\ 0 & z \end{pmatrix}$ has one of our standard forms, as in (10.41).

The idea is that the expression for the transformed v_c may be rearranged so that one can first evaluate $f((\tau' - 1)/(\tau' + 1))$, where $\tau' = \begin{pmatrix} x & y \\ 0 & z \end{pmatrix}\tau$; then using (10.51), we will show how to transform v_c into another v_d.

How can we find $\left(\begin{smallmatrix} \alpha & \beta \\ \gamma & \delta \end{smallmatrix}\right)$? If we multiply (10.53) by $\left(\begin{smallmatrix} 1 & 1 \\ -1 & 1 \end{smallmatrix}\right) = \left(\begin{smallmatrix} 1 & -1 \\ 1 & 1 \end{smallmatrix}\right)^{-1}$ on the left, then we obtain

$$\begin{pmatrix} x & y \\ 0 & z \end{pmatrix} = \begin{pmatrix} 1 & 1 \\ -1 & 1 \end{pmatrix} \begin{pmatrix} \alpha & \beta \\ \gamma & \delta \end{pmatrix} \begin{pmatrix} a & b \\ 0 & d \end{pmatrix} \begin{pmatrix} 1 & -1 \\ 1 & 1 \end{pmatrix},$$

and by comparing elements in the lower left component we find the conditions

$$(\gamma - \alpha)(a + b) + (\delta - \beta)d = 0, \quad \alpha\delta - \beta\gamma = 1.$$

We try the calculation for v_0, to begin with. For v_0 the condition (10.53) reads

$$\begin{pmatrix} \alpha & \beta \\ \gamma & \delta \end{pmatrix} \begin{pmatrix} 1 & 0 \\ 0 & 5 \end{pmatrix} \begin{pmatrix} 1 & -1 \\ 1 & 1 \end{pmatrix} = \begin{pmatrix} 1 & -1 \\ 1 & 1 \end{pmatrix} \begin{pmatrix} x & y \\ 0 & z \end{pmatrix}.$$

So we require $(\gamma - \alpha) + 5(\delta - \beta) = 0$ and $\alpha\delta - \beta\alpha = 1$. Try $\gamma - \alpha = -5$ and $\delta - \beta = 1$ and our first condition is satisfied. Using the second condition we find that $(\gamma + 5)\delta - \gamma(\delta - 1) = 1$, which gives $5\delta + \gamma = 1$. Try $\beta = 0$ (for simplicity and for reasons that will become apparent) and then we find $\alpha = 1, \beta = 0, \gamma = -4, \delta = 1$ and so

$$\begin{pmatrix} 1 & 0 \\ -4 & 1 \end{pmatrix} \begin{pmatrix} 1 & 0 \\ 0 & 5 \end{pmatrix} \begin{pmatrix} 1 & -1 \\ 1 & 1 \end{pmatrix} = \begin{pmatrix} 1 & -1 \\ 1 & 1 \end{pmatrix} \begin{pmatrix} x & y \\ 0 & z \end{pmatrix},$$

whence

$$\begin{pmatrix} x & y \\ 0 & z \end{pmatrix} = \begin{pmatrix} 1 & 4 \\ 0 & 5 \end{pmatrix}$$

and, by multiplying both sides on the left by $\left(\begin{smallmatrix} 1 & 0 \\ -4 & 1 \end{smallmatrix}\right)^{-1}$ we obtain

$$\begin{pmatrix} 1 & 0 \\ -4 & 1 \end{pmatrix} \begin{pmatrix} 1 & 0 \\ 0 & 5 \end{pmatrix} \begin{pmatrix} 1 & -1 \\ 1 & 1 \end{pmatrix} = \begin{pmatrix} 1 & -1 \\ 1 & 1 \end{pmatrix} \begin{pmatrix} 1 & 4 \\ 0 & 5 \end{pmatrix}. \qquad (10.54)$$

The relation for v_∞ may be obtained similarly (see Exercise 10.7.1) and reads

$$\begin{pmatrix} 1 & -4 \\ 0 & 1 \end{pmatrix} \begin{pmatrix} 5 & 0 \\ 0 & 1 \end{pmatrix} \begin{pmatrix} 1 & -1 \\ 1 & 1 \end{pmatrix} = \begin{pmatrix} 1 & -1 \\ 1 & 1 \end{pmatrix} \begin{pmatrix} 1 & -4 \\ 0 & 5 \end{pmatrix}. \qquad (10.55)$$

The relations for v_1 and v_4 are (see Exercises 10.7.2 and 10.7.3):

$$\begin{pmatrix} 3 & -1 \\ -2 & 1 \end{pmatrix} \begin{pmatrix} 1 & 1 \\ 0 & 5 \end{pmatrix} \begin{pmatrix} 1 & -1 \\ 1 & 1 \end{pmatrix} = \begin{pmatrix} 1 & -1 \\ 1 & 1 \end{pmatrix} \begin{pmatrix} 1 & 0 \\ 0 & 5 \end{pmatrix} \qquad (10.56)$$

and

$$\begin{pmatrix} 3 & 1 \\ 2 & 1 \end{pmatrix}\begin{pmatrix} 1 & -1 \\ 0 & 5 \end{pmatrix}\begin{pmatrix} 1 & -1 \\ 1 & 1 \end{pmatrix} = \begin{pmatrix} 1 & -1 \\ 1 & 1 \end{pmatrix}\begin{pmatrix} 5 & 0 \\ 0 & 1 \end{pmatrix}. \tag{10.57}$$

We shall show how (10.54) and (10.51) may be used to obtain the rule $v_0 \mapsto -\sqrt{2}/v_4$ under the transformation $\tau \mapsto (\tau - 1)/(\tau + 1)$.

First we use

$$f\left(\frac{\tau}{-4\tau + 1}\right) = f\left(\frac{4\tau - 1}{\tau}\right) = e^{-4\pi i/24} f(\tau)$$

and then

$$\frac{\sqrt{2}}{f\left(\dfrac{\tau + 4}{5}\right)} = e^{-4\pi i/24} f\left(\frac{1}{5}\frac{\tau - 1}{\tau + 1}\right).$$

Whence

$$f\left(\frac{1}{5}\frac{\tau - 1}{\tau + 1}\right) = \frac{-\sqrt{2}}{f\left(\dfrac{\tau - 96}{5}\right)},$$

which shows that $v_0 \mapsto -\sqrt{2}/v_4$.

By a similar argument (using (10.55) and (10.51)) we obtain $v_\infty \mapsto -\sqrt{2}/v_1$.

The calculations for v_1, v_4 and v_2, v_3 are similar though a little more complicated. We illustrate the essential ideas by considering v_4, and we shall show that v_4 is transformed into $-\sqrt{2}/v_\infty$; that is $v_4(\tau) \mapsto -\sqrt{2}/f(5\tau)$. To see that we argue as follows.

We recall that $v_4(\tau) = f((\tau - 96)/5)$. Write $\gamma = (\tau - 1)/(\tau + 1)$ and $\delta = (\gamma - 1)/5 = (2\tau/(5(\tau + 1)))$. Consider

$$f\left(\frac{\gamma - 96}{5}\right) = f\left(\frac{\gamma - 1 - 95}{5}\right) = f\left(\frac{\gamma - 1}{5} - 19\right)$$

$$= e^{19\pi i/24} f_1\left(\frac{\gamma - 1}{5}\right) \tag{10.58}$$

$$= e^{19\pi i/24} f_1(\delta)$$

(using Exercise 6.7.5)). We have to find $f_1(\delta)$ and to do that we begin by recalling that, on using (10.55),

$$f\left(\frac{3\delta + 1}{2\delta + 1}\right) = \frac{\sqrt{2}}{f(5\tau)}, \tag{10.59}$$

and we shall try to connect that with $f_1(\delta)$.

Now

$$f\left(\frac{3\delta+1}{2\delta+1}\right) = f\left(\frac{\delta}{2\delta+1}+1\right) = e^{-\pi i/24}f_1\left(\frac{\delta}{2\delta+1}\right)$$

$$= e^{-\pi i/24}f_2\left(-\frac{2\delta+1}{\delta}\right)$$

$$= e^{-\pi i/24}f_2\left(-2-\frac{1}{\delta}\right)$$

$$= e^{-5\pi i/24}f_1(\delta), \qquad (10.60)$$

on appealing to Exercises 6.7.5 and 6.7.6.

We refer to (10.58) and observe that

$$e^{19\pi i/24}f_1(\delta) = e^{\pi i}e^{-5\pi i/24}f_1(\delta) = -e^{-5\pi i/24}f_1(\delta)$$

$$= -\frac{\sqrt{2}}{f(5\tau)} = -\frac{\sqrt{2}}{v_\infty},$$

on using (10.59) and (10.68). That gives the desired result.

A similar argument (Exercise 10.7.6) shows that v_1 is turned into $-\sqrt{2}/v_0$.

Finally, we use the relation

$$\begin{pmatrix} -2 & 1 \\ 3 & -2 \end{pmatrix}\begin{pmatrix} 1 & 2 \\ 0 & 5 \end{pmatrix}\begin{pmatrix} 1 & -1 \\ 1 & 1 \end{pmatrix} = \begin{pmatrix} 1 & -1 \\ 1 & 1 \end{pmatrix}\begin{pmatrix} -1 & -2 \\ 0 & -5 \end{pmatrix} \qquad (10.61)$$

to show that v_2 is mapped to $-\sqrt{2}/v_2$ and a similar argument to show that v_3 is mapped to $-\sqrt{2}/v_3$ (see Exercises 10.7.6, 10.7.7 and 10.7.8).

We summarize our results as follows:

	u	v_∞	v_0	v_1	v_2	v_3	v_4
$\tau \mapsto \tau+2$	$e^{-\pi i/12}u$	εv_∞	εv_2	εv_3	εv_4	εv_0	εv_1
$\tau \mapsto \tau-\dfrac{1}{\tau}$	u	v_0	v_∞	v_4	v_2	v_3	v_1
$\tau \mapsto \dfrac{\tau-1}{\tau+1}$	$\dfrac{\sqrt{2}}{u}$	$-\dfrac{\sqrt{2}}{v_1}$	$-\dfrac{\sqrt{2}}{v_4}$	$-\dfrac{\sqrt{2}}{v_0}$	$-\dfrac{\sqrt{2}}{v_2}$	$-\dfrac{\sqrt{2}}{v_3}$	$-\dfrac{\sqrt{2}}{v_\infty}$

where we recall that $u = f(\tau)$ and we have written $\varepsilon = e^{-5\pi i/12}$.

Exercises 10.7

10.7.1 Obtain the relation (10.55), namely

$$\begin{pmatrix} 1 & -4 \\ 0 & 1 \end{pmatrix} \begin{pmatrix} 5 & 0 \\ 0 & 1 \end{pmatrix} \begin{pmatrix} 1 & -1 \\ 1 & 1 \end{pmatrix} = \begin{pmatrix} 1 & -1 \\ 1 & 1 \end{pmatrix} \begin{pmatrix} 1 & -4 \\ 0 & 5 \end{pmatrix}.$$

10.7.2 Obtain the relation for v_1, that is (10.56)

$$\begin{pmatrix} 3 & -1 \\ -2 & 1 \end{pmatrix} \begin{pmatrix} 1 & 1 \\ 0 & 5 \end{pmatrix} \begin{pmatrix} 1 & -1 \\ 1 & 1 \end{pmatrix} = \begin{pmatrix} 1 & -1 \\ 1 & 1 \end{pmatrix} \begin{pmatrix} 1 & 0 \\ 0 & 5 \end{pmatrix}.$$

10.7.3 Obtain the relation for v_4,

$$\begin{pmatrix} 3 & 1 \\ 2 & 1 \end{pmatrix} \begin{pmatrix} 1 & -1 \\ 0 & 5 \end{pmatrix} \begin{pmatrix} 1 & -1 \\ 1 & 1 \end{pmatrix} = \begin{pmatrix} 1 & -1 \\ 1 & 1 \end{pmatrix} \begin{pmatrix} 5 & 0 \\ 0 & 1 \end{pmatrix},$$

10.7.4 Check the details of the calculation leading to the result that the transformation $\tau \mapsto (\tau - 1)/(\tau + 1)$ turns v_0 into $-\sqrt{2}/v_4$.

10.7.5 Show that v_1 is turned into $-\sqrt{2}/v_0$.

10.7.6 Obtain the relation (10.61):

$$\begin{pmatrix} -2 & 1 \\ 3 & -2 \end{pmatrix} \begin{pmatrix} 1 & 2 \\ 0 & 5 \end{pmatrix} \begin{pmatrix} 1 & -1 \\ 1 & 1 \end{pmatrix} = \begin{pmatrix} 1 & -1 \\ 1 & 1 \end{pmatrix} \begin{pmatrix} -1 & -2 \\ 0 & -5 \end{pmatrix}.$$

10.7.7 Use the methods described in Section 10.7 and the result in Exercise 10.7.6 to show that v_2 is transformed into $-\sqrt{2}/v_2$.

10.7.8 Obtain the relation analogous to (10.61) for the case of v_3 and then use the method of 10.7.7 to prove that v_3 is transformed into $-\sqrt{2}/v_3$.

10.8 The modular equation and the solution of the quintic

In order to obtain the first form (10.26) of the modular equation and thence the solution of the quintic in Bring's form, (10.33), we shall use the ideas of Sections 10.4 to 10.7 and in particular the properties of functions of τ that are invariant under the transformations:

$$\tau \mapsto \tau + 2, \quad \tau \mapsto -\frac{1}{\tau} \quad \text{and} \quad \tau \mapsto \frac{\tau - 1}{\tau + 1}. \tag{10.62}$$

Before proceeding to the detailed calculations, it will be appropriate to make some general remarks about functions invariant under the transformations (10.62). The ideas are reminiscent of our treatment of the modular function, $j(\tau)$, and our construction of a fundamental region for the modular group and that function in Chapter 3, and of the problem of inversion discussed in Chapter 5 (and see also Chapter 7).

We recall that the modular functions are, by definition, invariant under the action of T and S; that is under the transformations $\tau \mapsto \tau + 1$ and $\tau \mapsto -1/\tau$. We are looking for a function invariant under (10.62) and one obvious candidate is our function $f(\tau)$ raised to the power 24, since $f^{24}(\tau)$ is transformed into itself under $\tau \mapsto \tau + 2$ and $\tau \mapsto -1/\tau$. However, the transformation $\tau \mapsto (\tau - 1)/(\tau + 1)$ sends $f(\tau)$ to $\sqrt{2}/f(\tau)$ and so $f^{24}(\tau) \mapsto 2^{12}/f^{24}(\tau)$. That suggests that we should consider the function

$$F(\tau) = f^{24}(\tau) + \frac{2^{12}}{f^{24}(\tau)} \tag{10.63}$$

and, on using the expression for $f(\tau)$ as an infinite product (see (10.35)), we obtain the expansion

$$F(\tau) = q^{-1} \prod_{n=1}^{\infty} (1 + q^{2n-1})^{24} + 2^{12} q \prod_{n=1}^{\infty} (1 + q^{2n-1})^{-24} \tag{10.64}$$
$$= q^{-1} + 24 + \cdots.$$

Clearly, $F(\tau)$ is invariant under the transformations (10.62). We already saw in Chapter 7, Section 7.7, as a consequence of Theorem 7.14 that a meromorphic function defined on H and of weight 0 is a rational function of $j = j(\tau)$. Moreover, for every c in the fundamental region of $SL_2(\mathbb{Z})$, one can find a value of τ such that $j(\tau) = c$ and indeed $j(\tau)$ takes every such value exactly once. Similar results are true for $F(\tau)$ (as we shall see): every function meromorphic in the upper half-plane and invariant under the transformation (10.62) is a rational function of $F(\tau)$ and the equation $F(\tau) = c$ has a solution for all $c \neq 0$ and in the domain defined by $\{\tau \in H || \tau | \geq 1, |\mathrm{Re}\,\tau| \leq 1/2\}$, which, is the fundamental domain for $\tau \mapsto \tau + 2$ and $\tau \mapsto -1/\tau$, the function $f^{24}(\tau)$ attains each non-zero value c exactly once.

For the present we assume the truth of all that and proceed to the deduction of the modular equation and the solution of the quintic.

Using the transformation table at the end of Section 10.7, we obtain the following transformation table for the functions uv and u/v where $u(\tau) = f(\tau)$ and $v_c(\tau) = f((\tau + c)/5)$, $c = 0, 1, \ldots, 4$, $v_\infty(\tau) = f(5\tau)$.

	uv	$\dfrac{u}{v}$
$\tau \mapsto \tau + 2$	$e^{-\pi i/2}uv$	$e^{\pi i/3}\dfrac{u}{v},$
$\tau \mapsto \tau - \dfrac{1}{\tau}$	uv	$\dfrac{u}{v},$
$\tau \mapsto \dfrac{\tau - 1}{\tau + 1}$	$-\dfrac{2}{uv}$	$-\dfrac{v}{u}$

As we have seen, the suffix c of v_c changes in the same way as τ. The idea now is to consider the function

$$\prod_c (A_c - B_c)^2, \quad c = \infty, 0, 1, 2, 3, 4, \tag{10.65}$$

where

$$A_c = \left(\frac{u}{v_c}\right)^3 + \left(\frac{v_c}{u}\right)^3, \quad B_c = (uv_c)^2 - \frac{4}{(uv_c)^2}. \tag{10.66}$$

On using the results for uv, u/v, we see that the functions A_c, B_c, as functions of τ, transform in accordance with the following table:

	A	B
$\tau \mapsto \tau + 2$	$-A$	$-B$
$\tau \mapsto -\dfrac{1}{\tau}$	A	B
$\tau \mapsto \dfrac{\tau - 1}{\tau + 1}$	$-A$	$-B$

And we deduce that $\prod_c (A_c - B_c)^2$ is invariant under those three transformations. We shall now show that the product (10.65) is a constant and that that constant is zero.

First we expand A_∞ and B_∞ as power series in $q = e^{\pi i \tau}$ and obtain

$$A_\infty = q^{-1/2}(1 - 2q + \cdots), \quad B_\infty = q^{-1/2}(1 - 2q + \cdots);$$

so $A_\infty - B_\infty$ vanishes as $\text{Im}\,\tau \to \infty$, that is at $q = 0$.

Now we show that $A_c - B_c$ is 0 at $q = 0$ for all c. First

$$u(5\tau - c) = f(5\tau) = v_\infty(\tau),$$

$$v_c(5\tau - c) = f\left(\frac{5\tau - c + c}{5}\right) = f(\tau) = u(\tau)$$

and A, B are invariant if u and v are interchanged. It follows that

$$A_c(5\tau - c) = A_\infty(\tau), \quad B_c(5\tau - c) = B_\infty(\tau) \tag{10.67}$$

and, since $\text{Im}(5\tau - c) \to \infty \Leftrightarrow \text{Im}(\tau) \to \infty$, we conclude that $A_c - B_c = 0$ at $q = 0$.

So the function in (10.65) is a constant that vanishes at $q = 0$ and therefore $A_c - B_c = 0$ for some c. But then (10.67) shows that $A_c - B_c = 0$ for all c and we conclude, from (10.66), that

$$\left(\frac{u}{v}\right)^3 + \left(\frac{v}{u}\right)^3 = (uv)^2 - \frac{4}{(uv)^2}, \tag{10.68}$$

that is,

$$u^6 + v^6 - u^5 v^5 + 4uv = 0,$$

which is the form of the modular equation quoted in (10.26). In particular (10.26) relates $u = f(\tau)$ and $v = f(5\tau)$, and if we think of it as a sextic equation for v, given u, its roots are

$$v_\infty(\tau) = f(5\tau), \quad v_c(\tau) = f\left(\frac{\tau + c}{5}\right), \quad c \equiv 0 (\text{mod } 48). \tag{10.69}$$

Using the relation between the product of the roots and the coefficient of the term independent of v, we obtain

$$\prod_c v_c = u^6. \tag{10.70}$$

Having expressed the roots of the sextic (10.26) in terms of $u = f(\tau)$, we shall use those roots to solve the general quintic in Bring's form. We consider the polynomial whose roots $\omega_0, \omega_1, \omega_2, \omega_3, \omega_4$ are given by

$$\omega_k = \frac{(v_\infty - v_k)(v_{k+1} - v_{k-1})(v_{k+2} - v_{k-2})}{\sqrt{5}u^3}, \quad k = 0, 1, 2, 3, 4. \tag{10.71}$$

The definition of w_k was originally suggested by Hermite (Hermite, 1861, *Sur la résolution de l'équation du cinquième degree*, and see Klein, 1884, pp. 162ff, and McKean & Moll, 1997, where the connection with Galois theory is very clearly explained; see also Briot and Bouqet, 1875). Using the transformation table at the end of Section 10.7, we construct the corresponding table for w_k:

	w_0	w_1	w_2	w_3	w_4
$\tau \mapsto \tau + 2$	$-w_2$	$-w_3$	$-w_4$	$-w_0$	$-w_1,$
$\tau \mapsto -\dfrac{1}{\tau}$	w_0	w_2	w_1	w_4	$w_3,$
$\tau \mapsto \dfrac{\tau - 1}{\tau + 1}$	$-w_0$	$-w_3$	$-w_4$	$-w_2$	$-w_1.$

$$(10.72)$$

(See Exercise 10.8.2; the last row uses (10.70).)

The polynomial whose roots are the w_k is given by

$$\prod_{k=0}^{4}(w - w_k) = w^5 + A_1 w^4 + A_2 w^3 + A_3 w^2 + A_4 w + A_5 \qquad (10.73)$$

and we note that the coefficients are finite provided that $u \neq 0, \infty$. Furthermore, the coefficients A_1^2, A_2, A_3^2, A_4 and A_5^2, considered as functions of τ, do not vary under the transformations on τ in (10.72) and so, given the discussion leading up to (10.63), whose claims are to be justified below, they are expressible as polynomials in:

$$u^{24} + \frac{2^{12}}{u^{24}} = q^{-1} + 24 + \cdots. \qquad (10.74)$$

We note that a polynomial in $u^{24} + 2^{12}/u^{24}$ is non-constant only if its Laurent expansion in q begins with a term in $q^r, r \leq -1$; so it has a pole of order at least 1. We are led therefore to consider the expansion of w_k.

On using the product, (10.35), for $f(\tau)$, we see that the first term in the expansion of v_c is $(q')^{-1/24}$, where $q' = e^{5\pi i \tau}$ if $c = \infty$ and $q' = e^{\pi i(\tau+c)/5}$, $c \neq \infty$. If we set $\alpha = e^{-4\pi i/5}$, and remembering that $c \equiv 0 \pmod{48}$, we find that $e^{-\pi i c/120} = \alpha^c$ and so the first term in the expansion of w_k is

$$\frac{q^{-5/24} q^{-1/120}(\alpha^{k+1} - \alpha^{k-1}) q^{-1/120}(\alpha^{k+2} - \alpha^{k-2})}{\sqrt{5} q^{-1/8}} = \lambda q^{-1/10},$$

where

$$\lambda = \frac{\alpha^{2k}(\alpha^3 - \alpha - \alpha^{-1} + \alpha^{-3})}{\sqrt{5}} = \alpha^{2k},$$

on using

$$\alpha^3 - \alpha - \alpha^{-1} + \alpha^{-3} = 2\left(\cos\frac{\pi}{10} + \cos\frac{\pi}{5}\right) = \sqrt{5}.$$

(See Exercise 10.8.3.)

It follows that the expansion of A_l begins with $q^{-l/10}$, or with a term of the form cq^p, $p \geq -l/10$. The coefficients A_1^2, A_2, A_3^2, A_4 are accordingly constants (because in those cases $q^{-l/10} > q^{-1}$), but the expansion of A_5^2 begins with the term $(\alpha^2 \alpha^4 \alpha^6 \alpha^8)^2 q^{-1} = q^{-1}$ and so A_5^2 is linearly dependent on

$$u^{24} + \frac{2^{12}}{u^{24}} = q^{-1} + 24 + \cdots.$$

On comparing the first terms in the expansions of the functions $u^{24} + 12u^{-24}$ and A_5^2, we obtain

$$A_5^2 = u^{24} + \frac{2^{12}}{u^{24}} + C$$

and we need to calculate C and the A_1, A_2, A_3 and A_4. To that end we calculate the value of $v_c(\tau)$ for one value of τ, and the most convenient for calculations is $\tau = i$. We note that $-1/i = i$ and so $f_1(i) = f_2(i)$ and also for $\tau = i$, $f(\tau)$, $f_1(\tau)$ and $f_2(\tau)$ are positive. We find that for $\tau = i$ the modular equation (10.26) takes the form

$$v^6 - a^5 v^5 + a^9 v + a^6 = 0, \quad a = 2^{1/4}, \tag{10.75}$$

and that the polynomial (10.73) has the form

$$w(w - i\sqrt{5})^2 (w + i\sqrt{5})^2 = w(w^2 + 5)^2. \tag{10.76}$$

(See Exercises 10.8.4, 10.8.5 and 10.8.6 for the details.)
It follows from (10.76) that $A_5^2(i) = 0$ and so

$$C = - \left(u^{24}(i) + \frac{2^{12}}{u^{24}(i)} \right) = -(2^6 + 2^6) = -2^7.$$

We deduce that

$$A_5^2 = \left(u^{24} + \frac{2^{12}}{u^{24}} - 2^7 \right) = \left(u^{12} - \frac{2^6}{u^{12}} \right)^2$$

and so

$$A_5 = \pm \left(u^{12} - \frac{2^6}{u^{12}} \right).$$

In order to remove the ambiguity in the sign, we consider the expansions in q of $A_5 = -w_0 w_1 w_2 w_3 w_4$ and $u^{12} - 2^6/u^{12}$ and note that they begin with $-q^{1/2}$ and $q^{1/2}$, respectively, whence $A_5 = -u^{12} + 2^6/u^{12}$ and we conclude that Equation (10.73) is of the form

$$w(w^2 + 5)^2 = u^{12} - 64u^{-12}. \tag{10.77}$$

On using the relations

$$f^8 = f_1^8 + f_2^8 \tag{10.78}$$

and

$$f f_1 f_2 = \sqrt{2}, \tag{10.79}$$

(see Exercise 6.7.3), we obtain

$$u^{12} - \frac{64}{u^{12}} = \frac{f^{24}(\tau) - 64}{f^{12}(\tau)} = \left(\frac{f_1^8(\tau) - f_2^8(\tau)}{f^2(\tau)} \right)^2,$$

whence, by (10.77),

$$\sqrt{w(\tau)} = \pm \frac{f_1^8(\tau) - f_2^8(\tau)}{f^2(\tau)(w^2(\tau) + 5)}.$$

If we now write

$$y(\tau) = \frac{f_1^8(\tau) - f_2^8(\tau)}{f^2(\tau)(w^2(\tau) + 5)}, \tag{10.80}$$

we obtain the equation

$$y^5 + 5y = y(w^2 + 5) = \frac{f_1^8 - f_2^8}{f^2}, \tag{10.81}$$

which is (10.34).

Suppose then that we want to solve (cf. (10.34) and (10.33)) the quintic in Bring's form

$$y^5 + 5y = a. \tag{10.82}$$

We have to find a value of τ such that

$$\frac{f_1^8 - f_2^8}{f^2} = a. \tag{10.83}$$

On using (10.78) and (10.79) and squaring (10.83) we obtain

$$f^{24} - a^2 f^{12} - 64 = 0$$

and that is a quadratic in f^{12} one of whose roots gives a solution of (10.80) (that is of (10.34)), the other a solution of that equation with a replaced by $-a$. As we shall see in Section 10.9, it is always possible to find a τ in H such that that holds.

Having in principle found τ we proceed to solve (10.81), as follows.

First, for that value of τ, find $v_c(\tau)$, $c = \infty, 0, 1, 2, 3, 4$ and then use those values in (10.71) to obtain $w_k(\tau)$ for $k = 0, 1, 2, 3, 4$. Then use (10.80) to find the five values $y_k(\tau)$. Those are the roots of (10.82) and our solution of the general quintic is completed.

Exercises 10.8

(These exercises are based on Prasolov & Solovyev, 1997, Section 7.15.)

10.8.1 Prove the relation (10.70):

$$\prod_c v_c = u^6.$$

10.8.2 Verify the entries in the table (10.72).

10.8.3 Let $\alpha = e^{-4\pi i/5}$. Prove that

$$\alpha^3 - \alpha - \alpha^{-1} + \alpha^{-3} = 2\left(\cos\frac{\pi}{10} + \cos\frac{\pi}{5}\right) = \sqrt{5}.$$

10.8.4 Using (10.78) and (10.79) and $f_1(i) = f_2(i)$, prove that

$$v_3 = f\left(\frac{i+48}{5}\right) = f\left(\frac{i-2}{5} + 10\right) = e^{-10\pi i/24} f\left(\frac{i-2}{5}\right)$$

$$= e^{-10\pi i/24} f(i+2) = -i\sqrt[4]{2}$$

and that

$$v_4 = i\sqrt[4]{2}.$$

10.8.5 Show that for $\tau = i$ the modular equation $u^6 + v^6 - u^5 v^5 + 4uv = 0$ takes the form $v^6 - a^5 v^5 + a^9 v + a^6 = 0$, where $a = \sqrt[4]{2}$. Observe that two roots of that equation are $v_3 = -ia$, $v_2 = ia$ and then divide by $v^2 + a^2$ to obtain

$$v^4 - a^5 v^3 + a^2 v^2 + a^7 v + a^4 = (v - \xi)^2 (v - \eta)^2,$$

where $\xi + \eta = a$ and $\xi\eta = -a^2$. Assume that $\xi > 0$ and $\eta < 0$ and hence show that $\xi = a(1 + \sqrt{5})/2$, $\eta = a(1 - \sqrt{5})/2$.
   Show further that

$$v_\infty = f(5i) = f\left(-\frac{1}{5i}\right) = f\left(\frac{i}{5}\right) = v_0$$

and note that $v_\infty > 0$, since $f(\tau) > 0$ for τ pure imaginary.
   Deduce that $v_0 = v_\infty = \sqrt[4]{2}(1+\sqrt{5})/2$, $v_1 = v_4 = \sqrt[4]{2}(1 - \sqrt{5})/2$.

10.8.6 Substitute $v_c(i)$ in (10.71) to obtain

$$w_0 = 0, \quad w_1 = w_2 = i\sqrt{5}, \quad w_3 = w_4 = -i\sqrt{5}.$$

Hence show that the polynomial (10.73) is of the form

$$w(w - i\sqrt{5})^2(w + i\sqrt{5})^2 = w(w^2 + 5)^2.$$

Deduce that $A_5^2(i) = 0$. (Cf. (10.75) and (10.76).)

10.9 The fundamental domain of the group generated by $\tau \mapsto \tau + 2$ and $\tau \mapsto -1/\tau$ and the function $f^{24}(\tau)$

In Chapter 3 we described the fundamental domain of the group generated by $\tau \mapsto \tau + 1$ and $\tau \mapsto -1/\tau$; that is, of the group $SL_2(\mathbb{Z})$ generated by T and S. Then in Chapter 7, Section 7.7, we solved the inversion problem for $j(\tau)$; namely that for every c in the fundamental region of $SL_2(\mathbb{Z})$, there is a unique τ in the fundamental region such that $j(\tau) = c$. Moreover, we showed that every function $g(\tau)$ defined in H and invariant under the action of $SL_2(\mathbb{Z})$ can be expressed as a rational function of $j(\tau)$.

In this section we obtain the basic results, similar to those referred to above, concerning functions invariant under the group generated by $\tau \mapsto \tau + 2$ and $\tau \mapsto -1/\tau$ (that is by T^2 and S). In particular, we shall obtain a description of the fundamental domain for that group; we shall show that the function $f^{24}(\tau)$ attains each non-zero value c in that domain exactly once; and we shall show if $g(\tau)$ is meromorphic in the fundamental region and invariant under the transformations $\tau \mapsto \tau + 2, \tau \mapsto -1/\tau, \ \tau \mapsto (\tau - 1)/(\tau + 1)$, then $g(\tau)$ is a rational function of $f^{24}(\tau) + 2/f^{24}(\tau)$ – the function $F(\tau)$ defined in (10.63).

(We follow the account given in Prasolov & Solovyev, 1997, Section 7.20.)

Theorem 10.1 *Let G_2 denote the group generated by $\tau \mapsto \tau + 2$ and $\tau \mapsto -1/\tau$; that is, by T^2 and S. Then:*

(a) *the fundamental domain D_2 of G_2 is defined by*

$$D_2 = \{\tau \in H \,||\tau| \geq 1, \ |\mathrm{Re}| \leq 1\};$$

(b) *if $c \in D_2, c \neq 0$, then the function of $f^{24}(\tau)$ attains the value c exactly once.*

(See Figure 10.1 for a sketch of D_2; not surprisingly, since $\tau \mapsto \tau + 1$ is replaced by $\tau \mapsto \tau + 2$, it is twice as wide as the fundamental region for $SL_2(\mathbb{Z})$.)

Proof Let τ' denote the image of $\tau \in H$ under the action of G_2 with the maximal value of $\mathrm{Im} \ \tau'$ (cf. the proof of Theorem 3.3). Since the transformation

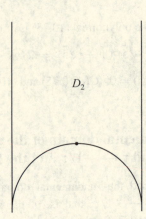

Figure 10.1 The fundamental region D_2 for G_2.

$\tau' \mapsto \tau \pm 2$ does not change $\operatorname{Im} \tau'$, we may assume that $|\operatorname{Re}(\tau')| \leq 1$. But then $|\tau'| \geq 1$, for, since $\operatorname{Im} \tau'$ is maximal, we have

$$\operatorname{Im} \tau' \geq \operatorname{Im}\left(-\frac{1}{\tau'}\right) = \operatorname{Im}\frac{\tau'}{|\tau'^2|}$$

and so $|\tau'| \geq 1$.

Any point $\tau \in H$ can be mapped to some $\tau' \in D_2$ under the action of G_2; in order to prove that distinct points of D_2 cannot be transformed into each other by G_2 it will suffice to prove (b).

We complete the domain D_2 (Figure 10.1) by including the points ± 1, corresponding to $q = -1$. For those points

$$f^{24}(\tau) = q^{-1} \prod_{k=1}^{\infty} (1 + q^{2k-1})^{24} = 0,$$

and for those points $\tau = \pm 1$ and the function

$$j(\tau) = \left(f(\tau)^{16} - \frac{16}{f(\tau)^8}\right)^3$$

has essential singularities (see Exercise 10.9.2). The function $j(\tau)$ has no poles in D_2 and therefore the function $f^{24}(\tau)$ does not vanish in D_2 and for $\operatorname{Im} \tau = \infty$ the values of the functions $j(\tau)$ and $f^{24}(\tau)$ are ∞. So we must consider the values of τ for which $f^{24}(\tau) \neq 0$.

Consider the region, D, in Figure 10.2, and denote by ∂D the boundary of D.

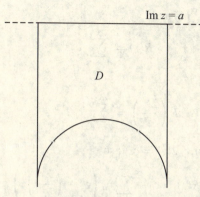

Figure 10.2 The fundamental region D for G.

For $c \neq 0$, we have

$$\int_{\partial D} d\ln(f^{24}(\tau) - c) = \int_{ia+1/2}^{ia-1/2} d\ln(f^{24}(\tau) - c).$$

Now $f^{24}(\tau) = q^{-1} + \cdots = e^{-\pi i \tau} + \cdots$, and so

$$d(\ln f^{24}(\tau) - c) = d(-\pi i \tau + \cdots) = -\pi i\, d\tau + \cdots$$

and so

$$\lim_{a \to \infty} \frac{1}{2\pi i} \int_{ia+1/2}^{ia-1/2} d\ln(f^{24}(\tau) - c) = 1.$$

We conclude that for each non-zero c in D_2 the function $f^{24}(\tau)$ assumes the value c exactly once, and that completes the proof of Theorem 10.1.

Finally, we look at the further condition requiring invariance under the transformation $\tau \mapsto (\tau - 1)/(\tau + 1)$. Denote by $g(\tau)$ a function invariant under the action of G_2 and also under $\tau \mapsto (\tau - 1)/(\tau + 1)$. From what we have proved,

$$g(\tau) = R(f^{24}(\tau)),$$

where R is a rational function subject to the further requirement that

$$R\left(f^{24}\left(\frac{\tau - 1}{\tau + 1}\right)\right) = R(f^{24}(\tau));$$

that is, on using

$$f(\tau)f\left(\frac{\tau - 1}{\tau + 1}\right) = \sqrt{2},$$

$$R\left(\frac{2^{12}}{f^{24}(\tau)}\right) = R(f^{24}(\tau)).$$

On writing $z = f^{24}(\tau)$ and

$$R(z) = \sum_{n=-\infty}^{\infty} c_n z^n,$$

we obtain

$$R\left(\frac{2^{12}}{z}\right) = \sum_{n=-\infty}^{\infty} c_n 2^{12n} z^{-n},$$

from which we obtain $c_n 2^{12n} = c_{-n}$; so

$$R(z) = c_0 + \sum_{n=1}^{\infty} c_n \left(z^n + \left(\frac{2^{12}}{z}\right)^n\right).$$

It follows (see Exercise 10.9.4) that

$$R(z) = P\left(z + \frac{2^{12}}{z}\right)$$

for some function P. Since R is rational, P has no singularities other than poles and so is a rational function.

We conclude that the function $g(\tau)$ is a rational function of $f^{24}(\tau) + 2^{12}/f^{24}(\tau)$.

Exercises 10.9

10.9.1 Prove the assertion made at the start of the proof of Theorem 10.1, namely that there exists τ' for which Im τ' is maximal.

10.9.2 Denote by $\lambda(\tau)$ the λ-function of Chapters 5, 6 and 7. Recall that (see Chapter 7, Section 7.7)

$$j(\tau) = 2^8 \frac{(1 - \lambda + \lambda^2)^3}{\lambda^2 (1 - \lambda)^2}$$

and hence obtain the result

$$j(\tau) = (2\pi)^8 \frac{\left(\theta_3^8 - \theta_2^4 \theta_0^4\right)^3}{\theta_1'^8}$$

$$= \left(f^{16} - f_1^8 f_2^8\right)^3$$

$$= \frac{(f^{24} - 16)^3}{f^{24}}$$

$$= \left(f^{16} - \frac{16}{f^8}\right)^3.$$

10.9.3 Show that the points $\tau = \pm 1$ are essential singularities for $j(\tau) = \left(f^{16} - 16/f^8 \right)^3$ and that for $\operatorname{Im} \tau = \infty$ the values of $j(\tau)$ and $f^{24}(\tau)$ are equal to ∞.

10.9.4 By considering the differences

$$\left(z + \frac{2^{12}}{z} \right)^n - \left(z^n + \left(\frac{2^{12}}{z} \right)^n \right),$$

use an induction argument to prove that

$$z^n + \left(\frac{2^{12}}{z} \right)^n$$

may be expressed as a polynomial in $z + 2^{12}/z$.

Solution of the quintic: the second method

This method is due to Hermite (1861) and Kronecker (1860), independently, and we shall follow (albeit very briefly, though one hopes not too superficially) the exposition by Klein (1884) and by Briot & Bouquet (1875); Dutta & Debnath (1965) offer an equally brief summary, based on Briot & Bouquet (1875). See also the very interesting account, in which the symmetries of the icosahedron are given prominence and afford considerable insight, in McKean & Moll, (1997).

We begin by deriving another form of the modular equation as in (10.32), where now $u = k^{1/4}(\tau)$ and $v = k^{1/4}(5\tau)$ (or $v = k^{1/4}(\tau/5 + 16n/5)$, $0 \le n < 5$), and where $k(\tau)$ denotes the modulus of the Jacobian elliptic function. The method used depends on a transformation of the elliptic integrals for the moduli $k(\tau)$ and $k(5\tau)$, respectively. Having done that we shall, as before, obtain the solution of the quintic in Bring's form, and we shall do so by using a variation of the ideas in (10.71) and (10.73).

At the final stage, the inversion problem for $\lambda(\tau) = k^2(\tau)$, as obtained in Chapter 5, plays a crucial role.

10.10 The modular equation for the quintic via elliptic integrals

The idea of this approach is to obtain a transformation of the form

$$y = \frac{a_0 + a_1 x + a_2 x^2 + \cdots + a_n x^n}{b_0 + b_1 x + b_2 x^2 + \cdots + b_n x^n}; \tag{10.84}$$

so that

$$\int \frac{dx}{\sqrt{(1-x^2)(1-k^2x^2)}} = M \int \frac{dy}{\sqrt{(1-y^2)(1-k_1^2y^2)}}, \qquad (10.85)$$

where k, k_1 are the moduli of the elliptic integrals (or of the elliptic functions they define by inversion) and M is a constant. The relationship between $u = k^{1/4}$ and $v = k_1^{1/4}$ is called the *modular equation* and we shall show that, for $n = 5$ and with $k = k(\tau)$, $k_1 = k(5\tau)$, we obtain the equation

$$u^6 - v^6 + 5u^2v^2(u^2 - v^2) + 4uv(1 - u^4v^4) = 0,$$

which is (10.32)

We begin with the equation

$$\frac{1-y}{1+y} = \frac{(1-x)(1-ax+bx^2)^2}{(1+x)(1+ax+bx^2)^2},$$

from which we obtain

$$y = \frac{x\{(2a+1) + (2ab+2b+a^2)x^2 + b^2x^4\}}{1 + (2b+2a+a^2)x^2 + (b^2+2ab)x^4}, \qquad (10.86)$$

which is of the form (10.84).

We seek to replace x, y by $1/(kx)$, $1/(k_1y)$ respectively, and the conditions for that are given by

$$k^2(2ab + 2b + a^2) = b^2(2a + 2b + a^2)$$
$$k^2(2a + 1) = b^2(b^2 + 2ab)$$
$$b^4k_1 = k^5. \qquad (10.87)$$

On writing $k = u^4$, $k_1 = v^4$, we obtain

$$b^4 = \frac{u^{20}}{v^4}, \quad b = \frac{u^5}{v},$$

and on substituting the latter value for b and $k = u^4$ in the second equation of (10.87) we find that

$$(2a + 1)u^{16} = \frac{u^{10}}{v^2}\left(\frac{u^{10}}{v^2} + \frac{2au^5}{v}\right),$$

that is

$$(2a + 1)uv^4 = u^5 + 2av$$

or, finally,

$$2av(1 - uv^3) = u(v^4 - u^4).$$

That last equation gives

$$2a = \frac{u(v^4 - u^4)}{v(1 - uv^3)}. \tag{10.88}$$

Using the substitutions $k = u^4, b = u^5/v$ again, the second relation in (10.87) may be written

$$(v^2 - u^2)(2b + a^2) = 2au^2(1 - u^3v) = \frac{u^3(v^4 - u^4)(1 - u^3v)}{v(1 - uv^3)},$$

or

$$a^2 = \frac{u^3}{v} \left\{ \frac{(v^2 + u^2)(1 - u^3v)}{1 - uv^3} \right\} 2u = \frac{u^3(v^2 - u^2)(1 + u^3v)}{v(1 - uv^3)}. \tag{10.89}$$

From (10.88) and (10.89) we obtain

$$2a = \frac{4u^2(1 + u^3v)}{u^2 + v^2};$$

that is

$$4u^2v(1 + u^3v)(1 - uv^3) = u(u^2 + v^2)(v^4 - u^4);$$

or, finally,

$$u^6 - v^6 + 5u^2v^2(u^2 - v^2) + 4uv(1 - u^4v^4) = 0,$$

which is the modular equation as in (10.32).

Exercises 10.10

10.10.1 Starting with the elliptic integrals

$$\int \frac{dx}{\sqrt{(1 - x^2)(1 - k^2x^2)}}, \quad \int \frac{dy}{\sqrt{(1 - y^2)\left(1 - k_1^2 y^2\right)}},$$

write

$$\frac{1 - y}{1 + y} = \left(\frac{1 - ax}{1 + ax}\right)^2 \frac{1 - x}{1 + x}$$

and hence

$$y = \frac{x(2a + 1 + a^2x^2)}{1 + a(a + 2)x^2}.$$

Show that the conditions for replacing x, y by $1/(kx)$, $1/(k_1y)$ are now given by

$$k^2(2a + 1) = a^3(a + 2), \quad k_1 = \frac{1}{a^2} = \frac{k^3}{a^4};$$

and then show that

$$k^2 = \frac{a^3(a+2)}{2a+1}, \quad k_1^2 = a\left(\frac{a+2}{2a+1}\right)^3.$$

Then show that

$$1 - y = \frac{(1-ax)^2(1-x)}{1+a(a+2)x^2},$$

$$1 + y = \frac{(1+ax)^2(1+x)}{1+a(a+2)x^2},$$

$$1 - k_1 y = \frac{\left(1-\dfrac{kx}{a}\right)^2(1-kx)}{1+a(a+2)x^2},$$

$$1 + k_1 y = \frac{\left(1+\dfrac{kx}{a}\right)^2(1+kx)}{1+a(a+2)x^2}.$$

Deduce that

$$\sqrt{(1-y^2)(1-k_1^2 y^2)} = \frac{(1-a^2 x^2)\left(1-\dfrac{k^2 x^2}{a^2}\right)}{(1+a(a+2)x^2)^2}\sqrt{(1-x^2)(1-k^2 x^2)}$$

and from that, by differentiation, show that

$$\int \frac{dy}{\sqrt{(1-y^2)(1-k_1^2 y^2)}} = (2a+1)\int \frac{dx}{\sqrt{(1-x^2)(1-k^2 x^2)}},$$

where

$$k^2 = \frac{a^3(a+2)}{2a+1}, \quad k_1^2 = a\left(\frac{a+2}{2a+1}\right)^3.$$

By putting $k^{1/4} = u$, $k_1^{1/4} = v$ obtain the equation

$$u^2 v^2 = \sqrt{kk_1} = \frac{a(a+2)}{(2a+1)} = \frac{u^3(u^3+2v)}{v(2u^3+v)}, \quad a = \frac{u^2}{v}$$

and thence the modular equation for the cubic transformation

$$u^4 - v^4 + 2uv(1-u^2 v^2) = 0.$$

10.11   Solution of the quintic equation

We begin with the modular equation (10.32), namely

$$u^6 - v^6 + 5u^2 v^2(u^2 - v^2) + 4uv(1-u^4 v^4) = 0, \tag{10.90}$$

where we recall that $u = k^{1/4}(\tau)$ and $v = k^{1/4}(5\tau)$; more precisely, for a given τ and $u = k^{1/4}(\tau)$, there are six roots of Equation (10.90), which we denote by

$$v_\infty = k^{1/4}(5\tau), \quad v_c = k^{1/4}\left(\frac{\tau + 16c}{5}\right), \quad 0 \le c < 5. \tag{10.91}$$

As we saw in Chapter 5, the modulus $k(\tau)$ is given by

$$u^8 = k^2(\tau) = \frac{\theta_2^4(0|\tau)}{\theta_3^4(0|\tau)} = \frac{e_3 - e_2}{e_1 - e_2}, \tag{10.92}$$

and so

$$u = k^{1/4}(\tau) = \frac{\theta_2^{1/2}(0|\tau)}{\theta_3^{1/2}(0|\tau)} = \sqrt{2}e^{i\pi\tau/8}\frac{\sum_{m\in\mathbb{Z}} e^{\pi i\tau(2m^2+m)}}{\sum_{m\in\mathbb{Z}} e^{\pi i\tau m^2}} = \phi(\tau), \tag{10.93}$$

say. Similarly

$$1 - u^8 = k'^2(\tau) = \frac{\theta_4^4(0|\tau)}{\theta_3^4(0|\tau)} = \frac{e_1 - e_3}{e_1 - e_2}, \tag{10.94}$$

and we shall denote that by $\psi^2(\tau)$.

In the notation of (10.93),

$$v_\infty = \phi(5\tau), \quad v_0 = \phi\left(\frac{\tau}{5}\right), \quad v_c = \phi\left(\frac{\tau + 16c}{5}\right), \quad c = 1, 2, 3, 4. \tag{10.95}$$

As in our 'first method' the idea now is to obtain a quintic equation in Bring's form from the sextic equation (10.90). (Hermite called it 'the depressed equation'; see McKean & Moll, 1997, p. 214.)

To that end we follow Hermite (see Klein, 1884, p. 163) and put

$$y = (v_\infty - v_0)(v_1 - v_4)(v_2 - v_3). \tag{10.96}$$

One can then show (see Exercises 10.11) that y is a root of the quintic

$$y^5 - 2^4 \cdot 5^3 \cdot u^4(1 - u^8)^2 y - 2^6 \cdot 5^{5/2} u^3(1 - u^8)^2(1 + u^8) = 0, \tag{10.97}$$

and that can be expressed in Bring's form

$$x^5 - x - A = 0, \tag{10.98}$$

by writing

$$y = 2 \cdot 5^{3/4} u\sqrt{1 - u^8} \cdot x, \tag{10.99}$$

and then A is given by

$$A = \frac{2}{5^{5/4}} \frac{1 + u^8}{u^2(1 - u^8)^{1/2}}. \tag{10.100}$$

So to solve the general quintic we first reduce it to Bring's form (10.98) and then we determine τ so that (10.100) holds. We do that by writing (10.100) as

$$A^2 5^{5/2} u^4 (1 - u^8) = 2^4 (1 + u^8)^2,$$

which is a quartic in u^4 and that may be solved, as we saw at the start of this chapter, in radicals, from which solution we obtain an expression for $u^4 = k(\tau)$ and, by the inversion theorem of Chapter 5, that may be solved to give a value of τ. Then we substitute that value of τ in the expression for y in (10.96), which gives us one root of our quintic.

The other roots are given by an appeal to Galois theory and the icosahedral group and are of the form:

$$(v_\infty - v_1)(v_2 - v_0)(v_3 - v_4),$$
$$(v_\infty - v_2)(v_1 - v_3)(v_0 - v_4),$$
$$(v_\infty - v_3)(v_2 - v_4)(v_1 - v_0),$$
$$(v_\infty - v_4)(v_0 - v_3)(v_1 - v_2). \tag{10.101}$$

(See McKean & Moll, 1997, p. 212, though the notation there is slightly different; for other related matters see Exercises 10.11, and for general reading see the books referred to in the introduction to the second method).

Exercises 10.11

10.11.1 Imitate the ideas that were used to derive the quintic (10.77) from the modular equation in Section 10.8 to obtain (10.97) from (10.90). See McKean & Moll, (1997), Section 5.5.

10.11.2 Verify that the substitutions (10.99) and (10.100) transform equation (10.97) into Bring's form, (10.98).

10.11.3 Verify the details concerning $u = k^{1/4}(\tau)$, given in (10.93).

10.11.4 Show that the roots of the quintic in (10.98) may be expressed in the form:

$$\frac{1}{2 \cdot 5^{3/4}} \frac{\Phi(\tau)}{\phi(\tau)\psi^{1/4}(\tau)}, \qquad \frac{1}{2 \cdot 5^{3/4}} \frac{\Phi(\tau + 16)}{\phi(\tau)\psi^{1/4}(\tau)},$$

$$\frac{1}{2 \cdot 5^{3/4}} \frac{\Phi(\tau + 2 \cdot 16)}{\phi(\tau)\psi^{1/4}(\tau)}, \quad \frac{1}{2 \cdot 5^{3/4}} \frac{\Phi(\tau + 3 \cdot 16)}{\phi(\tau)\psi^{1/4}(\tau)},$$

$$\frac{1}{2 \cdot 5^{3/4}} \frac{\Phi(\tau + 4 \cdot 16)}{\phi(\tau)\psi^{1/4}(\tau)},$$

where

$$\Phi(\tau) = \left\{ \phi(5\tau) + \phi\left(\frac{\tau}{5}\right) \right\} \left\{ \phi\left(\frac{\iota + 16}{5}\right) - \phi\left(\frac{\iota + 4 \cdot 16}{5}\right) \right\}$$

$$\times \left\{ \phi\left(\frac{\tau + 2 \cdot 16}{5}\right) - \phi\left(\frac{\tau + 3 \cdot 16}{5}\right) \right\}.$$

11

An arithmetic application of elliptic functions: the representation of a positive integer as a sum of three squares

11.1   Sums of three squares and triangular numbers

We have already seen that the elliptic functions enable us to obtain formulae for the number of representations of a positive integer, n, as a sum of two squares (and to identify those cases when such a representation is not possible) and also for the number of representations as a sum of four squares (in which case every positive integer is so representable). We turn now to the more difficult question of the representation as a sum of three squares.

To set the scene, we begin with the representation of n as a sum of three triangular numbers:

$$n = \frac{n_1(n_1 + 1)}{2} + \frac{n_2(n_2 + 1)}{2} + \frac{n_3(n_3 + 1)}{2}.$$

Given such a representation, we find that

$$8n + 3 = 4n_1^2 + 4n_1 + 4n_2^2 + 4n_2 + 4n_3^2 + 4n_3 + 3$$
$$= (2n_1 + 1)^2 + (2n_2 + 1)^2 + (2n_3 + 1)^2;$$

so $N = 8n + 3$ is representable as a sum of three odd squares. It is straightforward (by appealing to congruence conditions) to show that if n is representable as a sum of three squares then n cannot be of the form $4^a(8b + 7)$ and then, though this is not easy, there is an elementary proof (using quadratic forms and quadratic reciprocity) that that condition is also sufficient. But the truth lies much deeper than that, and we shall see that the elliptic functions and theta functions, together with the theory of binary quadratic forms, enable us to obtain a formula for the number of representations as a sum of three squares. We follow the ideas of Hermite (1862), Jacobi (1829), and

Kronecker (1860) to prove our main theorem (Theorem 11.1). The general background is to be found in Dickson (1934), and we begin by recalling some ideas, required in the enunciation of the theorem, from the theory of binary quadratic forms, which we shall have more to say about later. For the moment we follow the notation of Gauss and consider the *binary quadratic form*[1]

$$ax^2 + 2bxy + cy^2 \qquad (11.1)$$

of *discriminant* (Gauss uses 'determinant')

$$-\Delta = D = b^2 - ac. \qquad (11.2)$$

Two such forms are (properly) *equivalent* if one can be obtained from the other by a *unimodular* substitution $(x, y) \mapsto (x', y')$, $x = px' + qy'$, $y = rx' + sy'$, where p, q, r, s are integers with $ps - qr = 1$. Two such forms are said to be in the *same class* if one is transformed into the other by a unimodular substitution (see footnote 1).

Our main theorem is:

Theorem 11.1 *Denote by $G(n)$ the number of classes of binary forms of discriminant $-n$ and by $F(n)$ the number of classes of such forms in which at least one of the outer coefficients (a or c) is odd. Then the number of representations of n as a sum of three squares is $24F(n) - 12G(n)$.*

Example If $n \equiv 7 \pmod 8$ one can show that $G(n) = 2F(n)$ and so the number of representations as a sum of three squares is 0, which agrees with our introductory remarks.

The idea of the proof is to interpret the problem in terms of theta functions by showing that if we define $E(n)$ by

$$12\sum_{n=1}^{\infty} E(n)q^n = (\theta_3(0))^3 = (1 + 2q + 2q^4 + 2q^9 + \cdots)^3, \qquad (11.3)$$

then $12E(n)$ denotes the number of representations of n as a sum of three squares, and then we show that

$$E(n) = 2F(n) - G(n), \qquad (11.4)$$

---

[1] Gauss used the notation $ax^2 + 2bxy + cy^2$, as did Legendre and Dirichlet; the notation, $ax^2 + bxy + cy^2$, without the factor 2, was used by Lagrange, Kronecker and Dedekind. If one uses that notation, then the discriminant is $b^2 - 4ac$. For an excellent introduction, see Davenport (1952), and later editions.

by using a very involved argument from the theory of the Jacobi elliptic functions and from the theory of binary quadratic forms.

11.2   Outline of the proof

We follow the account given by Hermite, though Hermite uses Jacobi's earlier notation Θ, Θ_1, H, H_1 for the theta functions, whereas we shall use the notation summarized at the end of Chapter 4.

As usual, we write $q = e^{\pi i \tau}$, $\operatorname{Im} \tau > 0$, and $t = e^{2iz}$ and we shall prove the identities:

$$\theta_3(0)^2 \frac{\theta_4(x)\theta_2(x)}{\theta_3(x)\theta_1(x)} = \cot x + 4q \frac{\sin 2x}{1-q} - \frac{4q^2 \sin 4x}{1+q^2} + \frac{4q^3 \sin 6x}{1-q^3} - \cdots$$

$$(11.5)$$

and

$$\theta_3(0) \frac{\theta_3(x)\theta_1(x)}{\theta_2(x)} = \tan x + 2 \sum_{n=1}^{\infty} (-1)^{n-1} q^{n^2} \sin(2nx) \cdot B_n, \qquad (11.6)$$

where

$$B_n = 1 + 2q^{-1} + 2q^{-4} + \cdots + 2q^{-(n-1)^2}. \qquad (11.7)$$

We shall prove (11.5) and (11.6) in Sections 11.3 and 11.4, respectively.

Having done that, the idea, following Hermite, is to multiply (11.5) and (11.6) to obtain

$$\theta_3^3 \theta_4(x) = \left(\cot x + 4 \sum_{n=1}^{\infty} (-1)^{n-1} \frac{q^n \sin 2nx}{1+(-q)^n} \right)$$
$$\times \left(\tan x + 2 \sum_{n=1}^{\infty} \sin(2nx)(-1)^{n-1} q^{n^2} B_n \right)$$

and then, on integrating with respect to x from 0 to $\pi/2$, we obtain

$$\theta_3^3 = 1 + 4 \sum_{n=1}^{\infty} \frac{q^n}{1+(-q)^n} - 2 \sum_{n=1}^{\infty} (-1)^n q^{n^2} B_n + 4 \sum_{n=1}^{\infty} \frac{q^{n^2+n} B_n}{1+(-q)^n}.$$

$$(11.8)$$

As we shall see, in Section 11.5, the first two series in (11.8) may be expressed as power series in q with coefficients involving $d(m)$, the number of divisors of *m*. The third sum is expressible in the form

$$\sum_{n=1}^{\infty} \frac{q^{n^2+n}B_n}{(1+(-q)^n)} = \sum_{\substack{a,m=1 \\ 0 \le b^2 \le (n-1)^2}}^{\infty} (-1)^{a(n+1)} q^{n^2+n+an-b^2}. \tag{11.9}$$

Finally, in Section 11.5, we shall make use of the connection with the theory of binary quadratic forms. We shall write

$$A = n, \quad B = b, \quad C = n+1+a; \tag{11.10}$$

so that the exponent of q in (11.9) is

$$n^2 + n + an - b^2 = AC - B^2. \tag{11.11}$$

Then we consider the binary quadratic form

$$Ax^2 + 2Bxy + Cy^2$$

of discriminant $-\Delta$, where

$$\Delta = n^2 + an + n - b^2.$$

Now, as we have already remarked,

$$\theta_3^3 = \left(1 + 2\sum_{n=1}^{\infty} q^{n^2}\right)^3 = \sum_{N=0}^{\infty} r_3(N)q^N, \tag{11.12}$$

where $r_3(N)$ denotes the number of representations of N as a sum of three squares, or, as we wrote in (11.3), $r_3(N) = 12E(n)$.

A comparison between (11.8) and (11.12) suggests that we might be able to find a formula for $r_3(N)$ in terms of the number of classes of quadratic forms of discriminant $-\Delta$; perhaps

$$E(\Delta) = 2F(\Delta) - G(\Delta). \tag{11.13}$$

11.3   Proof of (11.5)

We begin with (11.5) and note that;

$$\theta_3^2 \frac{\theta_4(x)\theta_2(x)}{\theta_3(x)\theta_1(x)} = \theta_3^2 \frac{\dfrac{\theta_2(x)}{\theta_4(x)}}{\dfrac{\theta_3(x)}{\theta_4(x)} \dfrac{\theta_1(x)}{\theta_4(x)}}$$

$$= \theta_3^2 \frac{\dfrac{\theta_2}{\theta_4} cn\left(\dfrac{2Kx}{\pi}\right)}{\dfrac{\theta_3}{\theta_4} dn\left(\dfrac{2Kx}{\pi}\right) \dfrac{\theta_2}{\theta_3} sn\left(\dfrac{2Kx}{\pi}\right)}. \tag{11.14}$$

Recall that $\theta_3^2 = 2K/\pi$.

Consider

$$\frac{2K}{\pi} \frac{sn\left(\dfrac{2Kx}{\pi}\right)}{sn\left(K - \dfrac{2Kx}{\pi}\right)} = \frac{2K}{\pi} \frac{sn\left(\dfrac{2Kx}{\pi}\right)}{\left(cn\left(-\dfrac{2Kx}{\pi}\right) \bigg/ dn\left(-\dfrac{2Kx}{\pi}\right)\right)}$$

$$= \frac{2K}{\pi} \frac{sn\left(\dfrac{2Kx}{\pi}\right) dn\left(\dfrac{2Kx}{\pi}\right)}{cn\left(\dfrac{2Kx}{\pi}\right)}$$

$$= -\frac{d}{dx} \ln cn\left(\frac{2Kx}{\pi}\right). \tag{11.15}$$

If we can prove that

$$\frac{d}{dx} \ln cn\frac{2Kx}{\pi} = \tan x + 4q\frac{\sin 2x}{1-q} + 4q^2 \frac{\sin 4x}{1+q^2} + \cdots, \tag{11.16}$$

then, on replacing x by $\pi/2 - x$, we obtain

$$-\frac{d}{dx} \ln cn\left(K - \frac{2Kx}{\pi}\right) = \cot x + 4q\frac{\sin 2x}{1-q} - 4q^2 \frac{\sin 4x}{1+q^2}$$

$$+ \frac{4q^3 \sin 6x}{1-q^3} - \cdots,$$

which, by (11.14) and (11.15) gives (11.5).

We turn to the proof of (11.16).

We begin with the infinite product for the Jacobi function

$$cn\left(\frac{2Kx}{\pi}\right) = 2q^{1/4}k'^{1/2}k^{-1/2} \cos x \prod_{n=1}^{\infty} \frac{(1 + q^{2n}e^{ix})(1 + q^{2n}e^{-ix})}{(1 - q^{2n-1}e^{ix})(1 - q^{2n-1}e^{-ix})},$$

from which we obtain

$$\ln\left(cn\left(\frac{2Kx}{\pi}\right)\right) = \ln(2q^{1/4}k'^{1/2}k^{-1/2}) + \ln \cos x$$

$$+ \sum_{n=1}^{\infty} \{\ln(1 + q^{2n}e^{ix}) + \ln(1 + q^{2n}e^{-ix})$$

$$- \ln(1 - q^{2n-1}e^{ix}) - \ln(1 - q^{2n-1}e^{-ix})\},$$

and we note that the various series and products here, and in what follows, are absolutely convergent, since $q = e^{\pi i \tau}$ and $\operatorname{Im} \tau > 0$.

On expanding the logarithms as infinite series, we obtain

$$\ln\left(cn\left(\frac{2Kx}{\pi}\right)\right) = \ln(2q^{1/4}k'^{1/2}k^{-1/2}) + \ln\cos x$$

$$+ \sum_{n=1}^{\infty}\sum_{k=1}^{\infty}\frac{(-1)^{k-1}q^{2nk}}{k}2\cos 2kx$$

$$+ \sum_{n=1}^{\infty}\sum_{k=1}^{\infty}\frac{q^{(2n-1)k}}{k}2\cos 2kx. \qquad (11.17)$$

On differentiating with respect to x, we obtain from (11.17)

$$\frac{d}{dx}\ln\left(cn\left(\frac{2Kx}{\pi}\right)\right) = -\tan x - 4\sum_{n=1}^{\infty}\sum_{k=1}^{\infty}\frac{(-1)^{k-1}q^{2nk}}{k}k\sin 2kx$$

$$- 4\sum_{n=1}^{\infty}\sum_{k=1}^{\infty}\frac{q^{(2n-1)k}}{k}k\sin 2kx$$

$$= -\tan x - 4\sum_{k=1}^{\infty}\sum_{n=1}^{\infty}((-1)^{k-1}q^{2nk}\sin 2kx$$

$$+ q^{(2n-1)k}\sin 2kx),$$

whence

$$-\frac{d}{dx}\ln cn\frac{2Kx}{\pi} = \tan x + 4\left\{\frac{q\sin 2x}{1-q} + \frac{q^2\sin 4x}{1+q^2} + \cdots\right\}$$

and on replacing x by $\pi/2 - x$, as already observed,

$$\frac{d}{dx}\ln cn\left(\frac{2K}{\pi}\left(\frac{\pi}{2}-x\right)\right) = \cot x + 4\left\{\frac{q\sin 2x}{1-q} - \frac{q^2\sin 4x}{1+q^2} - \cdots\right\}$$

as required. So we have proved (11.5).

We turn to the proof of (11.16).

11.4 Proof of (11.6)

Hermite (1862) appears not to give the details in this case, so the reader is advised to work through the account offered here with care. See also Exercises 11.4.

We want to prove that

$$\theta_3\frac{\theta_3(x)\theta_1(x)}{\theta_2(x)} = \tan x + 2\sum_{n=1}^{\infty}\sin(2nx)(-1)^{n-1}q^{n^2}B_n,$$

where $B_n = 1 + 2q^{-1} + \cdots + 2q^{-(n-1)^2}$.

We write the left-hand side of (11.6) (as above) in the form

$$\theta_3\theta_3(x)\frac{\theta_1(x)/\theta_4(x)}{\theta_2(x)/\theta_4(x)} = \theta_3\theta_3(x)\frac{\dfrac{\theta_2}{\theta_3}\,sn(2Kx/\pi)}{\dfrac{\theta_2}{\theta_4}\,cn(2Kx/\pi)}$$

$$= \theta_4\theta_3(x)sc\left(\frac{2Kx}{\pi}\right). \tag{11.18}$$

Now expand $sc\,(2Kx/\pi)$ as a Fourier series (by integrating round a suitable period parallelogram – see Exercise 11.4.2) to obtain

$$\theta_3\frac{\theta_3(x)\theta_1(x)}{\theta_2(x)} = \theta_4\theta_3(x)\left\{\frac{\pi}{2Kk'}\tan x + \frac{2\pi}{Kk'}\sum_{n=1}^{\infty}\frac{(-1)^n q^{2n}}{1+q^{2n}}\sin 2nx\right\},$$
$$\tag{11.19}$$

and we also note that

$$\theta_3(x) = 1 + 2q\cos 2x + 2q^4\cos 4x + \cdots$$
$$= \sum_{m\in\mathbb{Z}}q^{m^2}e^{2imx}. \tag{11.20}$$

Now

$$\frac{\pi}{2K} = \frac{1}{\theta_3^2}, \quad k'^{1/2} = \frac{\theta_4}{\theta_3}, \tag{11.21}$$

and so the right-hand side of (11.19) reads

$$\theta_4\theta_3(x)\frac{1}{\theta_4^2}\left\{\tan x + 4\sum_{n=1}^{\infty}\frac{(-1)^n q^{2n}}{1+q^{2n}}\sin 2nx\right\}. \tag{11.22}$$

In order to obtain (11.22) in a suitable form, we begin with

$$\theta_3(x)\tan x = \left(1 + 2\sum_{k=1}^{\infty}q^{k^2}\cos 2kx\right)\tan x. \tag{11.23}$$

By using induction on k (for example) we can prove that

$$\cos 2kx\tan x = \sin 2kx - 2\sin(2k-2)x$$
$$+ 2\sin(2k-4)x + \cdots + (-1)^k\tan x \tag{11.24}$$

and then (11.24) and (11.23) give

$$\theta_3(x)\tan x = \left(1 + 2\sum_{n=1}^{\infty}q^{n^2}\cos 2nx\right)\tan x$$

$$= \tan x + 2\sum_{n=1}^{\infty}q^{n^2}(\sin 2nx - 2\sin(2n-2)x + \cdots + (-1)^n\tan x)$$

$$= \tan x \left(1 + 2\sum_{n=1}^{\infty}(-1)^n q^{n^2}\right)$$

$$+ 2q\sin 2x + 2q^4(\sin 4x - 2\sin 2x)$$
$$+ 2q^9(\sin 6x - 2\sin 4x + 2\sin 2x) + \cdots$$
$$+ 2q^{n^2}(\sin 2nx - 2\sin(n-2)x + \cdots + (-1)^{n-1}\sin 2x)$$
$$+ \cdots. \tag{11.25}$$

By rearranging (11.25) as a series in $\sin 2kx$, we obtain

$$\theta_3(x)\tan x = \theta_4 \tan x + 2\sin 2x(q - 2q^4 + 2q^9 - \cdots)$$
$$+ 2\sin 4x(q^4 - 2q^9 + 2q^{16} - \cdots)$$
$$+ 2\sin 6x(q^9 - 2q^{16} + 2q^{25} - \cdots) + \cdots$$
$$+ 2\sin 2Nx\left(q^{N^2} - 2q^{(N+1)^2} + 2q^{(N+2)^2} - \cdots\right)$$
$$+ \cdots. \tag{11.26}$$

Now we look at the remaining terms in the sum (11.22). We have

$$\frac{1}{\theta_4}\theta_3(x)4\sum_{m=1}^{\infty}(-1)^m\frac{q^{2m}}{1+q^{2m}}\sin 2mx$$

$$= \frac{1}{\theta_4}\left(1 + 2\sum_{n=1}^{\infty}q^{n^2}\cos 2nx\right)4\sum_{m=1}^{\infty}(-1)^m\frac{q^{2m}}{1+q^{2m}}\sin 2mx$$

$$= \frac{1}{\theta_4}\left[4\sum_{m=1}^{\infty}(-1)^m\frac{q^{2m}}{1+q^{2m}}\sin 2mx\right.$$
$$\left.+8\sum_{n=1}^{\infty}\sum_{m=1}^{\infty}\frac{q^{n^2}(-1)^m q^{2m}}{1+q^{2m}}\frac{1}{2}(\sin(2m+2n)x + \sin(2m-2n)x)\right]$$

$$= \frac{4}{\theta_4}\left[\sum_{m=1}^{\infty}\frac{(-1)^m q^{2m}}{1+q^{2m}}\sin 2mx\right.$$
$$\left.+ \sum_{n,m=1}^{\infty}\frac{q^{n^2}(-1)^m q^{2m}}{1+q^{2m}}(\sin 2(m+n)x + \sin 2(m-n)x)\right]$$

$$= \frac{4}{\theta_4}\left[\sum_{N=1}^{\infty}\frac{(-1)^N q^{2N}}{1+q^{2N}}\sin 2Nx + \sum_{N=1}^{\infty}\sum_{\substack{m,n \\ m+n=N}}\frac{q^{n^2}(-1)^m q^{2m}}{1+q^m}\sin 2Nx\right.$$

$$\left.+ \sum_{N=1}^{\infty}\sum_{\substack{m,n \\ m-n=\pm N}}\frac{q^{n^2}(-1)^m q^{2m}}{1+q^{2m}}\sin 2Nx\right]$$

$$= \frac{4}{\theta_4} \sum_{N=1}^{\infty} \left[\frac{(-1)^N q^{2N}}{1+q^{2N}} + \sum_{\substack{m,n \\ m+n=N}} \frac{q^{n^2}(-1)^m q^{2m}}{1+q^{2m}} + \sum_{\substack{m,n \\ m-n=N}} \frac{q^{n^2}(-1)^m q^{2m}}{1+q^{2m}} \right.$$

$$\left. - \sum_{\substack{m,n \\ n-m=N}} \frac{q^{n^2}(-1)^m q^{2m}}{1+q^{2m}} \right] \sin 2Nx. \tag{11.27}$$

Now the two sums over $m - n = \pm N$ may be written as

$$\sum_{m=1}^{\infty} \frac{q^{(m-N)^2}(-1)^m q^{2m}}{1+q^{2m}} - \sum_{m=1}^{\infty} \frac{q^{(m+N)^2}(-1)^m q^{2m}}{1+q^{2m}}$$

$$= \sum_{m=1}^{\infty} \frac{(-1)^m q^{2m}}{1+q^{2m}} \left(q^{(m-N)^2} - q^{(m+N)^2} \right)$$

$$= \sum_{m=1}^{\infty} \frac{(-1)^m q^{2m}}{1+q^{2m}} q^{(m-N)^2} \left(1 - q^{2m \cdot 2N} \right)$$

$$= \sum_{m=1}^{\infty} \frac{(-1)^m q^{2m}}{1+q^{2m}} q^{m^2+N^2} \left(q^{-2mN} - q^{2mN} \right). \tag{11.28}$$

(We note in passing the presence of series resembling Lambert series; see Grosswald, 1984.)

By combining (11.26), (11.27) and (11.28) and rearranging the terms in the sum for (11.28), we find that in order to prove (11.6) it suffices to prove that, for all $N \geq 1$,

$$2(1 - 2q + 2q^4 - 2q^9 + \cdots)(-1)^{N-1} q^{N^2} \left(1 + 2q^{-1} + \cdots + 2q^{-(N-1)^2} \right)$$

$$= 2 \left(q^{N^2} - 2q^{(N+1)^2} + 2q^{(N+2)^2} - \cdots \right)$$

$$+ 4 \sum_{m=1}^{\infty} \left(\frac{q^{(m-N)^2}(-1)^m q^{2m}}{1+q^{2m}} - \frac{q^{(m+N)^2}(-1)^m q^{2m}}{1+q^{2m}} \right) \tag{11.29}$$

$$= 2 \left(q^{N^2} - 2q^{(N+1)^2} + 2q^{(N+2)^2} - \cdots \right)$$

$$+ 4 \sum_{m=1}^{\infty} q^{(m^2+N^2)}(-1)^m \left(1 - q^{2m} + q^{4m} - \cdots - q^{2m(2N-1)} \right). \tag{11.30}$$

The proof is rather involved (the reader is invited to find a better one) and it will be convenient to state (11.30) formally as:

Propositon 11.1 *With the foregoing notation, for every integer $N \geq 1$ we have*

$$2\theta_4(-1)^{N-1} q^{N^2} \left(1 + 2q^{-1} + \cdots + 2q^{-(N-1)^2} \right)$$

$$= 2q^{N^2} + 4 \sum_{r=1}^{\infty} (-1)^r q^{(N+r)^2}$$

$$+ 4 \sum_{m=1}^{\infty} q^{(m^2+N^2)}(-1)^m \left(1 - q^{2m} + q^{4m} - \cdots - q^{2m(2N-1)} \right).$$

Proof We use induction on N.

The case $N = 1$ is straightforward. If $N = 1$, the sums on the right-hand side of (11.30) (and of (11.29)) read

$$2q - 4q^4 + 4q^9 + \cdots + 4 \sum_{m=1}^{\infty} (-1)^m q^{m^2+1} - 4 \sum_{m=1}^{\infty} (-1)^m q^{(m+1)^2}$$

$$= 2q + 4 \sum_{m=1}^{\infty} q^{(m+1)^2}(-1)^m + 4 \sum_{m=1}^{\infty} (-1)^m q^{m^2+1} - 4 \sum (-1)^m q^{(m+1)^2}$$

$$= 2q - 4q^2 + 4q^5 - \cdots$$

$$= 2q\theta_4,$$

which is the left-hand side, as required, and which gives the basis of the induction.

For the induction step, we suppose that (11.30) holds for N and we shall then deduce it for $N + 1$.

To simplify our notation, we recall that

$$\theta_4 = 1 - 2q + 2q^4 - 2q^9 + \cdots,$$

and we also recall that

$$B_N = 1 + 2q^{-1} + \cdots + 2q^{-(N-1)^2}.$$

Write

$$\phi_N = q^{N^2} - 2q^{(N+1)^2} + 2q^{(N+2)^2} - \cdots + (-1)^r 2q^{(N+r)^2} + \cdots, \qquad (11.31)$$

and

$$A_N(m) = q^{-2mN} - q^{-2m(N-1)} + \cdots + (-1)^{2N-1} q^{2m(N-1)}. \qquad (11.32)$$

We observe that

$$B_{N+1} = B_N + 2q^{-N^2} \qquad (11.33)$$

and

$$A_{N+1}(m) = q^{-2m(N+1)} - A_N(m) - q^{2mN}, \qquad (11.34)$$

results that are obviously related to the induction step. Finally, we define ξ_N by

$$\xi_N = 2q^{N^2-(N+1)^2}\phi_{N+1} + 2\phi_N; \tag{11.35}$$

so that

$$2\phi_{N+1} = q^{(N+1)^2-N^2}\xi_N - 2q^{(N+1)^2-N^2}\phi_N. \tag{11.36}$$

We note that

$$\xi_N = 2q^{N^2} + \sum_{r=1}^{\infty}(-1)^{r+1}q^{N^2+(N+r-1)(2N+r+1)}. \tag{11.37}$$

With the notation introduced above, the first stage of the induction step, for N, may be written

$$2\theta_4(-1)^{N-1}q^{N^2}B_N = 2\phi_N + 4q^{N^2}\sum_{m=1}^{\infty}c_m A_N(m), \tag{11.38}$$

where

$$c_m = q^{m^2+2m}(-1)^m. \tag{11.39}$$

The induction step reads

$$2\theta_4(-1)^N q^{(N+1)^2}B_{N+1} = 2\phi_{N+1} + 4q^{(N+1)^2}\sum_{m=1}^{\infty}c_m A_{N+1}(m); \tag{11.40}$$

so we have to prove that (11.38) implies (11.40).

Using (11.33), (11.34) and (11.37), we re-write (11.40) as

$$-2\theta_4(-1)^{N-1}q^{(N+1)^2}B_N + 4\theta_4(-1)^N q^{(N+1)^2}q^{-N^2}$$
$$= 2\phi_{N+1} + 4q^{(N+1)^2}\sum_{m=1}^{\infty}c_m\left(q^{-2m(N+1)} - A_N(m) - q^{2mN}\right).$$

On using (11.37) we have

$$-2\theta_4(-1)^{N-1}q^{(N+1)^2}B_N + 4\theta_4(-1)^N q^{(N+1)^2-N^2}$$
$$= q^{(N+1)^2-N^2}\xi_N - 2q^{(N+1)^2-N^2}\phi_N$$
$$+ 4q^{(N+1)^2}\sum_{m=1}^{\infty}c_m\left(q^{-2m(N+1)} - q^{2mN}\right)$$
$$- 4q^{(N+1)^2}\sum_{m=1}^{\infty}c_m A_m(N). \tag{11.41}$$

Finally, multiply both sides of (11.41) by $q^{N^2-(N+1)^2}$ to obtain

$$
- 2\theta_4(-1)^{N-1}q^{N^2}B_N + 4\theta_4(-1)^N
$$

$$
= \xi_N - 2\phi_N - 4q^{N^2}\sum_{m=1}^{\infty}c_m A_N(m) + 4q^{N^2}\sum_{m=1}^{\infty}c_m\left(q^{-2m(N+1)} - q^{2mN}\right).
$$

$$(11.42)$$

We now appeal to (11.38) to obtain some cancellation of the terms in (11.42), which leaves us with the following to be proved to give the induction step:

$$
4\theta_4(-1)^N = \xi_N + 4q^{N^2}\sum_{m=1}^{\infty}q^{m^2+2m}(-1)^m\left(q^{-2m(N+1)} - q^{2mN}\right). \qquad (11.43)
$$

That is,

$$
4(-1)^N\theta_4 = \xi_N + 4(-1)^N\sum_{m=1}^{\infty}q^{(m-N)^2}(-1)^{m-N}
$$

$$
+ 4(-1)^N\sum_{m=1}^{\infty}q^{(m+N+1)^2}q^{N^2-(N+1)^2}(-1)^{m+N+1}. \quad (11.44)
$$

Finally, (11.44) may be written in the form

$$
q^{(N+1)^2-N^2}\left\{4(-1)^N\theta_4 - 2\phi_N - 4(-1)^N\sum_{m=1}^{\infty}q^{(m-N)^2}(-1)^{m-N}\right\}
$$

$$
= 2\phi_{N+1} + 4(-1)^N\sum_{m=1}^{\infty}q^{(m+N+1)^2}(-1)^{m+N+1},
$$

or

$$
q^{(N+1)^2-N^2}\left\{4(-1)^N\theta_4 - 2\phi_N - 4(-1)^N\sum_{r=1-N}^{\infty}q^{r^2}(-1)^r\right\}
$$

$$
= 2\phi_{N+1} + 4(-1)^N\sum_{k=N+2}^{\infty}q^{k^2}(-1)^k. \qquad (11.45)
$$

Now it is straightforward, albeit tedious, to show that the two sides of (11.45) agree, and so we have proved (11.45) and so the induction step. Proposition 11.1 is proved.

Exercises 11.4

11.4.1 Verify the details in Equation (11.18).

11.4.2 Obtain the Fourier series expansion of $sc\,(2Kx/\pi)$, as used in (11.19).

(See Whittaker and Watson, 1927, Section 22.6, especially Section 22.61, pp. 511–512.)

11.4.3 Obtain the expression (11.24) for $\cos 2kx \tan x$.

(Hint: begin with the case $k = 1$, $\cos 2x \tan x = \sin 2x - \tan x$ and then use induction.)

11.4.4 By writing out each side of (11.45) in detail, show that the left-hand side equals the right-hand side and hence prove (11.45) and so complete the proof of Proposition 11.1.

11.5   Completion of the proof of Theorem 11.1

We refer to (11.8) in Section 11.2, which is obtained from the result of multiplying (11.5) by (11.6) and then integrating with respect to x from 0 to $\pi/2$. In other words,

$$\int_0^{\frac{\pi}{2}} \theta_3^3 \theta_4(x)\mathrm{d}x = \int_0^{\frac{\pi}{2}} \left(\cot x + 4\sum_{n=1}^{\infty}(-1)^{n-1}\frac{q^n \sin 2nx}{1+(-q)^n} \right)$$

$$\times \left(\tan x + 2\sum_{n=1}^{\infty} \sin 2nx(-1)^{n-1}q^{n^2}B_n \right) \mathrm{d}x \quad (11.46)$$

We use

$$\int_0^{\frac{\pi}{2}} \cot x \sin 2nx \, \mathrm{d}x = \frac{\pi}{2}$$

and

$$\int_0^{\frac{\pi}{2}} \tan x \sin 2nx \, \mathrm{d}x = (-1)^{n-1}\frac{\pi}{2}$$

and thence we obtain (11.8).

The idea now is to interpret the right-hand side of (11.46) in terms of the theory of binary quadratic forms, which we outlined in Section 11.1, but first we need to introduce a few more ideas.

A binary form $ax^2 + 2bxy + cy^2$ of negative discriminant $D = -\Delta$ is said to be *reduced* if it satisfies the general conditions:

$$|2b| \le |a|, \quad |2b| \le |c|, \quad |a| \le |c|; \quad (11.47)$$

to which one adds two special conditions:

$$\text{if } a = c, \quad b \ge 0; \quad \text{and if} \quad |2b| = |a|, \quad b \ge 0. \quad (11.48)$$

Those last two are included in order to enunciate the following theorem: *every class contains one and only one reduced form*. (See Davenport, 1952 for further details.)

It may happen that the two forms are both properly equivalent and improperly equivalent (see Section 11.1); a class consisting of such forms is said to be *ambiguous*. An *ambiguous form* is a form (a, b, c) in which $2b$ is divisible by a; if $2b = ma$, then the ambiguous form is transformed into itself by the substitution with matrix

$$\begin{pmatrix} 1 & m \\ 0 & -1 \end{pmatrix}$$

and determinant -1.

We recall the definition of the quadratic form

$$Ax^2 + 2Bxy + Cy^2$$

given by (11.10) and (11.11); we are looking to interpret the coefficients in the expansion (11.8) in terms of arithmetic properties of that quadratic form. We have already done so in the case of the third series, (11.9), and it will be convenient here to quote the results for the other two series in (11.8), obtained by expanding them as power series in q. To that end we use the following property of a Lambert series:

$$\sum_{n=1}^{\infty} a_n \frac{x^n}{1 - x^n} = \sum_{n=1}^{\infty} a_n \sum_{m=1}^{\infty} x^{mn} = \sum_{N=1}^{\infty} b_N x^N, \quad b_N = \sum_{n|N} a_n.$$

The results are:

$$\sum_{n=1}^{\infty} \frac{q^n}{1 + (-q)^n} = \sum_{m=1}^{\infty} d(m)q^m - \sum_{m=1,\sigma}^{\infty} (\sigma - 3)d(m)q^{2\sigma m} \tag{11.49}$$

and

$$\sum_{n=1}^{\infty} (-1)^n q^{n^2} B_n = \sum_{m=1}^{\infty} (-1)^{(m+1)/2} d(m)q^m$$

$$+ \sum_{m=1}^{\infty} d(m)q^{4m} + \sum_{m=1,\sigma}^{\infty} (\sigma - 3)d(m)q^{4 \cdot 2\sigma \cdot m}, \tag{11.50}$$

where σ runs through the values $1, 2, 3, \ldots$, and where $d(m)$ denotes the number of divisors of m.

For the third sum, we obtain (11.9) by expanding

$$\frac{1}{1 + (-q)^n} = \sum_{a=0}^{\infty} (-1)^{a(n+1)} q^{an}$$

and then substituting in the third series in (11.8). We recall the quadratic form derived from the exponent $n^2 + n + an - b^2$ in (11.9), which we denoted by (A, B, C) and $A = n$, $B = b$, $C = n + 1 + a$ (see (11.10) and (11.11)).

We begin with the interpretation of the component

$$q^{n^2+n+an-b^2} \tag{11.51}$$

of the nth term in (11.9) and defer consideration of the contribution of the part $(-1)^{a(n+1)}$ until later.

The number b in (11.51) takes the values $0, \pm 1, \pm 2, \cdots, \pm(n - 1)$ (cf. the definition (11.7) of B_n) and we note that for b with those values the form (A, B, C) is reduced if we restrict our attention to the values of b in that set as far as $\pm (n/2 - 1)$ if n is even and as far as $\pm ((n - 1)/2)$ if n is odd. That first set of values, on using

$$n^2 + n + an - b^2 = \Delta,$$

gives, exactly once for each choice of b, all the reduced forms of the discriminant $-\Delta$, provided that we exclude the cases of the ambiguous forms (A, B, C), where $2B = A$ and $C = A$ and, amongst the forms $(A, 0, C)$, the case in which $A = C$, which yields a square for Δ. In those cases, one is led to a (strictly) negative value of a.

We turn our attention to the second set of vales of b for which

$$\pm b = \frac{n}{2}, \frac{n}{2} + 1, \cdots, n - 1, \quad n \text{ even};$$

$$\pm b = \frac{n+1}{2}, \frac{n+1}{2} + 1, \cdots, n - 1, \quad n \text{ odd}.$$

Define $\varepsilon = \pm 1$ by $\varepsilon = b/|b|$ and make the substitution

$$x = \varepsilon X - Y, \quad y = \varepsilon Y$$

in the quadratic form

$$nx^2 + 2bxy + (n + 1 + a)y^2$$

to obtain

$$nX^2 - 2\varepsilon(n - b\varepsilon)XY + (2n - 2b\varepsilon + 1 + a)Y^2. \tag{11.52}$$

We shall denote the transformed form (11.52) by $(A, -\varepsilon B, C)$, where we set

$$A = n, \quad B = n - b\varepsilon, \quad C = 2n - 2b\varepsilon + 1 + a. \tag{11.53}$$

We observe that that definition satisfies the condition

$$2B < A, \quad 2B < C,$$

where the first term, A, is either always greater than or always less than the last, C.

That construction yields a sequence of forms having twice as many members as the set of reduced form, if one excludes the forms $(A, 0, C)$ which correspond to the choice $b = 0$. To see that, one writes Equations (11.53) in the form:

$$n = A, \quad b = \varepsilon(A - B), \quad a = C - 2B - 1, \tag{11.54}$$

and then, on interchanging A and C, we obtain

$$n = C, \quad b = \varepsilon(C - B), \quad a = A - 2B - 1. \tag{11.55}$$

Thus we see that each non-ambiguous reduced form gives two different systems (n, b, a), where n and a are positive and b is between the two assigned limits. However, with regard to the two ambiguous forms $(A, \varepsilon B, A)$, Equations (11.55) lead to a negative value for a and so one finally has a unique system of numbers n, b, a.

If one has the case $2B = A = C$, then one cannot appeal to Equations (11.54) and (11.55); so that case must be excluded, as was previously the case when $A = C$ and the ambiguous form was $(A, 0, C)$.

We summarize our conclusions as follows.

For a given discriminant Δ, denote by H the number of reduced, non-ambiguous forms, by h the number of ambiguous forms given by $(A, 0, C)$ and by h' the number of ambiguous forms corresponding to the cases $(2B, B, C)$ and (A, B, A). Then the expression

$$3H + h + h' \tag{11.56}$$

is the number of systems (n, b, a) which, under the given conditions, satisfy the equation

$$n^2 + n + an - b^2 = \Delta. \tag{11.57}$$

If Δ is a square or three times a square, that number, allowing for the exceptions arising from the forms given by $(1, 0, 1)$ and by $(2, 1, 2)$, must be reduced by one or two, and so one writes (11.56) in the form

$$3\mathfrak{H} - 2h - h', \tag{11.58}$$

where[2] $\mathfrak{H}$ denotes the total number of classes of forms of discriminant $-\Delta$, that is

$$\mathfrak{H} = H + h + h'. \tag{11.59}$$

---

[2] We follow the notation of Hermite's original paper, (1862).

We turn now to the factor $(-1)^{a(n+1)}$ and write, in the first place,

$$A = n, \quad B = n - b\varepsilon, \quad C = n + 1 + a,$$

and, in the second place,

$$A = n, \quad B = n - b\varepsilon, \quad C = 2n - 2b\varepsilon + 1 + a,$$

and then

$$C = n, \quad B = n - b\varepsilon, \quad A = 2n - 2b\varepsilon + 1 + a.$$

In all three cases one obtains

$$a(n + 1) = \Delta + A + B + C + 1 \pmod 2.$$

It follows that, for each discriminant, the factor $(-1)^{a(n+1)}$ is equal to $+1$ if one of the extreme coefficients A or C is odd, and is equal to -1 if both are even. On making that distinction in the reduced forms, call $\mathfrak{H}_0$ the total number of those forms and denote by h_0, h'_0, respectively, the number of ambiguous forms of the two kinds, as above. Let $\mathfrak{H}_1, h_1, h'_1$ have similar meanings in the case when the extreme coefficients are both even. Then one has

$$\sum_{n=1}^{\infty} (-1)^{a(n+1)} q^{n^2 + n + an - b^2}$$

$$= \sum_{\Delta} [(3\mathfrak{H}_0 - 2h_0 - h'_0) - (3\mathfrak{H}_1 - 2h_1 - h'_1)] q^{\Delta}$$

$$= 3 \sum_{\Delta} (\mathfrak{H}_0 - \mathfrak{H}_1) q^{\Delta} + \sum_{\Delta} (2h_1 + h'_1 - 2h_0 - h'_0) q^{\Delta},$$

$$(11.60)$$

which gives the expansion in powers of q of the series

$$\sum_{n=1}^{\infty} \frac{q^{n^2 + n} B_n}{1 + (-q)^n}.$$

The second sum in (11.60), namely

$$\sum_{\Delta} (2h_1 + h'_1 - 2h_0 - h'_0) q^{\Delta}, \qquad (11.61)$$

may be evaluated as follows.

Suppose first that $\Delta = m$, m odd, then

$$h_0 = \frac{1}{2} d(m), \quad h'_0 = \frac{1 - (-1)^{(m+1)/2}}{4} d(m),$$

$$h_1 = 0, \quad h'_1 = \frac{1 + (-1)^{(m+1)/2}}{4} d(m). \qquad (11.62)$$

Then, if $\Delta = 2m$, we have:

$$h_0 = d(m), \qquad h_0' = 0,$$
$$h_1 = 0, \qquad h_1' = 0. \tag{11.63}$$

If $\Delta = 4m$,

$$h_0 = d(m), \qquad h_0' = 0,$$
$$h_1 = \frac{1}{2}d(m), \qquad h_1' = \frac{1}{2}d(m), \tag{11.64}$$

and, finally, if $\Delta = 4 \cdot 2\sigma \cdot m$,

$$h_0 = d(m), \qquad h_0' = d(m),$$
$$h_1 = \frac{\sigma + 1}{2}d(m), \qquad h_1' = \frac{\sigma - 1}{2}d(m). \tag{11.65}$$

Equations (11.61) to (11.65) yield the result

$$\sum_\Delta (2h_1 + h_1' - 2h_0 - 2h_0')q^\Delta$$

$$= \sum_m \frac{(-1)^{(m+1)/2} - 2}{2}d(m)q^m - \sum_m 2d(m)q^{2m}$$

$$- \sum_m \frac{1}{2}d(m)q^{4m} + \sum_{m,\sigma} \frac{3\sigma - 5}{2}d(m)q^{4 \cdot 2\sigma \cdot m}. \tag{11.66}$$

Now in the basic equation (11.8), namely

$$\theta_3^3 = 1 + 4\sum_{n=1}^\infty \frac{q^n}{1 + (-q)^n} - 2\sum_{n=1}^\infty (-1)^n q^{n^2} B_n + 4\sum_{n=1}^\infty \frac{q^{n^2+n} B_n}{1 + (-q)^n},$$

substitute for each of the three series the results in (11.49), (11.50), (11.60) and (11.61) to (11.66); we find that the first two series are cancelled by those terms that arise from the ambiguous forms and the sum remaining gives

$$\theta_3^3 = 1 + \sum_\Delta 12(\mathfrak{H}_0 - \mathfrak{H}_1)q^\Delta. \tag{11.67}$$

We note that if Δ is a square or three times a square then the corresponding terms in the sum (11.67) may be replaced by

$$12\left[\mathfrak{H}_0 - \mathfrak{H}_1 + \frac{(-1)^n}{2}\right]q^{n^2} \tag{11.68}$$

or

$$12\left[\mathfrak{H}_0 - \mathfrak{H}_1 + \frac{2}{3}\right]q^{3n^2},$$

respectively.

However, one may avoid those two exceptional cases by combining the terms of the form q^{n^2} and q^{3n^2} and writing

$$\eta = \sum 2q^{3n^2} = \theta_3(q^3)$$ (11.69)

(that is the theta series θ_3 with q replaced by q^3). Whence

$$\theta_3^3 = \sum_\Delta 12(\mathfrak{H}_0 - \mathfrak{H}_1)q^\Delta + 30 + 4\eta - 6.$$ (11.70)

If we now introduce the notation used in the enunciation of Theorem 11.1 and write $\Delta = n$, a positive integer, we may write

$$\mathfrak{H}_0 = F(\Delta) = F(n), \quad \mathfrak{H}_1 = G(\Delta) - F(\Delta) = G(n) - F(n),$$ (11.71)

and so

$$\mathfrak{H}_0 - \mathfrak{H}_1 = 2F(n) - G(n) = E(n).$$ (11.72)

That completes the proof of Theorem 11.1.

Exercises 11.5

11.5.1 How should the arguments used in the foregoing in terms of the quadratic forms $ax^2 + 2bxy + cy^2$ be modified if the other notation $ax^2 + bxy + cy^2$ with discriminant $-\Delta = D = 4ac - b^2$ is used?

11.5.2 Verify the definite integrals

$$\int_0^{\frac{\pi}{2}} \cot x \sin 2nx \, dx = \frac{\pi}{2}$$

and

$$\int_0^{\frac{\pi}{2}} \tan x \sin 2nx \, dx = (-1)^{n-1}\frac{\pi}{2}$$

and then obtain (11.8) from (11.46).

11.5.3 Use the formula for the sum of a Lambert series to obtain the results (11.49) and (11.50).

11.5.4 Prove (see below) that the functions $F(n)$, $G(n)$ satisfy the equations

$$F(4n) = 2F(n), \quad G(4n) = F(4n) + G(n),$$
$$G(n) = F(n) \quad \text{if } n \equiv 1 \text{ or } 2(\mathrm{mod}\,4),$$
$$G(n) = 2F(n) \quad \text{if } n \equiv 7(\mathrm{mod}\,8),$$
$$G(n) = \frac{4}{3}F(n) \quad \text{if } n \equiv 3(\mathrm{mod}\,8).$$

(The proof of those results presupposes a knowledge of the theory of binary quadratic forms, and in turn they may be used to prove the theorem of Gauss concerning the number of representations of a positive integer as a sum of three squares. See Dickson, 1934, Chapter VII and Grosswald, 1984. See also Landau, 1947. For a general review of the application of elliptic and theta functions to problems in the theory of quadratic forms (in the Gauss notation), see the very interesting and still relevant account in H. J. S. Smith, 1965, *Report on the Theory of Numbers* originally published in a *Report of the British Association* for 1859.)

12

Applications in mechanics and statistics and other topics

Introduction

In the preceding chapters we have sought to introduce the reader to the rich and beautiful treasury of ideas originating in the work of Euler, Fagnano and others and developed by Gauss, Abel, Jacobi, Weierstrass – and others. But those astonishing results from pure mathematics have no less remarkable applications in mechanics, in probability and statistics (some of which we encountered in Chapter 1), in numerical analysis (again taking up a reference from Chapter 1) and in mathematical physics that, to quote Lawden (1989) in his preface, 'grace the topic with an additional aura'. In this final chapter we offer a brief introduction to some of those topics, though limitations of space prevent us from exploring them in the depth they deserve. Again, those limitations prevent us from describing such beautiful and fascinating subjects as the recent advances in the application of modular forms and functions and theta functions to superstring theory or to probability theory. For the former we recommend the book on *Superstring Theory* by Green, Schwarz and Witten (1987) and for the latter the reader is urged to consult the paper by Biane, Pitman and Yor (2001). Another remarkable application, for which, again, we must refer the reader elsewhere, is to the theory of sound and the design of organ pipes; see Rayleigh (1929).

Needless to say, our whole treatment has been based on a dynamical application, namely to the problem of the simple pendulum, and, like Greenhill, we derived more of our subsequent treatment from the insights so obtained.

We begin this chapter (Section 12.1) with a topic from dynamics, Euler's dynamical equations for the motion of a body about a fixed point. Then in Section 12.2 we shall consider the applications to the determination of

338

planetary orbits in the general theory of relativity. Our final example, in Section 12.3, from dynamics is to the generalization of a plane pendulum to a spherical pendulum in which the particle moves on a sphere rather than on a circle.[1]

Then (Section 12.4) we turn our attention to the classical problem of finding the Green's function for a rectangle.

In Section 12.5 we look at a statistical application concerning the theory of correlation, and in Section 12.6 we consider a problem in numerical analysis, which leads to Gauss' theory of the arithmetic/geometric mean.

Then in Section 12.7 we consider an application in analysis concerning the iteration of rational functions and rational maps whose Julia set is the entire sphere.

We shall conclude with an application to analysis and arithmetic, two of the 'Gaussian As', which is central to the theory of the Riemann zeta function.

12.1 Euler's dynamical equations

(We shall follow the applications of elliptic functions to the treatment of Euler's equations as given in Dutta & Debnath, 1965; Lawden, 1989; and Newboult, 1946.)

Consider the motion of a rigid body having one point fixed, at O. Let $Oxyz$ be the principal axes of inertia at O and A, B, C the corresponding moments of inertia, with ω_1, ω_2, ω_3 the component angular velocities. The axes are fixed in the body and ω_1, ω_2, ω_3 are accordingly the component angular velocities of the body; so the moment of momentum has components $A\omega_1$, $B\omega_2$, $C\omega_3$. The equations of motion of the body are

$$A\dot{\omega}_1 - (B - C)\omega_2\omega_3 = L,$$
$$B\dot{\omega}_2 - (C - A)\omega_3\omega_1 = M,$$
$$C\dot{\omega}_3 - (A - B)\omega_1\omega_2 = N, \tag{12.1}$$

where L, M, N are the moments about the axes of the forces under which the body is moving. These are *Euler's dynamical equations*.

In considering the motion of a rigid body relative to its centroid we may consider that point to be fixed in space, and so the forces acting on the body

---

[1] We shall rely on a number of valuable references which deserve to be read in full; details will be given in the appropriate section.

have no moment about O and so L, M, N are zero. Accordingly, Equations (12.1) now read

$$A\dot{\omega}_1 - (B - C)\omega_2\omega_3 = 0,$$
$$B\dot{\omega}_2 - (C - A)\omega_3\omega_1 = 0,$$
$$C\dot{\omega}_3 - (A - B)\omega_1\omega_2 = 0. \tag{12.2}$$

On multiplying those equations by $\omega_1, \omega_2, \omega_3$, respectively, adding and then integrating with respect to t we obtain

$$\frac{1}{2}(A\omega_1^2 + B\omega_2^2 + C\omega_3^2) = T, \tag{12.3}$$

where T is a constant, the kinetic energy. If we now multiply each equation in (12.2) in turn by $A\omega_1, B\omega_2, C\omega_3$, respectively, and integrate we obtain

$$A^2\omega_1^2 + B^2\omega_2^2 + C^2\omega_3^2 = H^2, \tag{12.4}$$

where H denotes the moment of momentum about O (a constant, since the forces have zero moment about O).

The problem is to express ω_1, ω_2 and ω_3 in terms of t, and if one looks at the form of, say, the first equation in (12.2) it is plausible that the Jacobian elliptic functions might provide the solution, for the first equation in (12.2) is of the form

$$\frac{d}{dt}\omega_1 = K\omega_2\omega_3, \quad K = (B - C)/A$$

and that is reminiscent, for example, of the equation

$$\frac{d}{du}dn(u) = -k^2 sn(u) cn(u),$$

and similarly for the others. That is indeed the required solution, and we shall adopt one approach suggested by classical mechanics, and we follow Newboult (1946) (for other related approaches see the books by Lawden, 1989, pp.130ff, and by Dutta and Debnath, 1965, pp.158ff).

The instantaneous axis of rotation describes a cone in the body, called the *polhode cone*, and a cone in space, the *herpolhode cone*. The actual motion of the body is produced by the polhode cone rolling on the herpolhode cone, and from (12.3) and (12.4) we see that the equation of the polhode cone is

$$A(2AT - H^2)\omega_1^2 + B(2BT - H^2)\omega_2^2 + C(2CT - H^2)\omega_3^2 = 0, \tag{12.5}$$

where the $\omega_1, \omega_2, \omega_3$ are to be thought of as rectangular coordinates. We shall suppose that $A > B > C$ and that $2AT - H^2 > 0, 2CT - H^2 < 0$, whilst $2BT - H^2$ may be positive or negative. When $2BT - H^2$ is positive the plane

$\omega_3 = 0$ meets the cone in imaginary lines, but the plane $\omega_1 = 0$ meets the cone in real lines. When $2BT - H^2$ is negative the situation is reversed: ω_1 never vanishes, but ω_3 does at certain times. In either case the plane $\omega_2 = 0$ meets the cone in real lines and $\omega_2 = 0$ at the corresponding times.

Whatever the sign of $2BT - H^2$, ω_2 will vanish at some times and we measure t in such a way that $t = 0$ gives one of those times. In accordance with that choice and with our intuition about the role of the elliptic functions, we write

$$\omega_2 = \beta \ sn(\lambda t), \tag{12.6}$$

where β, λ are constants to be determined.

Suppose first that $2BT - H^2 < 0$; so that ω_1 cannot vanish, but ω_3 does. Recall that $cn \ (\lambda t)$ has real zeros only, whereas the function $dn \ (\lambda t)$ has complex zeros only. Write

$$\omega_1 = \alpha_1 \ dn \ (\lambda t), \quad \omega_3 = \alpha_3 \ cn \ (\lambda t), \tag{12.7}$$

where $\alpha_1 = \omega_1(0)$, $\alpha_3 = \omega_3(0)$. A third constant is implicit in our choice of t and may be made explicit if desired by replacing t by $t - t_0$; so that t_0 gives one of the times at which ω_2 vanishes. Accordingly, we have obtained a recipe for the solution of our problem, but we have yet to determine β and λ and also the modulus, k, of our elliptic functions. To that end we use (12.6) and (12.7) and substitute in (12.2) to obtain

$$-k^2\lambda = \frac{B - C}{A} \frac{\beta\alpha_3}{\alpha_1}, \quad -\lambda = \frac{A - C}{B} \frac{\alpha_1\alpha_3}{\beta}, \quad -\lambda = \frac{A - B}{C} \frac{\beta\alpha_1}{\alpha_3} \tag{12.8}$$

(where we have used the results in (2.29) in the summary of results at the end of Chapter 2). From (12.8) we obtain

$$\beta^2 = \frac{(A - C)C}{(A - B)B}\alpha_3^2, \quad \lambda^2 = \frac{(A - B)(A - C)}{BC}\alpha_1^2, \quad k^2 = \frac{(B - C)C}{(A - B)A}\frac{\alpha_3^2}{\alpha_1^2}. \tag{12.9}$$

Since $A > B > C$ the values of β, λ are real, and we shall also require the modulus, k, such that $k^2 < 1$; that is

$$B\left(A\alpha_1^2 + C\alpha_3^2\right) < A^2\alpha_1^2 + C^2\alpha_3^2. \tag{12.10}$$

Since we are supposing that $2BT - H^2 < 0$, (12.10) holds.

If $2BT - H^2 > 0$, then ω_1 may vanish, but ω_3 does not and the solutions are then of the form

$$\omega_1 = \alpha_1 \ cn \ (\lambda t), \quad \omega_2 = \beta \ sn \ (\lambda t), \quad \omega_3 = \alpha_3 \ dn \ (\lambda t) \tag{12.11}$$

and we obtain

$$\beta^2 = \frac{(A - C)A}{(B - C)B}\,\alpha_1^2, \quad \lambda^2 = \frac{(A - C)(B - C)}{AB}\,\alpha_3^2,$$

$$k^2 = \frac{(A - B)A}{(B - C)C}\,\frac{\alpha_1^2}{\alpha_3^2}. \tag{12.12}$$

So we have obtained solutions to Euler's equations for the motion of a rigid body in terms of the Jacobi functions. For a version of the solution in terms of the Weierstrass functions, see Lawden (1989), pp. 187–191.

Exercises 12.1

12.1.1 Starting from (12.1) and (12.2), obtain the equations corresponding to (12.8), namely

$$-\lambda = \frac{B - C}{A}\,\frac{\beta\alpha_3}{\alpha_1}, \quad -\lambda = \frac{A - C}{B}\,\frac{\alpha_1\alpha_3}{\beta}, \quad -k^2\lambda = \frac{A - B}{C}\,\frac{\beta\alpha_1}{\alpha_3},$$

and hence obtain (12.12) and verify that $k^2 < 1$.

12.1.2 Show that either of the solutions (12.6) and (12.7) or (12.11) may be deduced from the other by using the formulae

$$sn\left(ku, \frac{1}{k}\right) = k\,sn(u, k), \quad cn\left(ku, \frac{1}{k}\right) = dn(u, k)$$

$$dn\left(ku, \frac{1}{k}\right) = cn(u, k).$$

12.2 Planetary orbits in general relativity

(See Lawden, 1989, Chapter 5, for the application of elliptic integrals to orbits under laws of attraction proportional to r^{-4} and r^{-5} as well as to the relativistic orbits discussed here.)

According to Newton's law, where the force on a particle of unit mass is μ/r^2, (r, θ) being polar coordinates at time t, a planet moves round the Sun in an ellipse and, if there are no other planets disturbing it, the ellipse remains the same for ever. According to Einstein's law, the Newtonian force is modified so that the gravitational force is of the form

$$\mu\left(\frac{1}{r^2} + \frac{3h^2}{c^2 r^4}\right) \tag{12.13}$$

per unit mass, where h denotes the angular momentum per unit mass of the planet about the Sun and c denotes the velocity of light. The prediction of Einstein's law is that the path is very nearly an ellipse, but it does not quite close up. In the next revolution the path will have advanced slightly in the direction in which the planet is moving and so the orbit is an ellipse which very slowly revolves. The advance of the orbit of Mercury (of the 'perihelion of Mercury') is indeed one of the tests of the theory of relativity. We shall show how elliptic functions help to explain all that.

One can show that the equations of motion are of the form

$$\frac{1}{2}(\dot{r}^2 + r^2\dot{\theta}^2) - \mu\left(\frac{1}{r} + \frac{h^2}{c^2 r^3}\right) = E, \quad r^2\dot{\theta} = h, \tag{12.14}$$

where E denotes the energy. On writing $u = 1/r$ we obtain

$$\left(\frac{du}{d\theta}\right)^2 = \frac{2\mu}{h^2}u - u^2 + \frac{2\mu}{c^2}u^3 + \frac{2E}{h^2} \quad , \tag{12.15}$$

for the equation of the orbit.

Now for the planets in the Solar System, the term $2\mu u^3/c^2$ is very small and so one introduces a parameter to take account of that; more precisely, we define a dimensionless variable v by

$$u = \mu v/h^2, \tag{12.16}$$

and in that notation (12.10) becomes

$$\left(\frac{dv}{d\theta}\right)^2 = 2v - v^2 + \alpha v^3 - \beta = f(v), \tag{12.17}$$

where

$$\alpha = 2(\mu/ch)^2, \quad \beta = -2Eh^2/\mu^2. \tag{12.18}$$

In order to prevent the planet escaping from the Sun's gravitational field we require $\beta > 0$. We also require $\beta \leq 1$, for otherwise if the term involving α is omitted and $\beta = 1 + \varepsilon$, $\varepsilon > 0$, then

$$\left(\frac{du}{d\theta}\right)^2 = -(v - 1)^2 - \varepsilon < 0.$$

The term α is very small, its largest value being that in the case of Mercury when $\alpha = 5.09 \times 10^{-8}$.

By considering the changes in sign of the derivative $f'(v)$ of $f(v)$, one can see that the zeros v_1, v_2, v_3 of $f(v)$ are real and positive and satisfy $0 < v_1 < 1 < v_2 < 2 < v_3$, where v_3 is large. Accordingly,

$$f(v) = \alpha(v - v_1)(v - v_2)(v - v_3) = \alpha(v - v_1)(v_2 - v)(v_3 - v) \tag{12.19}$$

and, since $f(v) \geq 0$, v must lie in the interval $v_1 \leq v \leq v_2$. The case $v \geq v_3$ would lead to $v \to \infty$ as $\theta \to \infty$ (so the planet would fall into the Sun) – and so must be excluded.

The zeros v_1, v_2, v_3 can be found by using the formula for the solution of a cubic in Chapter 10, and, given that α is small, we may use the relations between the roots and the coefficients of the cubic to expand the roots in ascending powers of α and so obtain

$$v_1 = 1 - e - \frac{\alpha}{2e}(1-e)^3 + O(\alpha^2),$$

$$v_2 = 1 + e + \frac{\alpha}{2e}(1+e)^3 + O(\alpha^2),$$

$$v_3 = \frac{1}{a} - 2 + O(\alpha), \qquad (12.20)$$

where $e^2 = 1 - \beta = 1 + 2Eh^2/\mu^2$.

Now from (12.17) we obtain

$$\alpha^{1/2}\theta = \int \frac{dv}{\sqrt{\{(v-v_1)(v_2-v)(v_3-v)\}}} \qquad (12.21)$$

and then the substitution $v = v_1 + 1/t^2$ yields the standard form

$$\alpha^{1/2}\theta = -\frac{2}{\sqrt{\{(v_2-v_1)(v_3-v_1)\}}} \int \frac{dt}{\sqrt{\{(t^2-a^2)(t^2-b^2)\}}}, \qquad (12.22)$$

where

$$a^2 = 1/(v_2-v_1), \quad b^2 = 1/(v_3-v_1).$$

Now refer to Chapter 8, Exercise 8.2.7 (d), to obtain

$$\alpha^{1/2}\theta = \frac{1}{\sqrt{(v_3-v_1)}}ns^{-1}(t\sqrt{(v_2-v_1)}), \qquad (12.23)$$

where the modulus, k, of the Jacobi functions is given by

$$k^2 = \frac{v_2-v_1}{v_3-v_1},$$

and finally we obtain for the equation of the orbit

$$v = v_1 + (v_2-v_1)sn^2\left(\frac{1}{2}\sqrt{\alpha(v_3-v_1)}\theta\right). \qquad (12.24)$$

Exercises 12.2

(The following exercises are based on Lawden, 1989, pp.128,129, where more details may be found.)

12.2.1 Verify the results in (12.20).

12.2.2 Use the results in (12.20) and (12.24) to obtain

$$\frac{1}{r} = \frac{\mu}{h^2}(A + B\,sn^2\eta\theta),$$

where

$$A = 1 - e - \frac{\alpha}{2e}(1 - e)^3 + O(\alpha^2),$$

$$B = 2e + \alpha \left(3e + \frac{1}{e} \right) + O(\alpha^2),$$

$$\eta = \frac{1}{2} - \frac{1}{4}(3 - e)\alpha + O(\alpha^2).$$

Show that the modulus is given by

$$k^2 = 2e\alpha + O(\alpha^2).$$

12.2.3 Show that if $\alpha = 0$, then $A = 1 - e$, $B = 2e$, $\eta = \dfrac{1}{2}$, $k = 0$ and the equation of the orbit reduces to the familiar

$$\frac{1}{r} = 1 - e \cos\theta,$$

where e now denotes the eccentricity of the ellipse.

12.2.4 In the orbit given by the equation in 12.2.2, perihelion occurs when $\theta = K/\eta$, where $K = K(k) = \int_0^{\pi/2} (1 - k^2 \sin^2 \theta)^{-1/2} d\theta$ (see Chapter 2, Section 2.10). Perihelion occurs again when $\theta = 3K/\eta$ and so θ increases by $2K/\eta$ (instead of 2π as in elliptic orbits and the equation in 12.2.3).

Show that the advance of the perihelion in each revolution is:

$$\frac{2K}{\eta} - 2\pi = \frac{\pi \left(1 + \frac{1}{4}k^2 + \cdots \right)}{\frac{1}{2} - \frac{1}{4}(3 - e)\alpha + \cdots} - 2\pi = 3\pi\alpha.$$

(See Lawden, 1989 for further details.)

12.3   The spherical pendulum

(See Dutta & Debnath, 1965, Chapter 5; Lawden, 1989, Chapter 7; and Greenhill, 1892, Chapter VII, for example, for further reading.)

A spherical pendulum is a generalization of the plane pendulum of Chapter 1 in which the 'heavy particle', P, is now constrained to move on a sphere rather than on a circle. We shall suppose that the particle has mass m and that the cylindrical polar coordinates of P with respect to the origin O, the centre of the sphere, are (r, θ, z). We denote by AOA' the vertical diameter (with A the

'South Pole') of the sphere on which P moves and by PN the perpendicular from P to that diameter; so that $NP = r$ and $ON = z$. If a is the radius of the sphere, then its equation is

$$r^2 + z^2 = a^2. \tag{12.25}$$

At time t the velocity of the particle has components $(\dot{r}, r\dot{\theta}, \dot{z})$ and the forces acting on it are:

 (i) its weight, mg, acting parallel to the z-axis;
(ii) the reaction passing through the centre of the sphere.

We shall denote the initial value of z, at time t_0, by z_0 and the initial velocity by v_0.

The sum of the moments of the forces about the z-axis is equal to 0 and, by the principle of the conservation of angular momentum, the moment of momentum about the z-axis is conserved; so for some constant, h,

$$r^2\dot{\theta} = h. \tag{12.26}$$

The energy is conserved, since there is no dissipative force on the particle, and so we may deduce from the energy equation that

$$\dot{r}^2 + r^2\dot{\theta}^2 + \dot{z}^2 = v_0^2 + 2g(z_0 - z) = 2g(l - z), \tag{12.27}$$

where

$$l = \frac{v_0^2 + 2gz_0}{2g}.$$

It follows from (12.25) that

$$r\dot{r} + z\dot{z} = 0,$$

and from that and (12.26) Equation (12.27) becomes

$$\dot{z}^2 \left(\frac{a^2}{a^2 - z^2} \right) + \frac{h^2}{(a^2 - z^2)} = 2g(l - z),$$

whence

$$\dot{z}^2 = \frac{1}{a^2}\{2g(l - z)(a^2 - z^2) - h^2\} = \frac{\theta(z)}{a^2}, \tag{12.28}$$

where

$$\theta(z) = 2g(l - z)(a^2 - z^2) - h^2. \tag{12.29}$$

It follows from (12.28) that at the point $z_0 = z(t_0)$, we must have $\theta(z_0) > 0$, since $\dot{z}$ is real when $t = t_0$.

We consider the cubic $\theta(z)$ in (12.29) and note that for real values of z it is positive as $z \to \infty$ and then $\theta(a)$ is negative, $\theta(z_0)$ is positive and $\theta(-a)$ is negative. It follows that the cubic $\theta(z) = 0$ has three roots, and if we denote those roots by z_1, z_2, z_3 we may suppose that

$$-a < z_1 < z_0 < z_2 < a < z_3 < \infty. \tag{12.30}$$

The root z_3 is not on the surface of the sphere and so the motion of the particle takes place on the surface of the sphere between the planes $z = z_1$ and $z = z_2$.

Now write $z = aw$ and $al' = l$; so that (12.28) becomes

$$a^2 \dot{w}^2 = 2ga(l' - w)(1 - w^2) - \frac{h^2}{a^2}. \tag{12.31}$$

If we now introduce another change of variable $t = cx$ then our equation becomes

$$\left(\frac{dw}{dx}\right)^2 = \frac{c^2}{a} 2g(l' - w)(1 - w^2) - \frac{c^2 h^2}{a^4}$$
$$= 4(l' - w)(1 - w^2) - h'^2, \tag{12.32}$$

where $gc^2 = 2a$ and $h' = \frac{ch}{a^2} = (\frac{2a}{g})^{1/2}\frac{h}{a^2} = (\frac{2}{ag})^{1/2}\frac{h}{a}$.

The idea now is to write (12.32) in such a way that its resemblance to the usual notation for the Weierstrass $\wp$-function becomes obvious, and to that end we first put $w = \eta + n$, where $n = l'/3$ and write

$$g_2 = \frac{4}{3}(l'^2 + 3), \quad g_3 = h'^2 + \frac{8}{27}l'^3 - \frac{8}{3}l', \tag{12.33}$$

to obtain

$$\left(\frac{d\eta}{dx}\right)^2 = 4\eta^3 - g_2\eta - g_3. \tag{12.34}$$

That suggests that we should set $\eta = \wp(u)$. Now

$$\left(\frac{d\eta}{du}\right)^2 = 4\eta^3 - g_2\eta - g_3$$

and

$$\frac{d\eta}{du} = \wp'(u)\frac{du}{dx},$$

where the dash denotes differentiation with respect to u. It follows that

$$\wp'(u)^2 \left(\frac{du}{dx}\right)^2 = 4\wp(u)^3 - g_2\wp(u) - g_3,$$

from which we deduce that

$$\frac{du}{dx} = \pm 1,$$

where $u = \pm(x + \alpha)$, α a constant.

Since $\wp(u)$ is an even function, we may choose the positive sign and then $u = x + \alpha$. It follows that

$$\eta = \wp(u) = \wp(x + \alpha). \tag{12.35}$$

Denote by e_1, e_2, e_3 the roots of $4\eta^3 - g_2\eta - g_3$, which are real, and we may suppose $e_1 > e_3 > e_2$ and then

$$e_1 = \frac{z_3}{a} - \frac{l'}{3}, \quad e_2 = \frac{z_1}{a} - \frac{l'}{3}, \quad e_3 = \frac{z_2}{a} - \frac{l'}{3},$$

where z_1, z_2, z_3 are given by (12.30).

We consider the integrals

$$\omega_1 = \int_{e_1}^{\infty} \frac{d\eta}{\sqrt{4\eta^3 - g_2\eta - g_3}},$$

where the integral is taken along the real axis, and

$$\omega_2 = \int_{e_2}^{\infty} \frac{d\eta}{\sqrt{4\eta^3 - g_2\eta - g_3}},$$

where we suppose the integral to be taken along the imaginary axis; so ω_1 is real and ω_3 is pure imaginary.

As z decreases from z_2 to z_1, η decreases from e_3 to e_2 and is real. It follows that $\alpha = \omega_2$ and that $2\omega_1$, $2\omega_3$ are periods of $\wp(u)$ and that ω_1 represents the time taken to reach z_1 from z_2.

So we have obtained

$$z = aw = a\wp\left\{\sqrt{\frac{g}{2a}}t + \omega_2\right\} + \frac{l}{3} \tag{12.36}$$

$$= a\wp\left\{\sqrt{\frac{g}{2a}}t + \omega_2\right\} + \frac{1}{6g}(v_0^2 + 2gz_0).$$

For the function θ we obtain

$$\dot{\theta} = \frac{h}{a^2 - z^2} = \frac{h}{2a}\left\{\frac{1}{a - z} + \frac{1}{a + z}\right\}$$

$$= \frac{h}{2a^2}\left\{\frac{1}{a - w} + \frac{1}{a + w}\right\}$$

and so

$$\frac{d\theta}{du} = \frac{ch}{2a^2} \left\{ \frac{1}{\wp(u) - \wp(\alpha)} - \frac{1}{\wp(u) - \wp(\beta)} \right\},$$

where

$$w = \wp(u) + \frac{l'}{3}, \quad -\wp(\alpha) = 1 + \frac{l'}{3}, \quad \wp(\beta) = 1 - \frac{l'}{3}. \qquad (12.37)$$

We obtain

$$\left(\frac{dw}{dx}\right)^2 = (\wp'(u))^2 = 4(l' - w)(1 - w^2) - h'^2 = -h'^2, \qquad (12.38)$$

if $w = \pm 1$ or $w = l'$.

The solution to the spherical pendulum problem may now be obtained as

$$r^2 = a^2 - z^2 = -a^2\{(\wp(u) - \wp(\alpha))(\wp(u) - \wp(\beta))\}. \qquad (12.39)$$

The details of that calculation are to be found in Exercise 12.3.1 (and in Dutta & Debnath, 1965, Chapter 5 with some changes of notation; see also Lawden, 1989, Chapter 7).

Exercises 12.3

12.3.1 Starting from Equation (12.38), show that

$$\wp'^2(\alpha) = \wp'^2(\beta) = -h'^2$$

and hence

$$\wp'(\alpha) = \wp'(\beta) = ih'.$$

Hence show that

$$2i\frac{d\theta}{du} = \frac{h}{h'}\sqrt{\frac{2a}{g}}\frac{1}{a^2}\left[\frac{\wp'(\alpha)}{\wp(u) - \wp(\alpha)} - \frac{\wp'(\beta)}{\wp(u) - \wp(\beta)}\right]$$
$$= \zeta(u + \beta) - \zeta(u - \beta) - 2\zeta(\beta) - \zeta(u + \alpha) + \zeta(u - \alpha) + 2\zeta(\alpha),$$

where ζ denotes the Weierstrass zeta function.

Integrate that last equation to obtain

$$e^{2i\theta} = -E^2 \frac{\sigma(u + \beta)\sigma(u - \alpha)}{\sigma(u + \alpha)\sigma(u - \beta)} \exp 2(\zeta(\alpha) - \zeta(\beta))u,$$

where the constant of integration, E, is determined by the initial conditions.

Using the fact that $r^2 = a^2 - z^2 = (a - z)(a + z)$, obtain the result in (12.39).

12.3.2 (This question recasts the results obtained above in terms of Cartesian coordinates x, y, z)

Use the result obtained in 12.3.1 for $e^{2i\theta}$ to show that

$$e^{2i\theta} = \frac{x + iy}{x - iy} = -E^2 \frac{\sigma(u + \beta)\sigma(u - \alpha)}{\sigma(u + \alpha)\sigma(u - \beta)} \exp 2(\zeta(\alpha) - \zeta(\beta))u,$$

and then use (12.39) to show that

$$r^2 = (x + iy)(x - iy) = -a^2 \frac{\sigma(u + \alpha)\sigma(u - \alpha)\sigma(u + \beta)\sigma(u - \beta)}{\sigma^4(u)\sigma^2(\alpha)\sigma^2(\beta)},$$

and deduce that

$$x + iy = Ea \frac{\sigma(u - \alpha)\sigma(u + \beta)}{\sigma^2(u)\sigma(\alpha)\sigma(\beta)} \exp(\zeta(\alpha) - \zeta(\beta))u,$$

$$x - iy = Ea \frac{\sigma(u + \alpha)\sigma(u - \beta)}{\sigma^2(u)\sigma(\alpha)\sigma(\beta)} \exp(-(\zeta(\alpha) - \zeta(\beta)))u.$$

12.4   Green's function for a rectangle

We begin by recalling briefly the essential properties of the Green's function (the following account is based on Courant & Hilbert, 1968, Chapter V, though many books on the applications of advanced calculus to physics offer insight into the essential ideas).

Denote by Δ the two-dimensional Laplace operator

$$\Delta u(x, y) = \frac{\partial^2 u}{\partial x^2} + \frac{\partial^2 u}{\partial y^2},$$

and consider the second-order partial differential equation

$$\Delta u(x, y) = -\varphi(x, y) \tag{12.40}$$

in some domain G of the (x, y)-plane, with homogeneous boundary conditions (for example $u = 0$ on the boundary of G). The solution of such an equation may be given in terms of *Green's function*, $K(x, y; \xi, \eta)$, where (ξ, η) is some isolated source point. The Green's function is related to the Dirac delta distribution $\delta(x - \xi, y - \eta)$ and possesses the essential property that

$$u(x, y) = \int\int_G K(x, y; \xi, \eta)\varphi(\xi, \eta)\mathrm{d}\xi \, \mathrm{d}\eta$$

and so provides a solution to Equation (12.40) subject to the given boundary conditions.

We illustrate the general idea and offer an application of the elliptic functions by taking G to be the interior of a rectangle, R, whose edges are parallel to the coordinate axes and whose vertices (taken in counter-clockwise order) are at the points with coordinates $(0, 0)$, $(a, 0)$, (a, b), $(0, b)$. We shall denote a general point of R by (x, y) and the source point by (ξ, η). For the boundary condition, we require $u = u(x, y) = 0$ on the edges of R, and we write

$$r = \sqrt{\{(x - \xi)^2 + (y - \eta)^2\}}. \tag{12.41}$$

We note that the solution $u(x, y)$ vanishes on the boundary of R and has a singularity at the point (ξ, η) determined by $-\log r/2\pi$.

Consider the reflections of the rectangle R in the coordinate axes, which produces four congruent rectangles and four points, the original (ξ, η) and the reflections $(-\xi, \eta)$, $(\xi, -\eta)$, $(-\xi, -\eta)$. We see from (12.41) that those four points are to be interpreted as 'sources' or 'sinks' according as they are obtained by an even or an odd number of reflections.

In order to obtain the potential, X, of the mass-distribution so obtained, we consider the analytic function $\varphi(x + iy) = X + iY$ and then

$$f(x + iy) = e^{2\pi(X+iY)} = e^{2\pi\varphi(x+iy)},$$

which has simple zeros or poles at the points $(\pm\xi, \pm\eta)$ according as they are obtained from (ξ, η) by an even or by an odd number of reflections. The four rectangles obtained from the original R define a new lattice with vertices at $(\pm a, \pm b)$ and the function $f(x + iy)$ has then two simple zeros and two simple poles which are symmetric with respect to the origin. The zeros are at (ξ, η) and $(-\xi, -\eta)$, the poles are at $(-\xi, \eta)$, $(\xi, -\eta)$.

At last we have introduced our elliptic functions, for the simplest analytic functions with that property are the elliptic functions for the period lattice $(-a, -b)$, $(a, -b)$, (a, b), $(-a, b)$ given in terms of the Weierstrass σ-functions by

$$f(z) = \frac{\sigma(z - \xi - i\eta)\sigma(z + \xi + i\eta)}{\sigma(z - \xi + i\eta)\sigma(z + \xi - i\eta)}, \tag{12.42}$$

where, in the notation of Chapter 7,

$$\sigma(z) = z \prod_{\omega}' \left\{ \left(1 - \frac{z}{2\omega}\right) \exp\left(\frac{z}{2\omega} + \frac{1}{8}\frac{z^2}{\omega^2}\right) \right\}, \tag{12.43}$$

$\omega = ka + lbi$, $k = 0, \pm 1, \ldots$, $l = 0, \pm 1, \ldots$ If we substitute from (12.43) in the expression for $f(z)$ and write 1 for $\exp(\xi\eta i/\omega^2)$ when $\omega = 0$, we obtain

$$ f(z) = \prod_{\omega = ka + lbi} \left[\frac{(z + \zeta - 2\omega)(z - \zeta - 2\omega)}{(z + \bar{\zeta} - 2\omega)(z - \bar{\zeta} - 2\omega)} \exp\left(\frac{\xi\eta i}{\omega^2} \right) \right], \qquad (12.44) $$

where $\zeta = \xi + i\eta$, $k, l \in \mathbb{Z}$. The verification of the boundary conditions is outlined in Exercise 12.4.1.

We conclude that the Green's function is given by

$$ K(x, y; \xi, \eta) = -\frac{1}{2\pi} Re\left(\log \frac{\sigma(z - \zeta, a, ib)\sigma(z + \zeta, a, ib)}{\sigma(z - \bar{\zeta}, a, ib)\sigma(z + \bar{\zeta}, a, ib)} \right), \qquad (12.45) $$

where $z = x + iy$, $\zeta = \xi + i\eta$.

The Green's function given in (12.45) has a Fourier expansion in terms of the eigenfunctions

$$ \frac{2}{\sqrt{ab}} \sin k\frac{\pi}{a}x \sin \frac{\pi}{b}y $$

of Δ, given by

$$ K(x, y; \xi, \eta) = \frac{4}{ab\pi^2} \sum_{m=1}^{\infty} \sum_{k=1}^{\infty} \frac{\sin k\frac{\pi}{a}x \sin m\frac{\pi}{b}y \sin k\frac{\pi}{a}\xi \sin m\frac{\pi}{b}\eta}{\frac{m^2}{b^2} + \frac{k^2}{a^2}}. $$

$$ (12.46) $$

Exercises 12.4

12.4.1 The boundary conditions for our problem require that the function $f(z)$ obtained in (12.44) should have absolute value 1 on the boundary of R. Verify that by carrying out the details of the method outlined as follows.

First suppose that $z = x$ and note that (by the convention introduced above) the factor with $\omega = 0$ has absolute value 1. Then for $\omega \neq 0$, pair corresponding factors in the infinite product so that the term in the numerator is paired with its complex conjugate in the denominator.

In the case $z = x + ib$ consider first the terms in l and then those in k. The product corresponding to the terms in l gives the exponential factor with a sum $\sum \frac{1}{\omega^2}$ over lattice points involving l, but with k fixed. Show that that sum converges absolutely and is real. The remaining terms can be paired so that $\omega = ka + l b$i corresponds to $\omega = ka - (l - 1)b$i and the product of such a pair has absolute value 1.

For $z = $iy, multiply first the terms in l and then pair the terms for $|k| > 0$ so that the term in k corresponds to the term in $-k$. Then show that the corresponding sums $\sum \frac{1}{\omega^2}$ over l converge absolutely and are real. The remaining terms are paired so that the factor with $\omega = ka + l b$i corresponds to the factor $\omega = -ka + l b$i. Again, the product has absolute value 1.

Finally, if $z = a + $iy, pair the factors with $\omega = ka + l b$i and $\omega = -(k - 1)a + l b$i and multiply over the l to obtain the result.

12.4.2 Obtain the Fourier expansion (12.46).

12.5 A statistical application: correlation and elliptic functions

There are many examples in which an analysis of the distribution of two variables seems to suggest a connection between the two; sometimes those connections are 'common-sense', in other case they are 'nonsense' – the apparent connections between the two variables are coincidence. In other cases the connection may be due to a third variable, which directly influences both. So the fundamental question is: how do we measure a possible link between two variables and how might we investigate the possibility that an apparent connection is due to a third (thus far unsuspected) variable? The theory of correlation seeks to do that (see Williams, 2001, for background and Kendall, 1941, for the connection with elliptic functions).

Consider two jointly distributed random variables, X and Y, and recall that the *variances* and *covariances* of X and Y are defined by

$$\text{var}(X) = E(X^2) - (E(X))^2,$$
$$\text{cov}(X, Y) = E(XY) - E(X)E(Y),$$

where $E(X)$ denotes the *expectation* of X. We shall suppose that the variances and covariances exist (and are not zero in the case of the variances). We then define the *correlation coefficient* (or simply the correlation) by

$$\rho = \frac{\text{cov}(X, Y)}{[\text{var}(X)\text{var}(Y)]^{1/2}}. \tag{12.47}$$

It follows from Cauchy's inequality that $\rho^2 \leq 1$, and one can show that ρ^2 provides a dimensionless measure of the degree of linear dependence between

X and Y. The more nearly the correlation approaches ± 1, the more nearly is the connection between X and Y perfectly linear.

In order to calculate ρ, we suppose that we have n paired data; say (x_1, y_1), $(x_2, y_2), \ldots, (x_n, y_n)$. Then it follows from the definitions quoted above that

$$\text{cov}(X, Y) = \left(\sum_{i=1}^{n} \frac{x_i y_i}{n} \right) - \left(\sum_{i=1}^{n} \frac{x_i}{n} \right) \left(\sum_{i=1}^{n} \frac{y_i}{n} \right)$$

and

$$\text{var}(X) = \left(\sum_{i=1}^{n} \frac{x_i^2}{n} \right) - \left(\sum_{i=1}^{n} \frac{x_i}{n} \right)^2,$$

with a similar formula for $\text{var}(Y)$. Accordingly, the definition (12.47) yields

$$\rho = \frac{\left(n \sum x_i y_i \right) - \left(\left(\sum x_i \right) \left(\sum y_i \right) \right)}{\left[\left(\left(n \sum x_i^2 \right) - \left(\sum x_i \right)^2 \right) \left(\left(n \sum y_i^2 \right) - \left(\sum y_i \right)^2 \right) \right]^{1/2}}.$$

So that is how to calculate ρ.

The correlation coefficient is used to measure the accuracy of an assumed linear relationship between the variables X and Y, which takes the form $y_i = m x_i + c$. In order to do that one introduces the concept of a *regression line*, and one way of doing that is minimize the sum

$$\sum (y_i - m x_i - c)^2,$$

and then the correlation coefficient measures how closely the two variables are related. The measure is determined by $|\rho|$; if $|\rho|$ is close to 1, then the more likely it is that there will be a linear relationship for the data as given. We consider a particular example in order to see how the Jacobi elliptic functions come in.

Suppose we are given three variables X_1, X_2, X_3 and three sets of data $(x_{1i}, x_{2i}), (x_{1i}, x_{3i}), (x_{2i}, x_{3i})$, $i = 1, \ldots, n$ and we consider the relationship between any two of the three variables. We denote the correlation coefficients corresponding to any two of our three variables by ρ_{12}, ρ_{13} and ρ_{23}, respectively. (Note that Kendall, 1941, uses r_{12} etc.) We also denote by $\rho_{ij \cdot k}$ the correlation between x_i and x_j, given that x_k is kept constant. (Kendall has $r_{ij \cdot k}$.) It can be shown that

$$\rho_{12 \cdot 3} = \frac{(\rho_{12} - \rho_{23} \rho_{13})}{\left(1 - \rho_{23}^2 \right)^{1/2} \left(1 - \rho_{13}^2 \right)^{1/2}}. \qquad (12.48)$$

Now we introduce the elliptic functions. Let

$$\rho_{12} = cn(u_3) = cn(u_3, k),$$
$$\rho_{13} = cn(u_2) = cn(u_2, k),$$
$$\rho_{23} = cn(u_1) = cn(u_1, k),$$

where $u_1 + u_2 + u_3 = 0$ and k will be obtained below. By the addition formula for the Jacobi functions, we know that

$$-dn(-(u_1 + u_2)) = -dn(u_1 + u_2) = \frac{cn(u_1 + u_2) - cn(u_1)cn(u_2)}{sn(u_1)sn(u_2)}$$

and now the formula (12.48) yields $-dn(u_3) = \rho_{12 \cdot 3}$, whence, by symmetry,

$$\rho_{12 \cdot 3} = -dn(u_3),$$
$$\rho_{23 \cdot 1} = -dn(u_1),$$
$$\rho_{13 \cdot 2} = -dn(u_2). \tag{12.49}$$

Since $dn^2 u = 1 - k^2 sn^2 u$, we can write $k^2 = (1 - dn^2 u)/(1 - cn^2 u)$ and so the modulus k satisfies

$$
\begin{aligned}
k^2 &= \frac{1 - dn^2 u_3}{1 - cn^2 u_3} \\
&= \frac{1 - \rho_{12 \cdot 3}^2}{1 - \rho_{12}^2} \cdot \frac{1 - \rho_{13}^2}{1 - \rho_{13}^2} \\
&= \frac{1 - R_{1(23)}^2}{\left(1 - \rho_{12}^2\right)\left(1 - \rho_{13}^2\right)},
\end{aligned} \tag{12.50}
$$

where $R_{1(23)}$ is defined as follows.

Let

$$
R = \begin{vmatrix} 1 & \rho_{12} & \rho_{13} \\ \rho_{12} & 1 & \rho_{23}, \\ \rho_{13} & \rho_{23} & 1 \end{vmatrix}, \tag{12.51}
$$

then $1 - R_{1(23)}^2 = R/(1 - \rho_{23}^2)$ and so (12.50) gives

$$
k^2 = \frac{R}{\left(1 - \rho_{12}^2\right)\left(1 - \rho_{23}^2\right)\left(1 - \rho_{13}^2\right)}; \tag{12.52}
$$

that is

$$
R = k^2 sn^2 u_1 \, sn^2 u_2 \, sn^2 u_3. \tag{12.53}
$$

(The number $R_{1(23)}$ is a correlation coefficient between expressions of the form $e_{1i} = x_{1i} - a x_{2i} - b x_{3i}$ and a variable X_i which is not the original X_i, but the original X_i minus its mean – a so-called *centred* random variable.)

On writing

$$R' = \begin{vmatrix} 1 & -r_{12\cdot3} & -r_{13\cdot2} \\ -r_{12\cdot3} & 1 & -r_{23\cdot1} \\ -r_{13\cdot2} & -r_{23\cdot1} & 1 \end{vmatrix},$$

we find (by direct calculation or by considerations of symmetry) that

$$R' = k^4 \, sn^2 u_1 \, sn^2 u_2 \, sn^2 u_3, \tag{12.54}$$

from which we obtain

$$k^2 = \frac{R'}{R}. \tag{12.55}$$

The matrix (12.51) whose determinant is R is called the correlation matrix. From (12.50) we see that

$$\begin{aligned} R^2_{1(23)} &= 1 - k^2 \, sn^2 u_2 \, sn^2 u_3 \\ &= sn^2 u_3 + dn^2 u_2 \, cn^2 u_3 \\ &= sn^2 u_2 + cn^2 u_2 \, dn^2 u_3, \end{aligned} \tag{12.56}$$

and there are similar expressions for the other two multiple correlations.

In the case of sampling from a *normal population* (that is the diagram is symmetric and bell-shaped – perhaps better to write Normal or Gaussian, since a distribution that is not normal is not abnormal!), the variances assume a simple form and, to order n^{-1}, we have

$$\begin{aligned} \mathrm{var}(\rho_{12}) &= \frac{1}{n}\left(1 - \rho_{12}^2\right)^2 \\ &= \frac{1}{n} \, sn^4 u_3; \end{aligned} \tag{12.57}$$

$$\begin{aligned} \mathrm{var}(\rho_{12\cdot3}) &= \frac{1}{n}\left(1 - dn^2 u_3\right)^2 \\ &= \frac{k^4}{n} \, sn^4 u_3; \end{aligned} \tag{12.58}$$

so the standard error of the partial correlation is k^2 times that of the total correlation. Kendall (1941), p. 283, also quotes an approximate result, due to Isserlis, which reads

$$\begin{aligned} \mathrm{var}(R_{1(23)}) &= \frac{1}{n}\left(1 - R^2_{1(23)}\right)^2 \\ &= \frac{k^4}{n} sn^4 u_2 \, sn^4 u_3 \\ &= n \, \mathrm{var}\rho_{13} \, \mathrm{var}\rho_{12\cdot3} \\ &= n \, \mathrm{var}\rho_{12} \, \mathrm{var}_{13\cdot2}. \end{aligned}$$

Exercises 12.5

12.5.1 Verify the formula (12.49).

12.5.2 Verify the results (12.52) and (12.53) as derived from (12.51).

12.5.3 Obtain the formulae similar to (12.56) for the other two multiple correlations.

12.6 Numerical analysis and the arithmetic-geometric mean of Gauss

The arithmetic-geometric mean (AGM) of two non-negative numbers a and b is defined in the following way. First, write $a = a_0$, $b = b_0$ and define two sequences $\{a_n\}$, $\{b_n\}$, $n = 0, 1, 2, \ldots$, by

$$a_{n+1} = \frac{1}{2}(a_n + b_n), \quad b_{n+1} = \sqrt{a_n b_n}, \tag{12.59}$$

where b_{n+1} denotes the positive square root. We shall assume that $a \geq b > 0$ and we recall the familiar inequality

$$\frac{a+b}{2} \geq \sqrt{ab}.$$

That inequality implies that $a_n \geq b_n$ for all $n \geq 0$, but the ideas here are much richer than that, and we shall see how there is a truly astonishing connection (discovered by Gauss) with the problem of the rectification of the lemniscate and the integral

$$\int_0^1 (1 - r^4)^{-1/2} dr, \tag{12.60}$$

whose connection with the lemniscate is given in Exercise 1.9.7. Our treatment will be based on the paper by Cox (1984) and we hope we shall whet the reader's appetite to explore that most interesting paper further (there are connections with theta functions and with $SL_2(\mathbb{Z})$, if one takes $\{a_n\}$, $\{b_n\}$ to be sequences of complex numbers, and with ideas of Abel and Gauss).

We begin by proving that the two sequences possess the properties

$$a \geq a_1 \geq \cdots \geq a_n \geq a_{n+1} \geq \cdots \geq b_{n+1} \geq b_n \geq \cdots \geq b_1 \geq b, \tag{12.61}$$

and

$$0 \leq a_n - b_n \leq 2^{-n}(a - b). \tag{12.62}$$

To prove (12.61), we use an induction argument. The induction step begins by observing that $a_n \geq b_n$ and $a_{n+1} \geq b_{n+1}$ imply, by the definition of the sequences,

$$a_n \geq \frac{a_n + b_n}{2} = a_{n+1} \geq b_{n+1} = (a_n b_n)^{1/2} \geq b_n. \qquad (12.63)$$

So $a_0 \geq a_1 \geq b_1 \geq b_0$ (which gives the basis of the induction) and then (12.63) shows that $a_n \geq b_n$ implies $a_{n+1} \geq b_{n+1}$.

To prove (12.62), we see that $b_{n+1} \geq b_n$ implies

$$a_{n+1} - b_{n+1} \leq a_{n+1} - b_n = 2^{-1}(a_n - b_n),$$

and we can now appeal to the induction argument to show that (12.62) holds.

From (12.61) we see that both of the limits $\lim_{n \to \infty} a_n$ and $\lim_{n \to \infty} b_n$ exist and then (12.62) implies that they are equal.

We have proved that the sequences $\{a_n\}$, $\{b_n\}$ converge to the common limit

$$M(a, b) = M(a_0, b_0) = \lim_{n \to \infty} a_n = \lim_{n \to \infty} b_n, \qquad (12.64)$$

which is the *arithmetic-geometric mean* (AGM) of Gauss, (see Gauss, 1799). For reasons that will emerge, Gauss was particularly interested in the case $M(\sqrt{2}, 1)$ and he calculated that to 19 decimal places, obtaining the value

$$M(\sqrt{2}, 1) = 1.198\ 140\ 234\ 473\ 559\ 220\ 74 \ldots$$

The amazing significance of $M(\sqrt{2}, 1)$ is that it is connected with the integral (12.60) by

$$2 \int_0^1 (1 - r^4)^{-1/2} dr = \frac{\pi}{2M(1, \sqrt{2})}, \qquad (12.65)$$

and Gauss discovered the generalization

$$\int_0^{\frac{\pi}{2}} \frac{d\phi}{\sqrt{a^2 \cos^2 \phi + b^2 \sin^2 \phi}} = \frac{\pi}{2M(a, b)}. \qquad (12.66)$$

We shall prove those two results.

First we show that

$$M(a, b) = M(a_1, b_1) = M(a_2, b_2) = \ldots,$$

which simply starts the sequence in (12.61) at the points (a_n, b_n) and then $M(a, b)$ satisfies the homogeneity condition

$$M(\lambda a, \lambda b) = \lambda M(a, b),$$

as is obvious from the definition.

Now we prove (12.66), namely:

Theorem 12.1 *Let $a \geq b > 0$ and define $M(a, b)$ by (12.64). Then*

$$M(a, b) \cdot \int_0^{\frac{\pi}{2}} (a^2 \cos^2 \phi + b^2 \sin^2 \phi)^{-1/2} d\phi = \frac{\pi}{2}.$$

Proof[2] Denote the integral in (12.66) by $I(a, b)$ and write $\mu = M(a, b)$; clearly we would like to prove that $I(a, b) = \pi/(2\mu)$. As a basis for an induction argument, we begin by proving that

$$I(a, b) = I(a_1, b_1). \tag{12.67}$$

Gauss introduced the variable ϕ' defined by

$$\sin \phi = \frac{2a \sin \phi'}{a + b + (a - b) \sin^2 \phi'}. \tag{12.68}$$

The inequality $0 \leq \phi' \leq \pi/2$ corresponds to $0 \leq \phi \leq \pi/2$, and Gauss then claimed that

$$(a^2 \cos^2 \phi + b^2 \sin^2 \phi)^{-1/2} d\phi = \left(a_1^2 \cos^2 \phi' + b_1^2 \sin^2 \phi'\right)^{-1/2} d\phi', \tag{12.69}$$

which is by no means obvious (see below), but from which (12.67) follows immediately.

Jacobi gives us some help, by first showing that

$$\cos \phi = \frac{2 \cos \phi' \left(a_1^2 \cos^2 \phi' + b_1^2 \sin^2 \phi'\right)^{1/2}}{a + b + (a - b) \sin^2 \phi'} \tag{12.70}$$

and then

$$(a^2 \cos^2 \phi + b \sin^2 \phi)^{1/2} = a \frac{a + b - (a - b) \sin^2 \phi'}{a + b + (a - b) \sin^2 \phi'} \tag{12.71}$$

(the reader is invited to follow Jacobi's hints in Exercise 12.6.1). Equation (12.69) then follows from (12.70) and (12.71) by differentiating (12.68).

Iteration of (12.67) now gives

$$I(a, b) = I(a_1, b_1) = I(a_2, b_2) = \cdots = I(a_n, b_n),$$

from which we deduce that

$$I(a, b) = \lim_{n \to \infty} I(a_n, b_n) = \frac{\pi}{2\mu},$$

---

[2] We follow the proof given in Cox (1984); which in turn uses the original proof of Gauss (1799), as amplified by Jacobi (1829).

since the functions

$$\left(a_n^2 \cos^2 \phi + b_n^2 \sin^2 \phi\right)^{-1/2}$$

converge uniformly to the constant function μ^{-1}. That completes the proof of Theorem 12.1.

We can use Theorem 12.1 to establish a connection with the complete elliptic integrals of the first kind, as introduced in Chapter 1 and developed later; that is, to the integrals of the form:

$$F\left(k, \frac{\pi}{2}\right) = \int_0^{\pi/2} \left(1 - k^2 \sin^2 \phi\right)^{-1/2} d\phi = \int_0^1 \{(1 - z^2)(1 - k^2 z^2)\}^{-1/2} dz.$$
$$(12.72)$$

To obtain (12.72), we define k by $k = (a - b)/(a + b)$ and then

$$I(a, b) = a^{-1} F\left(\frac{2\sqrt{k}}{1 + k}, \frac{\pi}{2}\right), \quad I(a_1, b_1) = a_1^{-1} F\left(k, \frac{\pi}{2}\right).$$

It follows that (12.67) is equivalent to

$$F\left(\frac{2\sqrt{k}}{1 + k}, \frac{\pi}{2}\right) = (1 + k) F\left(k, \frac{\pi}{2}\right),$$

which may be proved directly (see Exercise 12.6.3).

The substitution (12.70) can be written in the form

$$\sin \phi = \frac{(1 + k) \sin \phi'}{1 + k \sin^2 \phi'},$$

where k is defined as above, and which Tannery & Molk (1893–1902) p. 206, refer to as the 'Gauss transformation'.

Theorem 12.1 may be expressed in various ways; see Exercises 12.6.5, 12.6.6 and 12.6.7. We summarize the results as:

$$\frac{1}{M(1 + k, \ 1 - k)} = \frac{2}{\pi} \int_0^{\pi/2} (1 - k^2 \sin^2 \theta)^{-1/2} d\theta = \frac{2}{\pi} F\left(k, \frac{\pi}{2}\right);$$
$$(12.73)$$

$$\frac{1}{M(1 + k, \ 1 - k)} = \sum_{n=0}^{\infty} \left[\frac{1 \cdot 3 \cdots (2n - 1)}{2^n n!}\right]^2 k^{2n};$$
$$(12.74)$$

$$\frac{1}{M(1, \ k')} = \frac{2}{\pi} \int_0^{\pi/2} (1 - k^2 \sin^2 \theta)^{-1/2} d\theta; \quad k' = \sqrt{1 - k^2}.$$
$$(12.75)$$

A few words about numerical analysis are appropriate. The result proved in Exercise 12.6.7, namely

$$\frac{1}{M(1,\,k')} = \frac{2}{\pi} \int_0^{\pi/2} (1 - k^2 \sin^2 \theta)^{-1/2} d\theta,$$

where, in the usual notation for the complementary modulus, $k' = \sqrt{1 - k^2}$, shows that the average value of the function $(1 - k^2 \sin^2 \theta)^{-1/2}$ on the interval $[0, \pi/2]$ is the reciprocal of the AGM of the reciprocals of the maximum and minimum values of the function – an interpretation again due to Gauss (1799), p.371.

The result proved in Exercise 12.6.5 reads, if $0 \le k < 1$, then

$$\frac{1}{M(1 + k,\, 1 - k)} = \frac{2}{\pi} \int_0^{\pi/2} (1 - k^2 \sin^2 \theta)^{-1/2} d\theta = \frac{2}{\pi} F\left(k, \frac{\pi}{2}\right),$$
$$(12.76)$$

and that affords an efficient method for approximating the elliptic integral $F(k, \pi/2)$, and Cox, (1984), Section 3, shows that that problem led Lagrange independently to the algorithm for the AGM.

We saw in Chapter 1, Exercise 1.9.7, that Theorem 12.1 is related to the problem of finding the arc-length of the lemniscate, which leads one to investigate the efficiency of computational methods in connection with the evaluation of elliptic integrals.

One of the origins of the theory of elliptic integrals (and so of elliptic functions) is the observation, based on extensive computations, that if one denotes the elliptic integral (12.60) by ω, that is

$$\omega = 2 \int_0^1 (1 - r^4)^{-1/2} dr,$$

then the relation between the arc-length of the lemniscate and the AGM may be written

$$M(\sqrt{2}, 1) = \frac{\pi}{\omega}, \qquad (12.77)$$

and it was that connection that turned Gauss' attention to the subject and to one of the origins of the theory of elliptic functions. We conclude this section by showing how Gauss was led to prove (12.77).

We begin with the result (12.74) whose proof is outlined in Exercise 12.6.6. We look for another proof of that by showing, independently of our earlier arguments, that $(M(1 + k, 1 - k))^{-1} = y$, where

$$y = \sum_{n=0}^{\infty} \left(\frac{1 \cdot 3 \cdots (2n - 1)}{2^n n!} \right)^2 k^{2n}.$$

Gauss gave two proofs. The first begins by establishing the identity

$$M\left(1 + \frac{2t}{1+t^2},\ 1 - \frac{2t}{1+t^2}\right) = \frac{1}{1+t^2}M(1+t^2,\ 1-t^2) \qquad (12.78)$$

(see Exercise 12.6.8) and then he assumed that there is a power series expansion of the form

$$\frac{1}{M(1+k,\ 1-k)} = 1 + Ak^2 + Bk^4 + Ck^6 + \cdots.$$

Now write $k = t^2$ and also $k = 2t/(1+t^2)$ in that series and then substitute in the identity (12.78) to obtain

$$1 + A\left(\frac{2t}{1+t^2}\right)^2 + B\left(\frac{2t}{1+t^2}\right)^4 + C\left(\frac{2t}{1+t^2}\right)^6 + \cdots$$
$$= (1+t^2)(1 + At^4 + Bt^8 + Ct^{12} + \cdots).$$

On multiplying both sides by $2t/(1+t^2)$, that equation reads

$$\frac{2t}{1+t^2} + A\left(\frac{2t}{1+t^2}\right)^3 + B\left(\frac{2t}{1+t^2}\right)^5 + \cdots = 2t(1 + At^4 + Bt^8 + \cdots).$$

On comparing coefficients of like powers of t, we obtain a system of infinitely many equations in infinitely many unknowns, $A, B, C, \ldots$ Gauss then showed that that system is equivalent to the system of equations

$$0 = 1 - 4A = 9A - 16B = 25B - 36C = \ldots,$$

from which (12.74) follows.

His second proof depends on the fact that y satisfies the hypergeometric differential equation

$$(k^3 - k)\frac{d^2y}{dk^2} + (3k^2 - 1)\frac{dy}{dk} + ky = 0;$$

the details are to be found in Cox (1984), p. 382. Cox has some very interesting comments to make on the assumptions underlying Gauss' argument (for example, does $1/M(1+k, 1-k)$ possess a power series expansion?) and quotes Gauss' prediction that here was 'an entirely new field of analysis'.

The references to numerical analysis have been implicit rather than explicit, but we hope we have given some insight into how numerical calculations suggest a possible conclusion, then attempts to prove it point the way to new mathematical concepts and theories and, in this particular case, the calculations relating to the AGM led to the creation of the theory of elliptic functions.

It is natural to ask whether we can extend those ideas to the case when a and b are complex numbers. As Cox (1984) shows, that is not an easy problem but it leads to fascinating connections with theta functions, to the theory of modular forms and to elliptic integrals, so that

$$\frac{1}{M(1, k'(\tau))} = \frac{2}{\pi} \int_0^{\pi/2} (1 - k(\tau)^2 \sin^2 \phi)^{-1/2} d\phi.$$

See Cox (1984) for an elegant and complete account.

Exercises 12.6

12.6.1 Prove that $I(a, b) = I(a_1, b_1)$ (see (12.67)) by carrying out the details indicated in (12.68) and (12.69) and in the light of (12.70) and (12.71).

12.6.2 Prove that

$$I(a, b) = a^{-1} F\left(\frac{2\sqrt{k}}{1+k}, \frac{\pi}{2}\right), \quad I(a_1, b_1) = a_1^{-1} F\left(k, \frac{\pi}{2}\right).$$

12.6.3 Show that

$$I(a, b) = I(a_1, b_1)$$

is equivalent to the formula

$$F\left(\frac{2\sqrt{k}}{1+k}, \frac{\pi}{2}\right) = (1 + k)F\left(k, \frac{\pi}{2}\right),$$

and give a direct proof of that latter formula.

12.6.4 Prove that, in the notation of the AGM,

$$\int_0^{\pi/2} \frac{d\theta}{(a^2 \cos^2 \theta + b^2 \sin^2 \theta)^{1/2}} = \int_0^{\pi/2} \frac{d\phi}{(a_1^2 \cos^2 \phi + b_1^2 \sin^2 \phi)^{1/2}},$$

and deduce that, if $0 < k < 1$, then

$$K = \frac{\pi/2}{M(1, k')}, \quad K' = \frac{\pi/2}{M(1, k)}.$$

12.6.5 Show that Theorem 12.1 may be stated in the form: if $0 \le k < 1$, then

$$\frac{1}{M(1+k, 1-k)} = \frac{2}{\pi} \int_0^{\pi/2} (1 - k^2 \sin^2 \theta)^{-1/2} d\theta = \frac{2}{\pi} F\left(k, \frac{\pi}{2}\right).$$

12.6.6 Suppose that $0 < k < 1$. By expanding the integrand $(1 - k^2 \sin^2 \theta)^{-1/2}$ as a series of ascending powers of k^2, and noting that the series converges uniformly with respect to θ (by Weierstrass's M-test, since $0 \leq \sin^2 \theta \leq 1$), show that

$$\frac{1}{M(1+k, 1-k)} = \sum_{n=0}^{\infty} \left[\frac{1 \cdot 3 \cdots (2n-1)}{2^n n!}\right]^2 k^{2n}.$$

(The right-hand side is the hypergeometric function $\frac{1}{2}\pi F\left(\frac{1}{2}, \frac{1}{2}; 1, k^2\right)$.)

12.6.7 Using the homogeneity properties of M, show that (see (12.75))

$$\frac{1}{M(1, k')} = \frac{2}{\pi} \int_0^{\pi/2} (1 - k^2 \sin^2 \theta)^{-1/2} d\theta.$$

12.6.8 Use the homogeneity properties of $M(a, b)$ to derive the identity

$$M\left(1 + \frac{2t}{1+t^2}, \ 1 - \frac{2t}{1+t^2}\right) = \frac{1}{1+t^2} M(1 + t^2, \ 1 - t^2).$$

12.7   Rational maps with empty Fatou set

This example is taken from Beardon (1991), Section 4.3, where the reader will find further background material (as well as a summary of the basic properties of the Weierstrass $\wp$-function).

The idea is to consider rational functions, R, defined on a disc, D, in $\mathbb{C}$ and to show that for certain values of n, $R^n(D) = \mathbb{C}_\infty$. That means that for those values of n the functions R^n 'explode any small disc onto the whole (of the Riemann) sphere', from which one deduces that the family $\{R^n\}$ is not equicontinuous on any open subset of the complex sphere, and from that one deduces that the Julia set $J(R) = \mathbb{C}_\infty$ and hence one obtains that the Fatou set is empty.

We begin with the rational map, R, defined by

$$z \mapsto \frac{(z^2 + 1)^2}{4z(z^2 - 1)}, \tag{12.79}$$

and we shall show that its Julia set is the entire complex sphere.

Let $\lambda, \mu \in \mathbb{C}$ such that $(\lambda/\mu) \notin \mathbb{R}$ and let $\Lambda = \{m\lambda + n\mu : n, m \in \mathbb{Z}\}$ be the lattice with basis λ, μ. We denote a period parallelogram for the lattice Λ by Ω and we consider an elliptic function, f, for the lattice Λ. The function f maps $\mathbb{C}$ into $\mathbb{C}_\infty$ and, since $f(\mathbb{C}) = f(\Omega)$, $f(\mathbb{C})$ is a compact subset of $\mathbb{C}_\infty$. But, by the open mapping theorem, $f(\mathbb{C})$ is also an open subset of $\mathbb{C}_\infty$ and we deduce that

$$f(\Omega) = f(\mathbb{C}) = \mathbb{C}_\infty.$$

We shall use the properties of the Weierstrass elliptic functions, and in particular the addition theorem for $\wp$ in the form

$$\wp(2z) = R(\wp(z)), \tag{12.80}$$

where R denotes the rational function

$$R(z) = \frac{z^4 + g_2 z^2/2 + 2g_3 z + (g_2/4)^2}{4z^3 - g_2 z - g_3} \tag{12.81}$$

(See Exercise 12.7.1).

Let D denote an open disc in $\mathbb{C}$ and define U by $U = \wp^{-1}(D)$. Let $\varphi(z) = 2z$. Since U is open and since $\varphi^n(U)$ is the set U expanded by a factor 2^n, we may suppose that, for n sufficiently large, $\varphi^n(U)$ contains a period parallelogram, Ω, of $\wp$. From that and from (12.80) we see that, for such a value of n,

$$R^n(D) = R^n(\wp(U)) = \wp(2^n U) = \mathbb{C}_\infty. \tag{12.82}$$

Since D is an arbitrary open set, (12.82) implies that the family $\{R^n\}$ is not equicontinuous on any open subset of the complex sphere, and we obtain the result that the Julia set $J(R) = \mathbb{C}_\infty$.

The preceding argument yields a family of rational maps whose Julia set is the complex sphere. Moreover (as we saw in Chapter 7),

$$g_2^3 - 27g_3^2 \neq 0 \tag{12.83}$$

and we can always find a lattice Λ such that the pair $(g_2(\Lambda), g_3(\Lambda))$ provides a pair g_2, g_3 satisfying (12.83). In particular, there is a lattice Λ such that $g_2 = 4$ and $g_3 = 0$ and then R as defined in (12.81) gives the function in (12.79).

There is an easier construction, based on the properties of the square lattice for $\wp$, as given in Chapter 7. Denote by x a positive real number and consider the square lattice with basis $\lambda = x$, $\mu = \mathrm{i}x$. As we saw in Chapter 7, that choice gives $g_3 = 0$ and leads thence to a simplification of (12.81), given by

$$R(z) = \frac{16z^4 + 8g_2 z^2 + g_2^2}{16z(4z^2 - g_2)}. \tag{12.84}$$

From (12.83), $g_2 \neq 0$, and by defining $h(z) = 2z/\sqrt{g_2}$, we see that the function hRh^{-1} (which also has $\mathbb{C}_\infty$ as its Julia set) is the function in (12.79).

Finally, we observe that it follows from the properties of the Weierstrass function that it is reasonable to expect that a formula of the type $\wp(2z) = R(\wp(z))$ should hold. For, given $w \in \mathbb{C}_\infty$, there are exactly two solutions z, modulo Λ, of the equation $\wp(z) = w$, and we may take those to be of the form u and $\lambda + \mu - u$. Now

$$\wp(2(\lambda + \mu - u)) = \wp(2u), \tag{12.85}$$

and so we can define a map $w \mapsto \wp(2u)$ of $\mathbb{C}_\infty$ onto $\mathbb{C}_\infty$ which is independent of the choice of u. That map must be analytic (see Exercise 12.7.5) and so must be a rational map, R, and accordingly there is a relation of the type

$$R(\wp(u)) = \wp(2u).$$

(Note that one can replace 2 by any integer; see Exercise 12.7.6.)

Exercises 12.7

12.7.1 Obtain the expression for $R(\wp(z))$ given in (12.81). (Hint: use the duplication formula for $\wp(2z)$ and then replace z by $\wp(z)$ in (12.81).)

12.7.2 Obtain the simplified form of $R(z)$ in the case of the square lattice with basis $(\lambda, \mu) = (x, ix)$ in (12.84).

12.7.3 Show that the function hRh^{-1}, where $h(z) = 2z/\sqrt{g_2}$, has $\mathbb{C}_\infty$ as its Julia set and check that it is the function defined in (12.79).

12.7.4 Revise the details leading up to the result in (12.85) and the definition of the map $w \mapsto \wp(2u)$.

12.7.5 Prove that the map $w \mapsto \wp(2u)$ (as recalled in Exercise 12.7.4) is analytic and deduce (as is claimed in the text) that it must be a rational map, R.

12.7.6 Prove that the formula $R(\wp(u)) = \wp(2u)$ may be replaced by one of the type $R(\wp(u)) = \wp(nu)$, where n is a positive integer, and thence obtain the result for any integer.

12.8   A final, arithmetic, application: heat-flow on a circle and the Riemann zeta function

Consider the temperature in a circular wire, of length 1, or of a periodic distribution of temperature in an infinite wire, from $-\infty$ to $+\infty$ (that is in $\mathbb{R}^1$). The flow of heat preserves such periodicity and we have to solve the differential equation

$$\frac{\partial u}{\partial t} = \frac{1}{2}\frac{\partial^2 u}{\partial x^2}$$

on the circle $0 \leq x < 1$, where $u = u(x, t)$ is the temperature at time t and at the point x. One conjectures that the solution is determined by the initial distribution $u(x, 0) = f(x)$ at time $t = 0$.

One can show (see, for example, Dym & McKean, 1972, pp.63–66) that a solution is given by

$$u(x, t) = \sum_{n \in \mathbb{Z}} \hat{f}(n) \exp(-2\pi^2 n^2 t) e_n(x)$$

$$= \sum \exp(-2\pi^2 n^2 t) e_n(x) \int_0^1 f(y) e_n^*(y) dy$$

$$= \int_0^1 \left[\sum_{n \in \mathbb{Z}} \exp(-2\pi^2 n^2 t) e_n(x - y) \right] f(y) dy$$

$$= p \circ f, \tag{12.86}$$

where $p \circ f$ denotes the convolution

$$p \circ f = \int_0^1 g(x - y) f(y) dy,$$

$e_n(x) = e^{2\pi i n x}$ and $e_n^*(x) = e^{-2\pi i n x}$.

In the context of the ideas introduced in Section 12.5, the function

$$p = p_t(x) = \sum_{n \in \mathbb{Z}} \exp(-2\pi^2 n^2 t) e_n(x) \tag{12.87}$$

is the Green's function for the problem, and we note that, when $t = 0$,

$$p = \sum_{n \in \mathbb{Z}} e_n(x),$$

which is the Fourier expansion of the periodic Dirac distribution $\delta(x)$, which is defined to be 1 when $x = 0$ and 0 if $x \neq 0$.

Those ideas relate to Jacobi's identity for the theta function,

$$\sum_{n \in \mathbb{Z}} e^{-2\pi^2 n^2 t} e^{2\pi i n x}. \tag{12.88}$$

Consider the sum

$$q = q_t(x) = \frac{1}{\sqrt{2\pi t}} \sum_{n \in \mathbb{Z}} \exp(-(x - n)^2 / 2t). \tag{12.89}$$

Then the convolution $q \circ f$ solves the heat flow problem on $\mathbb{R}^1$, that is periodic of period 1 and $q_t(x) \to f(x)$ as $t \to 0$. But (given that the solution is unique), that means that $q \circ f = p \circ f$ and we conclude that $q = p$. (That last is an example of Kelvin's 'method of images', in which the integers in the sum defining q are understood to be fictitious sources of heat at the integers on the universal covering surface $\mathbb{R}^1$ of the circle.)

The fact that $q = p$ gives us another proof of the Jacobi identity for the theta function

$$\theta_3(x) = \sum_{n=-\infty}^{\infty} e^{-\pi n^2 x} \tag{12.90}$$

(in which we have made the change of notation in which t is replaced by x and there are other minor changes to conform to the usage in the theory of numbers[3]).

Let $s = \sigma + it$ be a complex variable, with $\sigma > 0$, and recall that the gamma function, $\Gamma(s/2)$, is given by

$$\int_0^\infty x^{s/2-1} e^{-n^2 \pi x} \, dx = \frac{\Gamma(s/2)}{n^s \pi^{s/2}}. \tag{12.91}$$

The Riemann zeta function, $\zeta(s)$, is defined for $\sigma > 1$ by

$$\zeta(s) = \sum_{n=1}^{\infty} \frac{1}{n^s}$$

and then (12.91) gives

$$\frac{\Gamma(s/2)\zeta(s)}{\pi^{s/2}} = \sum_{n=1}^{\infty} \int_0^\infty x^{s/2-1} e^{-n^2 \pi x} \, dx = \int_0^\infty x^{s/2-1} \sum_{n=1}^{\infty} e^{-n^2 \pi x} \, dx, \tag{12.92}$$

the interchange of summation and integration being justified by absolute convergence.

Now write

$$\psi(x) = \sum_{n=1}^{\infty} e^{-\pi n^2 x}, \tag{12.93}$$

and then (12.92) is

$$\zeta(s) = \frac{\pi^{s/2}}{\Gamma(s/2)} \int_0^\infty x^{s/2-1} \psi(x) \, dx, \qquad \sigma > 1. \tag{12.94}$$

Now Jacobi's identity (as obtained above or in Chapter 4) gives

$$\sum_{n=-\infty}^{\infty} e^{-\pi n^2 x} = \frac{1}{\sqrt{x}} \sum_{n=-\infty}^{\infty} e^{-\pi n^2 / x}, \qquad x > 0,$$

---

[3] One of the standard references is to Titchmarsh, 1986. We refer the reader to pp. 21–22, in Chapter II.

and so

$$2\psi(x) + 1 = \frac{1}{\sqrt{x}}\left(2\psi\left(\frac{1}{x}\right) + 1\right). \tag{12.95}$$

It follows from (12.94) that

$$\pi^{-s/2}\Gamma(s/2)\zeta(s) = \int_0^1 x^{s/2-1}\psi(x)dx + \int_1^\infty x^{s/2-1}\psi(x)dx$$

$$= \int_0^1 x^{s/2-1}\left\{\frac{1}{\sqrt{x}}\psi\left(\frac{1}{x}\right) + \frac{1}{2\sqrt{x}} - \frac{1}{2}\right\}dx + \int_1^\infty x^{s/2-1}\psi(x)dx$$

$$= \frac{1}{s-1} - \frac{1}{s} + \int_0^1 x^{s/2-3/2}\psi\left(\frac{1}{x}\right)dx + \int_1^\infty x^{s/2-1}\psi(x)dx$$

$$= \frac{1}{s(s-1)} + \int_1^\infty \left(x^{-s/2-1/2} + x^{s/2-1}\right)\psi(x)dx. \tag{12.96}$$

That last integral is convergent for all s and so the formula, proved originally for $\mathrm{Re}\,s > 1$, holds for all s by analytic continuation. The right-hand side of (12.96) is unchanged if s is replaced by $1 - s$ and so

$$\pi^{-\frac{s}{2}}\Gamma\left(\frac{s}{2}\right)\zeta(s) = \pi^{-1/2+s/2}\Gamma\left(\frac{1}{2} - \frac{s}{2}\right)\zeta(1 - s), \tag{12.97}$$

which is the functional equation of the Riemann zeta function and which plays a crucial role in the formulation of the Riemann Hypothesis and its application to the problem of the distribution of prime numbers.

Exercises 12.8

12.8.1 Check the details of the proof given above that $q = p$ and so of Jacobi's identity for $\theta_3(x)$.

12.8.2 In the usual notation, $q = e^{\pi i\tau}$,

$$\theta_3(z|\tau) = 1 + 2\sum_{n=1}^\infty q^{n^2}\cos 2nz,$$

show that

$$\frac{\partial^2\theta_3(z|\tau)}{\partial z^2} = -\frac{4}{\pi i}\frac{\partial\theta_3(z|\tau)}{\partial\tau}.$$

Check that the other three theta functions also satisfy that equation.

Appendix

A.1 Introduction

In this appendix we present an unorthodox (in some respects) sequence of simple propositions that update and make rigorous Euler's elementary derivation of the product formulae for the sine and cosine and some related identities, to which we have appealed in the main text and which are also related to the approach to elliptic functions given by Eisenstein and Kronecker, see Weil [1975], where, indeed, another proof affording insight into the elliptic functions is given. We also include, for convenience of reference and to save overburdening other parts of the book, some propositions concerning the Benoulli numbers and related topics.

A.2 The formula for arg (z)

We begin by recalling the properties of the argument (amplitude) of a complex number, $z = x + iy$. The principal value of the argument, θ, is that which satisfies the inequality $-\pi < arg z \leq \pi$, and in this introductory section we shall denote that by Arg z.

Cut the complex plane along the negative axis. Then if $z = x + iy$ lies in the cut plane Ω remaining, one can draw Figure A.1.

Clearly, one can take $-\pi/2 < \theta/2 < \pi/2$ and $\tan(\theta/2) = y/(r+x)$, whence $\theta/2 = \arctan y/(r+x)$ or

$$\theta = 2 \arctan \frac{y}{r+x} \quad (z \in \Omega). \tag{A.1}$$

Formula (A.1) has been available since Lambert in 1765; it ought to appear in every calculus text. Note that (A.1) becomes a purely analytical expression upon setting

$$\arctan t = \int_0^t \frac{du}{1+u^2} \quad (-\infty < t < \infty). \tag{A.2}$$

370

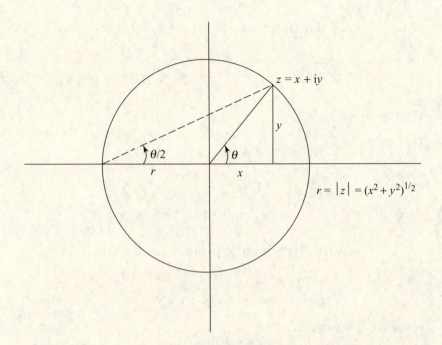

Figure A.1 The formula for arg z.

It follows immediately that $\arg(z) \equiv \theta = \theta(x, y)$ is a continuous function of z, a proposition usually either overlooked in the literature or established by a compactness argument. Hence, if one sets

$$\text{Log} z = \log r + i\theta \ (z \in \Omega), \qquad (A.3)$$

where θ is given by (A.1), Log$z$ is continuous in Ω, the crucial fact one needs for the inverse function argument that establishes $\text{Log}'z = z^{-1} \ (z \in \Omega)$, where the dash$'$ denotes differentiation with respect to z. Although it is preferable to avoid the problems caused by the multiple valued logarithm, and so Log z is in some respect preferable to log z, we prefer to use the latter and to follow the notation of Whittaker and Watson (1927, p.589) where there is no risk of ambiguity, or where, as is the case here, the argument is restricted to some interval like $-\pi < \theta < \pi$.

Exercises A.2

A.2.1 Show directly that $\text{Log}'z = z^{-1}$ in Ω, via the Cauchy–Riemann equations.

A.2.2 Prove that when $x < 0$, $\lim_{y \to 0\pm} \theta = \pm\pi$. Here by definition $\pi/4 = \int_0^1 \left(1 + u^2\right)^{-1} du$. Compare with (A.2) above.

A.3 The formula $\exp z = \lim_n (1 + z/n)^n$

That some authors make heavy weather of this proposition may be due to their desire to avoid the morass of multiple-valuedness of the logarithm; but one can successfully skirt that swamp.

Proof When $n > |z|$ set

$$z_n = n\mathrm{Log}\,(1 + z/n) = n\left[\mathrm{Log}1 + \left(\mathrm{Log}'1\right) z/n + o\,(1)\,z/n\right]$$
$$= z\left[1 + o\,(1)\right] \to z \text{ as } n \to \infty.$$

Hence

$$\exp z = \lim_n \exp z_n = \lim_n \exp\left[n\mathrm{Log}\,(1 + z/n)\right]$$
$$= \lim_n \left[\exp \mathrm{Log}\,(1 + z/n)\right]^n \lim_n (1 + z/n)^n.$$

A.4 Euler's infinite product for the sine

Euler's beautiful formula reads

$$\sin w = w \prod_{k=1}^{\infty} \left(1 - \frac{w^2}{k^2\pi^2}\right) (w \in \mathbb{C}), \tag{A.4}$$

or equivalently (on setting $z = iw$)

$$\sinh z = z \prod_{k=1}^{\infty} \left(1 + \frac{z^2}{k^2\pi^2}\right) (z \in \mathbb{C}). \tag{A.4'}$$

That classical identity is usually proved by relatively deep methods of complex analysis. The elementary proof of (A.4′) that follows is a simplification of an earlier proof by one of us (Eberlein (1966), (1977)).

We have

$$\sinh z = \lim_n p_n(z),$$

where

$$p_n(z) = \frac{1}{2} \left[(1 + z/n)^n - (1 - z/n)^n\right]. \tag{A.5}$$

Our procedure will be to factor p_n and then pass (judiciously) to the limit as $n \to \infty$. Now, $p_n(z) = 0$ if and only if $1 + z/n = \omega(1 - z/n)$, where $\omega^n = 1$. Then $z = n(\omega - 1)/(\omega + 1)$. But $\omega = e^{i\theta}$, whence

$$\frac{\omega - 1}{\omega + 1} = \frac{e^{i\theta} - 1}{e^{i\theta} + 1} = \frac{e^{i\theta/2} - e^{-i\theta/2}}{e^{i\theta/2} + e^{i\theta/2}} = i \tan \frac{\theta}{2}.$$

Now simplify by assuming n is odd. Then $n = 2m + 1$ and $\theta = k2\pi/(2m + 1)$ $(-m \le k \le m)$. Hence the zeros of p_{2m+1} are $z_k = (2m + 1)\,i\tan\,[k\pi/(2m + 1)]$ $(-m \le k \le m)$. Since p_{2m+1} is a polynomial of degree $2m + 1$, for some constant A we must have

$$p_{2m+1}(z) = Az \prod_{k=-m, k\neq 0}^{m} \left[1 - \frac{z}{(2m + 1)\,i\tan\left[k\pi/(2m + 1)\right]}\right]$$
$$= Az \prod_{k=1}^{m} \left[1 + \frac{z^2}{(2m + 1)^2 \tan^2\left(k\pi/(2m + 1)\right)}\right].$$

But it is obvious from the definition of $p_n(z)$ that the coefficient of z is 1, whence $A = 1$ and we have established

$$p_{2m+1}(z) = z \prod_{k=1}^{m} \left[1 + \frac{z^2}{(2m+1)^2 \tan^2 (k\pi/(2m+1))} \right]. \tag{A.6}$$

If we now let $m \to \infty$ and appeal to the fact that $\lim_{\theta \to 0} (\tan \theta)/\theta = 1$ we would *formally* obtain (A.4′) and have updated Euler's original 'proof'. But there is a double limit involved, and one must proceed more cautiously.

That (A.4′) holds in the sense of uniform convergence on bounded sets is an immediate consequence of the following.

Lemma A.1 $|z| \le M$ implies

$$\left| \sinh z - z \prod_{k=1}^{l} \left(1 + \frac{z^2}{k^2 \pi^2} \right) \right| \le \sinh M - p_{2l+1}(M)$$

$$\to 0 \, (l \to \infty).$$

Proof Let $m > l \ge 1$ and set $n = 2m + 1$. Then

$$\left| p_n(z) - z \prod_{k=1}^{l} \left[1 + \frac{z^2}{n^2 \tan^2 (k\pi/n)} \right] \right|$$

$$= \left| z \prod_{k=1}^{l} \left[1 + \frac{z^2}{n^2 \tan^2 (k\pi/n)} \right] \right| \left| \prod_{k=l+1}^{m} \left[1 + \frac{z^2}{n^2 \tan^2 (k\pi/n)} \right] - 1 \right|$$

$$\le |z| \prod_{k=1}^{l} \left[1 + \frac{|z|^2}{n^2 \tan^2 (k\pi/n)} \right] \left[\prod_{k=l+1}^{m} \left\{ 1 + \frac{|z|^2}{n^2 \tan^2 (k\pi/n)} \right\} - 1 \right]$$

$$\le M \prod_{k=1}^{l} \left[1 + \frac{M^2}{n^2 \tan^2 (kl/n)} \right] \left[\prod_{k=l+1}^{m} \left\{ 1 + \frac{M^2}{n^2 \tan^2 (k\pi/n)} \right\} - 1 \right]$$

$$= p_n(M) - M \prod_{k=1}^{l} \left(1 + \frac{M^2}{n^2 \tan^2 (k\pi/n)} \right).$$

Let $m \to \infty$ to obtain

$$\left| \sinh z - z \prod_{k=1}^{l} \left(1 + \frac{z^2}{k^2 \pi^2} \right) \right| \le \sinh M - M \prod_{k=1}^{l} \left(1 + \frac{M^2}{k^2 \pi^2} \right).$$

The inequality $\tan \theta \ge \theta \, (0 \le \theta < \pi/2)$ implies that

$$p_{2l+1}(M) \le M \prod_{k=1}^{l} \left(1 + \frac{M^2}{k^2 \pi^2} \right).$$

Hence $\left| \sinh z - z \prod_{k=1}^{l} \left(1 + \frac{z^2}{k^2 \pi^2} \right) \right| \le \sinh M - p_{2l+1}(M)$, as was to be proved.

A.5 Euler's infinite product for the cosine

This reads

$$\cos w = \prod_{n=0}^{\infty} \left(1 - \frac{4w^2}{(2n+1)^2 \pi^2} \right) \ (w \in C), \tag{A.7}$$

or equivalently (on setting $z = iw$)

$$\cosh z = \prod_{n=0}^{\infty} \left(1 + \frac{4z^2}{(2n+1)^2 \pi^2} \right) \ (z \in C), \tag{A.7'}$$

One could derive $(A.7')$ from the formula $\cosh z = \lim_n \left[(1 + z/n)^n + (1 - z/n)^n \right]/2$, but it is simpler to derive (A.7) from (A.4). Set $w = \pi/2$ in the latter to obtain the Wallis product

$$1 = \frac{\pi}{2} \prod_{n=1}^{\infty} \left(1 - \frac{1}{2^2 n^2} \right).$$

Hence

$$\sin w = \frac{\sin w}{1} = \frac{2w}{\pi} \prod_{n=1}^{\infty} \frac{1 - w^2/n^2}{1 - (2n)^{-2}}$$

$$= \frac{2w}{\pi} \prod_{n=1}^{\infty} \frac{(2n\pi)^2 - (2w)^2}{(2n\pi)^2 - \pi^2} = \frac{2w}{\pi} \prod_{n=1}^{\infty} \frac{(2n\pi - 2w)(2n\pi + 2w)}{(2n-1)\pi (2n+1)\pi}.$$

Now replace w by $\pi/2 - w$ to obtain

$$\cos w = \sin \left(\frac{\pi}{2} - w \right) = \frac{\pi - 2w}{\pi} \prod_{n=1}^{\infty} \frac{[(2n-1)\pi + 2w][(2n+1)\pi - 2w]}{(2n-1)\pi (2n+1)\pi}$$

$$= (1 - 2w/\pi) \prod_{n=1}^{\infty} [1 + 2w/(2n-1)\pi][1 - 2w/(2n+1)\pi]$$

$$= \lim_{N \to \infty} (1 - 2w/\pi) \prod_{n=1}^{N} [1 + 2w/(2n-1)\pi][1 - 2w/(2n+1)\pi]$$

$$= \lim_{N \to \infty} [1 - 2w/(2N+1)\pi] \prod_{n=0}^{N-1} \left[1 - 4w^2/(2n+1)^2 \pi^2 \right]$$

$$= \prod_{n=0}^{\infty} \left[1 - 4w^2/(2n+1)^2 \pi^2 \right].$$

Exercises A.5

A.5.1 Show that the convergence above is uniform on bounded sets.

A.5.2 Use the Euler product for $\sin w$ and logarithmic differentiation to obtain
the formula

$$w \cot w = 1 + 2 \sum_{n=1}^{\infty} \frac{w^2}{w^2 - n^2\pi^2}$$

$$= 1 - 2 \sum_{n=1}^{\infty} \sum_{k=1}^{\infty} \frac{w^{2k}}{n^{2k}\pi^{2k}}.$$

A.5.3 By taking $w = \pi z$ in the previous exercise, obtain the formula

$$\pi \cot \pi z = \frac{1}{z} + \sum_{n=1}^{\infty} \left(\frac{1}{z+n} + \frac{1}{z-n} \right).$$

A.5.4 Starting from the result in A.5.3, write $q = e^{2\pi i z}$ to obtain

$$\pi \cot \pi z = i\pi \frac{q+1}{q-1}$$

$$= i\pi - \frac{2i\pi}{1-q}$$

$$= i\pi - 2i\pi \sum_{n=1}^{\infty} q^n.$$

Deduce that

$$\frac{1}{z} + \sum_{n=1}^{\infty} \left(\frac{1}{z+n} + \frac{1}{z-n} \right) = i\pi - 2i\pi \sum_{n=1}^{\infty} q^n.$$

By differentiating the previous result, prove that, for $k \geq 2$,

$$\sum_{n\in\mathbb{Z}} \frac{1}{(n+z)^k} = \frac{1}{(k-1)!} (-2i\pi)^k \sum_{n=1}^{\infty} n^{k-1} q^n.$$

A.6 Evaluation of $\zeta(2n)$

Set

$$\zeta(s) = \sum_{n=1}^{\infty} n^{-s} \quad (s > 1). \tag{A.8}$$

Euler's motivation for developing his product formulae was to evaluate $\zeta(2)$.
To see this, compare the Maclaurin series

$$x^{-1} \sin x = 1 - \frac{x^2}{3!} + \frac{x^4}{5!} \cdots.$$

with the infinite product

$$x^{-1}\sin x = \prod_{n=1}^{\infty}\left(1 - \frac{x^2}{n^2\pi^2}\right)$$

$$= 1 - \left\{\pi^{-2}\sum_{1}^{\infty}n^{-2}\right\}x^2 + \pi^{-4}\left\{\sum_{m<n}^{\infty}m^{-2}n^{-2}\right\}x^4 - \cdots$$

(The formal multiplication is easily justified.)

Equate the coefficients of x^2 to obtain

$$\zeta(2) = \sum_{1}^{\infty}n^{-2} = \frac{\pi^2}{6}. \tag{A.9}$$

To evaluate $\zeta(4)$ note that

$$\sum_{m<n}m^{-2}n^{-2} = \frac{1}{2}\left[\sum m^{-2}n^{-2} - \sum_{1}^{\infty}n^{-4}\right]$$

$$= \frac{1}{2}\left[\left\{\sum_{1}^{\infty}m^{-2}\right\}\left\{\sum_{1}^{\infty}n^{-2}\right\} - \sum_{1}^{\infty}n^{-4}\right]$$

$$= \frac{1}{2}\left[\frac{\pi^4}{36} - \zeta(4)\right].$$

Now equate the coefficients of x^4 to obtain

$$\zeta(4) = \sum_{1}^{\infty}n^{-4} = \frac{\pi^4}{90}. \tag{A.10}$$

A modification of this analysis evaluates $\zeta(2n)$ ($n = 1, 2, \ldots$). Those results may also be obtained by an appeal to the theory of residues in complex analysis. Now we express them in terms of the Bernoulli numbers, which play a prominent part in our evaluation of the Eisenstein series in connection with $g_2(\tau)$ and $g_3(\tau)$ (see Chapter 7).

We begin by defining the Bernoulli numbers, B_k, in terms of the power series expansion for $x/(e^x - 1)$; thus

$$\frac{x}{e^x - 1} = 1 - \frac{x}{2} + \sum_{k=1}^{\infty}(-1)^{k+1}B_k\frac{x^{2k}}{(2k)!} \tag{A.11}$$

One can show that $B_1 = 1/6$, $B_2 = 1/30$, $B_3 = 1/42$, $B_4 = 1/30$ (see, for example Serre, 1970, Chapter VII for further examples).

We can now show that

$$\zeta(2k) = \frac{2^{2k-1}}{(2k)!}B_k\pi^{2k} \tag{A.12}$$

(from which our evaluations in (A.9) and (A.10) may be read off), as follows.
First, on writing $x = 2iz$ in the definition (A.11), we obtain

$$z \cot z = 1 - \sum_{k=1}^{\infty} B_k \frac{2^{2k} z^{2k}}{(2k)!} \qquad \text{(A.13)}$$

On the other hand, Exercises A.5.2 and A.5.3 show that

$$z \cot z = 1 + 2 \sum_{n=1}^{\infty} \frac{z^2}{z^2 - n^2 \pi^2}$$

$$= 1 - 2 \sum_{n=1}^{\infty} \sum_{k=1}^{\infty} \frac{z^{2k}}{n^{2k} \pi^{2k}} \qquad \text{(A.14)}$$

and, on comparing (A.14) and (A.13) we obtain (A.12).

References

Abel, N. H., 1881, *Oeuvres Complètes*, 2 vols., Christiana, Grndahl and Son.

Agnew, R. P., 1960, *Differential Equations*, second edn., New York, McGraw-Hill.

Ahlfors, L. V., 1979, *Complex Analysis*, third edn., New York, McGraw-Hill.

Apostol, T. M., 1969, *Mathematical Analysis*, Reading, Mass, Addison-Wesley.

Appell, P., 1878, Sur une interprétation des valeurs imaginaires du temp en mécanique, *C. R. Acad. Sci. Paris*, **87**, 1074–1077.

Baker, H. F., 1992, *Principles of Geometry*, 6 vols., Cambridge, Cambridge University Press.

Bateman, H., 1953, *Higher Transcendental Functions*, 3 vols., (eds. Erdélyi, Magnus, Oberhettinger and Tricomi) New York, McGraw-Hill.

Beardon, A. F., 1991, *Iteration of Rational Functions*, Graduate texts in Mathematics, **132**, Section 4.3, New York, Springer.

Biane, P., Pitman, J. and Yor, M., 2001, Probability laws related to the Jacobi theta and Riemann zeta functions, and Brownian excursions, *Bulletin American Math. Soc.*, **38**, 435–465.

Birch, B. J. and Swinnerton-Dyer, H. P. F., 1965, Notes on elliptic curves, II, *J. Reive Angew. Math.*, **218**, 79–108.

Bowman, F., 1961, *Introduction to Elliptic Functions with Applications*, New York, Dover.

Briot, C. and Bouquet, J. C., 1875, *Théorie des fonctions élliptiques*, Paris, Gauthier Villars.

Burnside, W. S. and Panton, A. W., 1901, *Theory of Equations* (especially Vol. II) Dublin, Dublin University Press.

Cassels, J. W. S., 1991, *Lectures on Elliptic Curves*, London Mathematical Society Student Texts, 24, Cambridge, Cambridge University Press.

Cayley, A., 1895, *An Elementary Treatise on Elliptic Functions*, London, George Bell and Sons.

Coddington, E. A., 1961, *An Introduction to Ordinary Differential Equations*, Englewood Cliffs, N. J., Prentice-Hall.

Coddington, E. A. and Levinson, N., 1955, *Theory of Differential Equations*, New York, McGraw-Hill.

Copson, E. T., 1935, *An Introduction to the Theory of Functions of a Complex Variable*, Oxford, Oxford University Press.

Cornell, G., Silverman, J. H. and Stevens, G., 1997, *Modular Forms and Fermat's Last Theorem*, Berlin, Springer.

Courant, R. and Hilbert, D., 1968, *Methoden der mathematischen Physik*, 2 Vols., Berlin, Springer.

Cox, D. A., 1984, The Arithmetic-Geometric Mean of Gauss, *L'Enseig. Math.*, **30**, 275–330.

Coxeter, H. S. M., 1961, *Introduction to Geometry*, New York and London, Wiley, (second edn., 1969).

Davenport, H., 1952, *The Higher Arithmetic*, Hutchinson's, University Library, London, Hutchinson.

Dickson, L. E., 1934, *History of the Theory of Numbers* (especially Vol. 2, Chapter 9), New York, G. E. Stechert & Co.

Durell, C. V., 1948, *Projective Geometry*, London, Macmillan.

Dutta, M. and Debnath, L., 1965, *Elements of the Theory of Elliptic and Associated Functions with Applications*, Calcutta, The World Press Private Ltd.

Dym, H. and McKean, H. P., 1972, *Fourier Series and Integrals*, New York and London, Academic Press.

Eberlein, W. F., 1954, The Elementary Transcendental Functions, *Amer. Math. Monthly*, **61**, 386–392.

1966, The Circular Function(s), *Math. Mag.*, **39**, 197–201.

Erdos, P., 2000, Spiraling the earth with C. G. J. Jacobi, *American Journal of Physics*, **68** (10), 888–895.

Evelyn, C. J. A., Money-Coutts, G. B. and Tyrrell, J. A., 1974, *The Seven Circles Theorem and Other New Theorems*, London, Stacey International.

Faltings, G., 1995, The proof of Fermat's Last Theorem by R. Taylor and A. Wiles, *Not. Amer. Math. Soc.*, **42**, No.7, 743–746.

Fuchs, L., 1870, Die Periodicitäz moduln der hyperelliptischen Integrale als Funktionen eines Parameters aufgefasst, *J. reine angew. Math.*, **71**, 91–136.

Gauss, C. F., 1799, Arithmetisch Geometrisches Mittel, *Werke* **III**, 361–432.

1801, Disquisitiones Arithmeticae, Leipzig, Fleischer, reprinted by Impression Anastaltique, Culture et Civilisation, Bruxelles, 1968. English translation by A. A. Clarke, Yace University Press, 1966. Also in Gauss, C. F., 1870, *Werke*, **I**, Leipzig, Teubner.

Glaisher, J. W. L., 1881, On some elliptic functions and trigonometrical theorems, *Messenger of Mathematics*, **X**, 92–97.

1882, Systems of formulae in elliptic functions, *Messenger of Mathematics*, **XI**, 86.

Glaisher, J. W. L., 1907, On the representation of a number as the sum of two, four, six, eight, ten and twelve squares, *Quart. J. Pure Appl. Math.*, **XXXVIII**, 1–62.

Green, M. B. Schwarz, J. M. and Witten, E., 1987, *Superstring Theory*, Cambridge, Cambridge University Press.

Greenhill, A. G., 1892, *The Applications of Elliptic Functions*, London, Macmillan.

Grosswald, E., 1984, *Representations of Integers as Sums of Squares*, New York, Springer-Verlag.

Halphen, G. H., 1886–91, *Traité des fonctions élliptiques et de leurs applications*, 3 vols., Paris, Gauthier-Villars.

Hancock, H., 1958, *Lectures on the Theory of Elliptic Functions*, New York, Dover.

Hardy, G. H., 1944, *A Course of Pure Mathematics*, Cambridge, Cambridge University Press.

Hardy, G. H. and Wright, E. M., 1979, *An Introduction to the Theory of Numbers*, fifth edn., Oxford, Oxford University Press.

Hermite, C., 1862, Sur les theorems de M. Kronecker relatifs aux formes quadratiques, *Comptes Rendus de l'Academie des Sciences*, **LV**, 11–85, (See *Oeuvres*, **II**, Paris, Gauthier-Villars, 1908, pp. 241–263.)

   1861, Lettre addressée à M. Liouville, Theorie des Fonctions Elliptiques et ses Applications Arithmétiques, *Comptes Rendees* **LIII**; *Oeuvres*, 109–124.

Hurwitz, A. and Courant, R., 1964, *Vorlesungen ber allgemeine Funktionentheorie und elliptische Funktionen*, Berlin, Springer.

Jacobi, C. G. J., 1829, *Fundamenta Nova Theoriae Functionarum Ellipticarum*, reprinted in *Gesammelte Werke*, **I**, (1882), 49–239.

   1835, *Journal für Math.*, **XIII**, 54–56; *Gesammelte Werke*, **II**, (1882), 25–26.

   1838, Theorie der elliptischen Funktionen aus der Eigenschaften der Thetareihen abgeleitet, *Gesammelte Werke*, **I**, (1882), 497–538.

Jones, G. A. and Singerman, D., 1987, *Complex Functions*, Cambridge, Cambridge University Press.

Jordan, M. C., 1893, *Cours d'analyse de l'Ecole Polytechnique*, Paris, Gauthier-Villars.

Kendall, M. G., 1941, Correlation and elliptic functions, *J. Royal Statistical Society*, **104**, 281–283.

Klein, F., 1884, *Vorlesungen über das Ikosaeder und die Auflösung der Gleichungen von fünften Grade*, Leipzig, Teubner; English translation, *The Icosahedron and the Solution of Equations of the Fifth Degree*, trans. G. G. Morrice, New York, Dover.

Klein, F. and Fricke, R., 1890, *Vorlesungen ber die Theorie der elliptischen Modulfunctionen*, Leipzig, Teubner.

Knopp, M., 1970, *Modular Forms in Analytic Number Theory*, Chicago, Markham.

Koblitz, N., 1991, *A Course in Number Theory and Cryptography*, New York, Berlin and Heidelberg, Springer.

Kronecker, L., 1860, Über die Anzahl der verschiedenen Classen quadratische Formen von negative Determinante, *J. reine. angew. Math.* **57**; (reprinted in *Werke* (1895–1931), 5 vols., Leipzig, Teubner, 248–255, reprinted New York, Chelsea, 1968.)

Landau, E., 1947, *Vorlesungen über Zahlentheorie*, Part 1, Chelsea reprint, New York, Chelsea, 114–125.

Lang, S., 1987, *Elliptic Functions*, second edn., Berlin, Springer.

Lawden, D. F., 1989, *Elliptic Functions and Applications*, Applied Mathematical Sciences, vol. 80, New York, Springer-Verlag.

Lowan, A. N., Blanch, G. and Horenstein, W., 1942, On the inversion of the *q*-series associated with Jacobian elliptic functions, *Bull. Amer. Math. Soc.*, **48**, 737–738.

McKean, H. and Moll, V., 1997, *Elliptic Curves, Function Theory, Geometry, Arithmetic*, Cambridge, Cambridge University Press.

Nevanlinna, R. and Paatero, V., 1969, *Introduction to Complex Analysis* (tr. T. Kövari and G. S. Goodman), Reading, Mass, Addison-Wesley.

Neville, E. H., 1944, *The Jacobian elliptic functions*, Oxford, Oxford University Press.

Newboult, H. O., 1946, *Analytical Methods in Dynamics*, Oxford, Oxford University Press.

Ore, O., 1957, *Niels Henrik Abel: Mathematician Extraordinary*, Minneapolis, University of Minnesota Press.

Osgood, W. F., 1935, *Advanced Calculus*, New York, Macmillan.

Polya, G., 1921, Über eine Aufgabe der Wahrscheinlichkeitsrechnung betreffend die Irrfahrt im Strassennetz, *Math. Ann.*, **84**,149–60. Reprinted in *Collected Papers*, vol. 4, 69–80, Cambridge, Mass, MIT Press, 1974–84.

  1927, Elementarer Beweis einer Thetaformel, *Sitz. der Phys. Math. Klasse*, 158–161, Berlin, Preuss. Akad. der Wiss. Reprinted in *Collected Papers*, Vol. 1, 303–306, Cambridge, Mass MIT Press, 1974–84.

  1954, *Induction and Analogy in Mathematics and Patterns of Plausible Inference* (2 vols.), New Jersey, Princeton.

Prasolov, V. and Solovyev, Y., 1997, *Elliptic Functions and Elliptic Integrals*, Translations of Mathematical Monographs, Vol. 170, Providence, Rhode Island, American Mathematical Society.

Rademacher, H., 1973, *Topics in Analytic Number Theory*, Berlin, Springer.

Lord Rayleigh, 1929, *The Theory of Sound*, 2 vols., second edn., London, Macmillan.

Rigby, J. F., 1981, On the Money-Coutts configuration of 9 anti-tangent cycles, *Proc. London Mathematical Society*, **43**, (1), 110–132.

Serre, J.-P., 1970, *Cours d'Arithmétique*, Paris, Presses Universitaires de France. (English translation, 1973, *A Course in Arithmetic,* GTM 7, Berlin, Springer-Verlag.)

Siegel, C. L., 1969, *Topics in Complex Function Theory*, Vol. 1, New York, Wiley-Interscience.

Smith, H. J. S., 1965, *Report on the Theory of Numbers*, New York, Chelsea.

Stewart, I. N., 2003, *Galois Theory*, third edn. Buca Raton, Florida, Chapman and Hall, CRC Press.

Tannery J. and Molk, J., 1893–1902, *Elements de la théorie des Functions Elliptiques*, 4 vols., Paris, Gauthier-Villars. Reprinted New York, Chelsea, 1972.

  1939, *The Theory of Functions*, Oxford, Oxford University Press.

Titchmarsh, E. C., 1986, *The Theory of the Riemann Zeta-functions*, revised by D. R. Heath-Brown, Oxford, Oxford University Press.

Tyrrell, J. A. and Powell, M. T., 1971, A theorem in circle geometry, *Bull. London Mathematical Society*, **3**, 70–74.

du Val, P., 1973, *Elliptic Functions and Elliptic Curves*, London Mathematical Society Lecture Note Series, 9, Cambridge, Cambridge University Press.

Watson, G. N., 1939, Three triple integrals, *Oxford Quart. J. Math.*, **10**, 266–276.

  1944, *A Treatise on the Theory of Bessel Functions*, Cambridge, Cambridge University Press.

Weber, H. 1908, *Lehrbuch der Algebra* (especially Vol. 3), second edn. Braunschweig, Vieweg. Reprinted New York, Chelsea, 1961.

Weierstrass, K., 1883, Zur Theorie der elliptischen Funktionen, *Sitzungberichte der Akadamie des Wissenschaften zu Berlin*, 193–203. (*Math. Werke*, **2**, 1895, 257–309) Berlin, Mayer and Müller.

Weil, A., 1975, *Elliptic Functions according to Eisenstein and Kronecker*, Berlin, Springer.

Whittaker, E. T., 1937, *A Treatise on the Analytical Dynamics of Particles and Rigid Bodies*, fourth edn., Cambridge, Cambridge University Press.

Whittaker, E. T., and Watson, G. N., 1927, *A Course of Modern Analysis*, fourth edn., Cambridge, Cambridge University Press.

Williams, D., 2001, *Weighing the Odds – a Course in Probability and Statistics*, Cambridge, Cambridge University Press.

Zagier, D., 1991, The Birch–Swinnerton-Dyer conjecture from a naïve point of view, *Prog. Math.*, **89**, 377–389.

Further reading

The book by P. du Val (1973) presents the theory of elliptic functions very attractively and with an emphasis on the classical applications. It includes an account of the root functions, which throws light on those functions as used in this book. It concludes with a very interesting historical note (pp. 238–240), which is recommended and which refers the reader to the much fuller account given in the book by Fricke (1913).

The references that follow are strongly recommended. It will be seen that several of them (for example, the book by Newboult, 1946) are on mechanics or dynamics, and they are the sources of the examples of the applications of elliptic functions given in Chapter 12 (and of the elliptic integrals in Chapter 8).

Alling, N. L., 1981, *Real Elliptic Curves*, Amsterdam, North-Holland.

Bell, E. T., 1939, *Men of Mathematics*, London, Victor Gollancz.

Birkhoff, G. and MacLane, S., 1948, *A Survey of Modern Algebra*, New York, Macmillan.

Eagle, A., 1958, *The Elliptic Functions as They Should Be*, Cambridge, Cambridge University Press.

Eberlein, W. F., 1977, On Euler's Infinite Product for the Sine, *J. Math. Anal.*, **58**, 147–151.

Fagnano, G. C. di, 1911–12, *Opera Mathematiche*, ed. D. Alighieri di Albright, Milan, Segati e c.

Fricke, R., 1913, Elliptische Funktionen, *Encyk. Math. Wiss.*, **II.b.3.2**. (2), 177–348, Leipzig.

1916, 1992, *Die elliptischen Funktionen und ihre Awendungen*, 2 vols., Leipzig, Teubner.

1926, *Vorlesungen über die Therie der Automorphen Funktionen*, Leipzig, Teubner.

Gauss, C. F., 1797, Elegantiores integralis $\int (1 - x^4)^{1/2}\, \mathrm{d}x$ proprietates et de curva lemniscata. *Werke*, **III**, 404–432.

Goldstein, H., 1980, *Classical Mechanics*, second edn., Reading, Mass., Addison Wesley.

Hille, E., 1962, *Analytic Function Theory*, Vol. II, Boston, Ginn and Company.

1976, *Ordinary Differential Equations in the Complex Domain*, New York, Wiley-Interscience.

Jahnke, E. and Emde, F., 1945, *Tables of Functions with Formulae and Curves*, New York, Dover.

Kilmister, C. W. and Reeve, J. E., 1966, *Rational Mechanics*, London, Longman's, Green & Co. Ltd.

Landsberg, M., 1893, Zur Theorie der Gauss'schen Summen und der linearen Transformation der Theta funktionen, *J. reine angew. Math.*, **111**, 234–253.

Mordell, L. J., 1917, On the representation of numbers as sums of 2r squares, *Quarterly Journal of Mathematics*, **48**, 93–104.

Neville, E. H., 1944, *The Jacobian Elliptic Functions*, Oxford, Oxford University Press.

Newboult, H. O., 1946, *Analytical Methods in Dynamics*, Oxford, Oxford University Press.

Rauch, H. E. and Lebowitz A., 1973, *Elliptic Functions, Theta Functions and Riemann Surfaces*, Baltimore, Williams and Wilkins.

Schwarz, H. A., 1893, *Formelen und Lehrsätze zum Gebrauch der elliptischen Funktionen nach Vorlesungen und Aufzeichnungen des Herren Prof. K. Weierstrasse*, second edn., Berlin, Springer-Verlag.

Silverman, J., 1986, *The Arithmetic of Elliptic Curves,* GTM 106, New York, Springer-Verlag.

Spenceley, G. W. and Spenceley, R. M., 1947, *Smithsonian Elliptic Functions Tables*, Washington, Smithsonian Institution.

Synge, J. L. and Griffiths, B. A., 1959, *Principles of Mechanics*, third edn., New York, McGraw-Hill.

Vladut, S. G., 1991, *Kronecker's Jugendtraum and Modular Functions*, studies in the Development of Modern Mathematics, volume 2, New York, Gordon and Breach Science Publications.

Index